Flow Assurance Solids in Oil and Gas Production

T0136116

Flow Assurance Solids in Oil and Gas Production

Jon Steinar Gudmundsson

CRC Press
Taylor & Francis Group
Boca Raton London New York

CRC Press is an imprint of the
Taylor & Francis Group, an **informa** business

A BALKEMA BOOK

CRC Press
Taylor & Francis Group
6000 Broken Sound Parkway NW, Suite 300
Boca Raton, FL 33487-2742

First issued in paperback 2020

ISBN-13: 978-1-138-73784-6 (hbk)
ISBN-13: 978-0-367-78192-7 (pbk)

Typeset by MPS Limited, Chennai, India

Library of Congress Cataloging-in-Publication Data

Names: Gudmundsson, Jon Steinar, author.
Title: Flow assurance solids in oil and gas production / Jon Steinar Gudmundsson.
Description: London, UK : CRC Press/Balkema, [2017] | Includes bibliographical
   references and index.
Identifiers: LCCN 2017028024 (print) | LCCN 2017036290 (ebook) | ISBN
   9781315185118 (ebook) | ISBN 9781138737846 (hardcover : alk. paper)
Subjects:LCSH:Flowassurance(Petroleumengineering)|Petroleumchemicals—Analysis.|
   Petroleumpipelines—Fluiddynamics.|Descaling.
Classification: LCC TN879.555 (ebook) | LCC TN879.555 .G83 2017 (print) |
   DDC665.5/44—dc23
LC record available at https://lccn.loc.gov/2017028024

Visit the Taylor & Francis Web site at
http://www.taylorandfrancis.com

and the CRC Press Web site at
http://www.crcpress.com

Dedication

Dedicated to my families in Norway, Iceland and the United States

Table of contents

Preface

The book was written for engineers and natural scientists working in the petroleum industry and related industries. The engineers and scientists that are faced with the challenges and problems of flow assurance solids, from reservoirs to processing facilities. The book is intended for non-specialists and specialists in the petroleum industry, throughout the lifetime of a hydrocarbon asset, from development to abandonment. Any important issue in the petroleum industry is also of relevance in academic institutions. The materials in the book are relevant for students taking advanced degrees at universities. Furthermore, the contents of the book are relevant for overview and specific courses in the petroleum and related industries.

In teaching, writing and otherwise the main challenge is not to be understood, but not to be misunderstood. This idea has always been in my mind when teaching and advising students. And it has been an important reason for writing this book on flow assurance solids. The desire to understand a subject matter is a powerful driving force. Asking questions to yourself and others, reflects our desire to understand how and why a particular result was obtained. One can use an equation in calculations, but knowing its background and derivation provides the knowledge to make better decisions.

A friend of mine told me once that he was always happy to exchange his ideas for better ones. The same applies to everything in the book in your hands. Any and all comments are welcome to the address flow.assurance.solids@gmail.com. A feed-back from you on what is wrong, what can be improved, what is missing and even what may be useless, will be received with thanks.

Trondheim, May 2017
Jón Steinar Guðmundsson

Acknowledgements

The writing of a book is an endeavour that needs the advice and support from trusted colleagues and friends. Fortunately, one of my former doctoral students has always be ready to help and to advice; my sincere thanks to Benjamin Pierre. Two masters of hydrocarbon thermodynamics, Abbas Firoozabadi and Tony Moorwood, have kindly commented on parts of the book. Baard Kaasa and Sébastien Simon were very helpful with detailed comments on a couple of chapters. Kari Ramstad, Odd Ivar Levik, Kevin Su and Javier Dufour, have generously commented on materials in the early stages of writing. I thank them all. Now that the book has been completed, any and all mistakes are mine and mine alone.

Chapter 1

Introduction
The view of solids

The transport of hydrocarbons from reservoirs to processing facilities is the essence of flow assurance. The concept of flow assurance has emerged as the all-inclusive term for low-risk and cost-effective delivery of multiphase mixtures in the oil and gas industry. In an industry where considerable risks exist at all levels of activities, flow assurance has become the new panacea for a whole range of technical decisions. Professional engineers and natural scientists stand at the centre of the flow assurance activities in oil companies, service companies, engineering companies and the cluster of small and large companies serving the industry. Financial, insurance and regulatory institutions look to flow assurance expertise to secure investments, safety and the environment.

What flow assurance encapsulates is not new. Oilfield operators and their employees, partners and supporting industries have toiled for decades with the technical, economical and safety issues needed in bringing hydrocarbons from upstream and midstream to downstream. The advent of substantial offshore operations has brought together *traditional* and *new* disciplines to serve the industry. Multiphase flow is the workhorse of the combined disciplines. The integration of hydrocarbon phase behaviour and the chemistry of precipitating solids into multiphase flow is where the new discipline of flow assurance solids stands. The problem of paraffin wax precipitation and deposition is as old as the oil industry. The newcomer to the flow assurance solids field of study is the formation and deposition of naphthenate. Timewise between the old and the new are the formation of natural gas hydrate and asphaltene. These organic deposits are joined by the old foe of water-based industries, namely inorganic scale. The four organic and one inorganic solids are here called the *big five* flow assurance solids.

1.1 OILFIELD PRODUCTION

The production of oil and gas (and water) from hydrocarbon resources, can be shown with time as illustrated in Figure 1.1 for an oil reservoir. For a non-associated gas reservoir, the plateau period would be relatively longer. The volume rates of oil, associated gas and water at *standard conditions*, are shown versus time. The development of a typical oil-dominated asset can be divided into several major phases, here exemplified by the following: (A) exploration, (B) development 0–20%, (C) plateau 20–50%, (D) mature 50–95% and (E) tail-end 95–100%. The recovery factors shown in percentages,

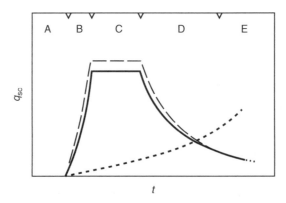

Figure 1.1   Production profiles of oil (solid line), gas (dashed line) and water (dotted line) versus time, at standard conditions. The periods A, B, C, D and E represent the following phases: exploration, development, plateau production, mature production and tail-end production, respectively.

refer to recovered reserves, not the resources in-place. Reserves are generally defined as the percentage of the *in-situ* resources that can be recovered, based on current cost-effective technology and economics (price of oil). The recoverable reserves are typically 40% of the resources in-place for oil-dominated reservoirs and 60% for natural gas reservoirs (non-associated gas). Professional engineers and natural scientists, work hard to increase the recovery factor of mainly oil reservoirs. Enhanced and improved recovery methods are high on the agenda in the oil industry. The plateau period of oil reservoirs can typically be extended by water injection for pressure support.

The solid line in Figure 1.1, illustrates the oil production rate with time. The dashed line illustrates the gas production with time. The dotted line illustrates the non-desired water production with time. In petroleum engineering, volume rates are conventionally reported at standard conditions, being 15°C and 100 kPa. The watercut at the start-up of new fields may be negligible, increasing with time to considerable volumes. Large oil companies world-wide, are likely to produce more water than oil in the mature phase. Where the solid oil-line and the dotted water-line cross, the watercut will be 50%. After the mature phase comes the tail-end phase. It is not unusual for large companies to hand-over the operation of oil fields in this phase to smaller companies. The argument being that the overhead of large companies cannot justify the operation of tail-end production. An important consideration in the life-time of a development is the field abandonment. The abandonment of exhausted oil and gas fields requires fulfilling regulatory requirements for environmentally acceptable and long-term and risk-free closure.

The precipitation of the big five solids occurs due to changes in temperature, pressure and composition (concentration). These intensive properties change with time during the life-time of a hydrocarbon reservoir. Petroleum engineering is the field of study that measures and models the changes in pressure and temperature from reservoir conditions to the surface, and along flowlines to processing facilities.

The big five solids are illustrated in Figure 1.2. The figure is a conventional pressure-temperature diagram, used to shown the phase behaviour of hydrocarbons.

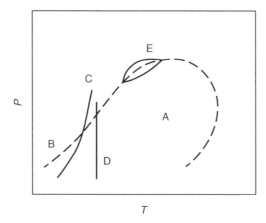

Figure 1.2 Hydrocarbon phase envelope and the equilibrium lines of the four organic flow assurance solids: A = two-phase region, B = envelope limit, C = gas hydrate, D = paraffin wax and E = asphaltene.

The composition of produced hydrocarbons changes *within* the dashed-line phase envelope, identified by A in Figure 1.2. Within the envelope, produced oil fluids exist in two phases, liquid and gas. The dashed line identified by B, is the boundary between single-phase and two-phase mixtures; called *bubble line* between oil and two-phase and *dew-point* line between gas and two-phase. Simply stated, oil exists at pressures above the boundary line and gas exists at temperatures to the right of the boundary line. The details of phase envelopes are beyond what can be presented in the present text. The critical point of the hydrocarbon will be located at high pressure and temperature, as illustrated for black oil, volatile oil, gas condensate and natural gas in *Appendix X1 – Crude Oil Composition*.

The line C in Figure 1.2 illustrates the dissociation line (equilibrium line) of natural gas hydrate. Gas hydrate forms when the line is crossed from right-to-left, from higher temperature to lower temperature (for example, cooling of subsea flowline). The temperature drop (cooling) in flowlines is more important than the pressure drop. The equilibrium temperature increases with pressure. Gas hydrates are typically formed at 10 MPa and 20°C in flowlines (fluid composition is not defined). The line D in Figure 1.2 illustrates the solubility of paraffin wax (the line is practically vertical). The solubility is not much affected by pressure, but depends strongly on temperature. Paraffin wax precipitates typically at temperature in the range 30–40°C. The small envelope E at the top of the phase envelope, illustrates the region of asphaltene formation, at high pressure and high temperature. The precipitation of asphaltene starts slightly above the bubble point of the oil and stops slightly below the bubble point. The precipitation depends on the density of the crude oil. The density decreases as the bubble pressure is approached. Inside the phase envelope, the light gases dissolved in the crude oil are liberated. The gas liberation increases the density of the crude oil and thereby stops the precipitation of asphaltenes.

1.2 FLOW ASSURANCE

Flow assurance encapsulates the technical aspects of transporting hydrocarbon mixtures from reservoir conditions to processing conditions. The challenges involved arise because the materials involved are natural fluids: crude oils, *saturated gases* and formation brines. The crude oils range in density from light-to-heavy (volatile oil and black oil). Extra heavy oil is also produced, increasingly in recent years. The saturated gases range from water vapour saturated gas to gas condensate. The formation brines range a lot in salinity and are often comingled (different formations and breakthrough). The challenge for the operator of a hydrocarbon asset, is to *produce and process* the natural fluids to transferable and saleable products. Herein lie the challenges of the professional engineers and natural scientists in the world-wide petroleum industry, the E&P (exploration and production) industry.

The *big five* flow assurance *solids* are the primary *subject matters* of the present book, constituting a *field of study*. The five solids are *each* presented and explained in dedicated chapters. Flow assurance related and solid-specific materials are presented in more than several *appendices*. For clarity, it may be useful to identify what subject matters are *not* included in the present book. For this purpose, Table 1.1 was made. The big five solids will not be commented on presently, but the excluded subject matters will be briefly addressed.

Multiphase flow represents a major strand of research and development in many industries. Early work in the nuclear energy industry, provided the groundwork for the modelling of multiphase R&D in the petroleum industry. The problem statements in the petroleum industry are more complex than in the nuclear industry. Unlike the nuclear industry, dealing with two alike fluids (water and steam), the petroleum industry must deal with three fluids simultaneously, two alike and one different (oil, gas, water). Another important difference lies in the distances and orientation of the flow channels. In the nuclear industry and similarly in the power generation (boiler) industry, the flow channels (boiler tubes) are short and vertical (usually). Power cycles are relatively compact. Flow channels in the petroleum industry are all but compact. Wellbores are 2–4 km deep and flowlines 5–15 km long. Of course, other depths and lengths are found. An extraordinary challenge in the petroleum (oil and gas) industry, are the phenomena of intermittent flow, meaning hydrodynamic slugging, terrain slugging, operational slugging and riser (severe) slugging (see *Section 2.4 Two-Phase Flow in Pipelines*).

Table 1.1 Subject matters included and excluded in the present book.

Included	Excluded
Asphaltene	Multiphase flow
Paraffin wax	Emulsions*
Gas hydrate	Sand and erosion
Inorganic scale	Corrosion
Naphthenate	Capex and opex

*Foaming is also a problem area.

Emulsions of oil and water are a major problem in the petroleum industry. The increase in produced water/brine is illustrated in Figure 1.2. In the early-years in the lifetime of an oil field, the watercut (WC = percentage of volumetric flowrate of water over the total liquid flowrate of oil and water) tends to be low. The watercut increases with time, in many cases reaching 80–90 percent in the late-years in the lifetime of an oil field. Oil-continuous emulsion will dominate in the early-years and water-continuous emulsions in later years. Emulsions of oil and water are stabilized by the big five solids, in particular asphaltenes and naphthenates, but also by other solids and precipitates. Stabilized emulsions are difficult to separate in processing facilities, calling for the use of specialty chemicals, called emulsion-breaker and demulsifiers. Specialty chemicals are also used break foams (gas bubbles) in separators. *Appendix X2 – Emulsions of Crude Oil and Brine*, may be consulted for details.

Sand and erosion are problems affecting the *integrity* of oilfield installations. The prevention and mitigation of sand in oil and gas production, is a major field of study with an overlap with the deposition of flow assurance solids. The collapse of oil and gas producing formations, leads to sand production. The sand can *settle out* in flowlines and various pieces of equipment, commonly in tank separators where the flow is more quiescent than elsewhere in the total flowstream. Reservoir formations range from consolidated to non-consolidated. The consolidation of sandstone reservoir formations varies. Production wells are fitted with sand screens and other devices to hinder sand production. The operation of wells with sanding-tendencies needs to follow specific operational procedures related to near-wellbore flow conditions. Too high a flowrate will exuberate sand production problems. The literature on sand production in oil and gas wells is considerable. In the present context, the production of formation sand and fines (much smaller mineral grains), may lead to co-deposition with the big five flow assurance solids. The presence of sand and fines may also provide seeding-sites for the nucleation of flow assurance solids. The behaviour of sand and fines in flow assurance situations are considered in *Section 2.7 – Particle Mass Transfer*.

The presence of sand particles in oil and gas flows, commonly lead to metal erosion, affecting the integrity of wellhead equipment and installations. Guidelines are offered in the oil and gas industry, specifying maximum flow velocities to avoid serious erosion problems. Without going into the details of a complicated subject matter, an example of recommended maximum fluid velocities for sizing of *carbon steel* flowlines and pipelines, are shown in Table 1.2. Values for other pipeline materials such as stainless steel, are found in the international standards literature, including NORSOK (1997).

Table 1.2 Recommended maximum fluid velocities for sizing of carbon steel flowlines and pipelines (NORSOK, 1997).

Fluid	Velocity
Liquids	6 m/s
Liquids with sand*	5 m/s
Slurry (mud and silt)*	4 m/s
Untreated seawater	2 m/s
Deoxygenated seawater	6 m/s

*Minimum velocity 0.8 m/s.

In some parts of the present book, fluid velocity in the range 2–4 m/s have been stated as being typical in liquid-carrying pipelines. In flowlines and pipelines carrying natural gas (gas-dominated flow), the maximum recommended velocity is much higher and is given either by the expression

$$u_{max} = 175\rho_g^{-0.43}$$

or 60 m/s, whichever is lower. The velocity is given in m/s and the *in-situ* gas density in kg/m^3. That is, the gas density at operating conditions, not standard conditions. The term gas-dominated flow is used to indicate that the recommended maximum gas velocity can perhaps also be used for mist flow (gas with some liquid droplets). The table and expression above are given here to underline the importance of erosion in flow assurance management, without going into the details of a complicated subject.

Corrosion is found in all oil and gas pipelines and equipment. Of course, it is a major field of study in all industrial practice. Corrosion products are commonly found in flow assurance solids. Unfortunately, corrosion and erosion are two problems that commonly appear together. The composition and pH of produced water are the determining factors for the degree of corrosion in wells (production *and* injection) and flowlines. Care is made in treating injection water (often seawater) by filtering, degassing and dosing of specialty chemicals (corrosion and scaling inhibitors and biocides). Of direct relevance in dealing with corrosion, is the injection of basic chemicals to increase the pH value of produced water. However, increased pH value affects both the formation of calcium carbonate scale and the formation of naphthenates. The operators of oil and gas assets need to *find the right balance* between the level of corrosion and the risk of solids precipitation and deposition. Corrosion problems affect the integrity of wells and flowlines and processing facilities.

Capex and opex of field development and life-time operations, involve most aspect of the flowstream from reservoir to processing facilities. However, cost issues are far beyond the purpose of the present book. At the same time it must be recognized, that flow assurance can have a considerable impact on both capex (capital expenditure) and opex (operational expenditure). *Money saved* during front-end studies and field development, can easily be *money lost* during field operations and maintenance. That is why flow assurance professionals need to be involved in the planning and engineering of new field development projects, notwithstanding their participation in the operation of hydrocarbon assets, throughout the life-time of commercial production to eventual abandonment.

1.3   CHAPTERS AND APPENDICES

The book is organized into *seven* chapters and 33 appendices. The first chapter is the current introductory chapter, setting the scene for the following main chapters. The second chapter presents the fundamental of transport phenomena, as relevant for the deposition of the big five flow assurance solids. Classical transport phenomena, heat transfer, mass transfer and momentum transfer, are presented and accompanied by several appendices. In the second chapter the argument is put forward, that the literature on flow assurance solids is dominated by the chemistry of precipitation. Indeed, a very important subject. The transport phenomena related to flowrate, on the other hand, have received much less attention. The deposition process is controlled by flowrate related diffusive and convective effects (turbulence). The terms chemistry-dominated and flowrate-dominated is used. The stated imbalance is addressed in the second chapter.

The flowrate-dominated material in the second chapter includes particle mass transfer and a presentation of deposition-release models. Particle mass transfer depends greatly on the particle diameter, expressed by the particle relaxation time. Particle mass transfer has not yet been applied to flow assurance solids deposition. The combination of deposition-release models and particle mass transfer places focus on the necessity to know more about the size of particles in the flow assurance of solids. The intention being that realistic modelling of solids deposition needs to take into consideration both *molecular transport* and *particle transport*. Furthermore, the release of deposit materials emphasizes the need for considering carefully the factors that control the *adhesion* and *cohesion* of flow assurance deposits. An improved understanding of these processes has the potential to improve deposition modelling. Additionally, such improved understanding may contribute to the *mitigation* of deposition and facilitate *cleaning/removal* operations and work-overs.

The details of the big five solids chapters will not be presented and discussed in the current chapter. Each chapter can be described as *stand-alone* for the particular flow assurance solid. To learn about a particular solid, only that particular chapter needs to be studied and understood. In a way, the chapters are modular. However, there are some materials in particular chapters that may be relevant for better understanding of other chapters. For example, the regular solution model used to calculate the precipitation of asphaltenes, is also relevant for the precipitation of paraffin wax (see *Appendix L2 – Regular Solution Model*). The monitoring of pipelines in the paraffin wax chapter should be equally applicable to other flowline and pipeline solids, perhaps most directly to inorganic scaling. The presence of water in reservoirs is dealt with in several chapters. Produced water is of paramount importance in inorganic scaling, but also in the formation/precipitation of naphthenate solids. Common threads are to be found in the primary chapters on the big five flow assurance solids.

The specialized appendices serve several purposes. They provide extra materials of relevance to the flow assurance solids field of study. Some of the appendices are of general relevance; some of the appendices are relevant to a particular solid; some of the appendices have a common relevance to two or more of the solids. The details presented in the appendices need not necessary to be known to understand the primary chapters on the big five solids. The details are there to provide *background knowledge* and an opportunity to explore how the main equations are derived, and to provide information that gives potentially a deeper understanding of a particular problem statement. Often in the literature, the reasons for using a particular equation and/or assumption are not available/known. That can sometimes be frustrating when studying something new. Not necessarily new in the general sense, but new for the person needing/wanting to understand.

1.4 SUPPORTING CONSIDERATIONS

The field of study of *flow assurance solids* does not exist in isolation. Similar fields of study provide supporting knowledge and techniques. To appreciate what these may

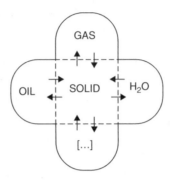

Figure 1.3 The strands of flow assurance solids. The fluid phases interacting to precipitate solids. The square-bracket represents the engineering and natural sciences.

entail, it should be useful to consider the fluid phases involved and their interactions resulting in precipitation and deposition. The *strands* of flow assurance solids are shown in Figure 1.3. The fluids involved are gas, oil and water. The fluids interact to form/make the solids, subject to changes in temperature and pressure and composition. Changes in composition occur due to evaporation/flashing of the constituent fluids and by the mixing/comingling of fluids having different concentrations. The strand, shown by [...] in the figure, are the chemical and physical *laws and models* that describe precipitation and deposition. The major engineering and natural sciences that provide the laws and theories are chemistry, fluid mechanics and thermodynamics. The workings of the laws and theories are based on the rules of mathematics. To make theoretical results useful, the constants, coefficients, numbers and factors must be determined through laboratory experiments and field testing.

Engineering and natural sciences are used to find answers to flow assurance solids challenges and problem. The results are often formulated and implemented in proprietary computer programs; commercial simulation software. The use of simulation software requires many skills and the running of sophisticated simulations does not guarantee correct results. The various theoretical assumptions underlying any and all software must be understood to ensure proper application. This holds equally true for flow assurance solids calculations. Engineers and scientists working on flow assurance solids projects need to be skilled in the fundamentals; they need to be skilled in the art of the field of study.

The above statement on the use of sophisticated simulation tools, is often expressed in the phrase *garbage in, garbage out* (=GIGO). In general terms, the adage that arguments are unsound if their premises are flawed, holds true in the flow assurance field of study. On the same subject, it may be worth noting that *correlations* do not necessarily mean *causation*. Understanding how the world holds together, can often be tortuous. The resulting reflection must be that a solid (!) knowledge of the four strands of flow assurance solids, is a prerequisite for sound decisions.

Flow assurance professionals work to either maximize or minimize a variable in field development and all phases of operations and maintenance. The *deliverability* of an asset must be maximized, meaning trouble free *reservoir performance*, trouble free *inflow performance*, trouble free production *tubing performance* and trouble free wellhead and flowline performances. The maximum *availability* of all equipment and facilities secures minimum down-time and minimum lost production. Asset operators are very focused on maintaining steady production, to secure steady monetary income. In the case of the most critical components (equipment) in a production and processing train, it is common to double-up, meaning that two components are installed in parallel when only one component is sufficient for the required service. A typical situation might be the use of rotating equipment (pumps and compressors) in parallel. In the case of planned maintenance or break-down, that the production stream need not be shut-down.

The *operability* of wells, flowlines and process facilities, makes adjustments to meet changing conditions with time possible. Installed systems, flowlines and process facilities, should have the appropriate *turn-down* capacity, for start-up, shut-down and transient (slugging) conditions. Maximum deliverability, maximum availability and maximum operability (providing flexibility in operations), contribute largely to the overall success of particular oil and gas assets. The success must be reflected in excellent health-and-safety practices and environmental compliance. The operation of field facilities and processing facilities must be reliable and exhibit first-class technical integrity.

The Society of Petroleum Engineers (SPE), has published a whole range of books of interest and relevance to the E&P industry. Books on flow assurance solids are Frenier and Ziauddin (2008) and Frenier and Ziauddin (2010). The first book has the title *Organic Deposits in Oil and Gas Production* (362 pp) and the second book has the title *Formation, Removal, and Inhibition of Inorganic Scale in the Oilfield Environment* (230 pp). Based on the titles, the two books complement much of what is presented in the present text. Apparently, the material presented stems from industrial experience and technical consulting. The current author has not had an opportunity to study the first book. The deposition of paraffin wax has recently been presented by Huang *et al.* (2016), with the title *Wax Deposition: Experimental Characterizations, Theoretical Modeling, and Field Practices* (184 pp). The title signals clearly direct relevance to parts of the present book; the chapter on paraffin wax. Apparently, the material presented stems from long-term academic R&D. The current author has not had an opportunity to study the book.

The phase behaviour of petroleum fluids is central in any and all studies of flow assurance solids. A book on the topic is that of Pedersen *et al.* (2014), with the title *Phase Behavior of Petroleum Reservoir Fluids* (465 pp). The book is comprehensive and should be a must-read for all flow assurance solids professionals. The contents show that the book covers three of the big five organic solids (not naphthenate) and the basics of inorganic scale. The current author has not had an opportunity to study the mentioned second edition of the book, but is familiar with the first edition (Pedersen *et al.*, 1989). A book on the thermodynamics of hydrocarbons is that of Firoozabadi (2016). The book has the title *Thermodynamics and Applications in Hydrocarbon Energy Production* (467 pp). The essence of the book is the presentation of a united view of classical thermodynamics, and how these are utilized in the engineering and physics of hydrocarbon production.

The subject of oilfield chemicals is not covered in the present book. The choice was made that it would be too voluminous and outside the focus of the book, just as sand production has not been included. Fortunately, a most useful book on the subject is that of Kelland (2014). The title of the book is *Production Chemicals for the Oil and Gas Industry* (454 pp). Three of the big five flow assurance solids are dealt with in the book: naphthenate, gas hydrate and paraffin wax. The operative word is *control*. Asphaltene control and scale control was included in the first edition of the book (Kelland, 2009), but apparently not in the second edition. The second edition, in additions to dealing with the mentioned flow assurance solids, has chapters on demulsifiers, foam control, flocculants, biocides, scavengers (hydrogen sulphide and oxygen) as well as drag reduction agents (DRA's). Clearly, much of the material covered in the book is relevant to flow assurance. The current author has not had an opportunity to study the second edition, but has benefitted from studying the first edition. Topics of relevance not mentioned above, are pour point depressants, asphaltene dispersants and scale inhibitors.

1.5   SYMBOLS AND UNITS

Traditions. Different fields of study have traditions in the use of symbols and units. In the present text, the symbols and units are based on the authors education background in chemical engineering and as a petroleum engineering educator, with work experience in geothermal engineering. The symbols and units used in the engineering disciplines, probably have a greater commonality than in relation to the natural sciences. The point being that great care needs to be exercised when using data and correlations across fields of study. The *symbols* used in the present book are author-background-based and the *units* used are the International System of Units (SI = *Système international d'unités*), sprinkled by units specific to the petroleum industry.

Moles. Concentrations can be expressed in terms of moles using two units, called molal m mol/kg and molar M mol/L. The mole is a SI unit expressed by mol. The difference between molal and molar concentrations may appear insignificant, but it is not in chemistry calculations. Mass does not change with temperature while volume changes with temperature. The symbol M is popular and has many names and functions. M stands for molecular weight g/mol (=kg/kmol) in the present text. It may occasionally be used for the extensive property mass. Lower-case m kg/s is used for mass rate. In general, lower case letters stand for a specific property (intensive property) of an extensive property. V stands for volume m^3 and v for specific volume m^3/mol.

Named units. The symbol for pressure is p and the symbol for temperature is T. The SI unit for pressure is pascal Pa (=N/m^2) and the unit for temperature is kelvin K, without the degree symbol. N stands for newton, the unit of force F = mass × acceleration. In passing, the symbol and unit of energy is the joule J (=Nm). Note that the SI units for temperature, pressure, force, energy and power are given names of the scientists, Kelvin, Pascal, Newton, Joule and Watt, but are written using lower-case for the first letter. It is common in the literature to use P (capital p) for pressure; not here, because the author uses P for power with the unit watt W (=J/s).

Temperature and pressure. Practical units of temperature and pressure are the Celsius degree °C and bar. The unit bar is understood to stand for absolute pressure, written bara to underscore that it does not stand for gauge pressure barg. The difference is the same as the engineering unit psia and psig (absolute and gauge). Not yet mentioned, is pressure in atmosphere. Standard atmosphere pressure 1 atm = 101.325 kPa. It was used as standard reference pressure for many years; older tabulations of thermodynamic properties are for this pressure. The standard pressure is now 1 bar (=1 bara = 10^5 Pa). However, for most engineering calculations, the difference has negligible consequence (Smith *et al.*, 1996).

Energy. The symbols A, G, H, Q and U express the extensive properties of Helmholtz free energy, Gibbs free energy, enthalpy and heat (thermal energy) and internal energy, all having the unit joule J, presented in *Appendix S – Gibbs Free Energy*. The capital letter A being a popular symbol, stands also for area m^2. The rate of heat flow is given by the lower-case symbol q J/s (=W). The symbol for specific enthalpy is given by the lower-case h kJ/kg, with the subscripts f for fluid (liquid), g for gas (vapour) and fg for latent heat (of vaporization) in Steam Tables (Rogers and Mayhew, 1980). In multiphase flow situations, the subscript L and G are used for liquid and gas, respectively. The liquid phase may be a mixture of oil and water. In some situations, the subscript M is used for oil-water mixtures and in other situations it is used for gas-liquid mixtures. Once again, the subscript is also used for mixing of solutions (see *Appendix L2 – Regular Solution Model*). Care needs to be used when working with specific units. For example, specific enthalpy may be given based on a mole of substance; the molar enthalpy being h J/mol (=kJ/kmol). When extensive properties are divided by mass or mole, they become intrinsic properties. For example, chemical potential μ J/mol, corresponds to *Gibbs Free Energy* per mole.

Flowrate. In the present text, volumetric flowrates are given by the lower-case letter q, as in petroleum engineering. Actual rates are at the *in-situ* temperature and pressure, while standard rates are at standard temperature and pressure. The subscripts used are the lower-case o for oil, g for gas and w for water. The SI flowrate unit would be m^3/s. The same subscripts are used for such properties and density and viscosity. Oilfield units are widely used in the industry, usually based on the volume unit of barrel bbl ($\simeq$159 L). Somewhere it was said that bbl stood for blue barrel, to distinguish from small oil producers. In the case of gas, the oilfield unit is cubic foot, given the symbol scf and occasionally written ft^3. The s means standard; that is, standard oilfield conditions (1 atm. and 60°F). What may be confusing is the use of prefixes for oilfield units. The prefix MM, as in MMscf/d, is not $10^6 \times 10^6$, but the Roman numeral for 1000 twice. In other words, MM in the oilfield stands for $10^3 \times 10^3 = 10^6$. In the SI system of units, time is given in seconds, by the symbol s. In oilfield units, time is commonly expressed in hours h and sometimes in days d.

Prefixes and period. The SI system of units is blessed with the use of prefixes: μ for micro, m for milli and M for mega, *et cetera*. It makes it easier to work with numbers. A common property in engineering is dynamic viscosity, here and elsewhere given the symbol μ mPa·s. Viscosity is a combined unit involving pascal and second. In the present text, it has been chosen to use the period to separate the basic units. The same applies to the gas constant $R = 8.314 \, J/mol \cdot K$. This use of period avoids problems when expressing thermal conductivity $W/m \cdot K$, instead of W/mK where mK would properly stand for millikelvin. The symbol $\cdot$ is the proper SI way of separating units in combined units. In the case of concentration, the unit mmol ($=10^{-3}$) is commonly used. The practical concentration unit ppm means parts-per-million expressing mg/kg.

Abbreviations. The abbreviations aq., wt. and vol. stand for aqueous (as in chemistry), wt. stands for weight (as in % wt.) and vol. stands for volume (as in % vol.). The abbreviation atm. (with a period) should perhaps be added to this list. The abbreviation for *mole* has the unit mol in the SI system, without a period. The author has avoided using the abbreviations *e.g.* = for example (*exempli gratia*), *i.e.* = that is (*id est*) and *etc.* = and so forth (*et cetera*). The main reason being that the author believes that written texts should be as close to spoken language as possible. However, one Latin abbreviation is used throughout the book is *et al.* for and others (*et alia*), when giving references.

Standard conditions. The chemical and physical properties of the hydrocarbons and inorganic minerals encountered in flow assurance studies, are tabulated in the literature at standard reference conditions. However, there is no universally accepted definition of the standard reference conditions of temperature and pressure. In chemistry, physics and engineering, the pressure can be either 1 atm. ($=101.325 \, Pa$), alternatively 1 bara ($=100 \, kPa$). The difference in pressure is not significant in engineering calculations. For the standard reference conditions of temperature, several alternatives exist: 0°C, 15°C, 20°C and 25°C. The *official* Society of Petroleum Engineers standard conditions are 15°C and 100 kPa. The *practical* oilfield standard state is 60°F ($=15.6$°C) and 1 atm. pressure.

Oil and gas. In Norway, the standard conditions are 15°C and 100 kPa in the oil and gas industry. The temperature and pressure are particularly important when dealing with natural gas. Standard gas volumes are expressed by Sm^3 and the standard gas volumetric flowrate by $q_g \, Sm^3/h$. The rate is based on an hour instead of second for practical reasons; the unit is therefore a hybrid SI unit. Standard volumes of oil are given at the same standard conditions, such that GOR (gas-oil-ratio) becomes Sm^3/Sm^3. Note that the GOR in oilfield units will have a different numerical value. Oilfield standard gas rate would be q_g MMscf/d. In oilfield units, the lower-case *s* means standard. The *actual* flowrate of oil in oilfield units is typically barrels per hour ($=bbl/h$). The standard flowrate is given stock-tank barrels per hour ($=stb/h$). The GOR in oilfield units would therefore be MMscf/bbl.

Standard and normal. It has no meaning to state that such and such flow of oil and gas is at standard conditions, unless the temperature and pressure are defined. STP stands for *standard* temperature and pressure, given at 0°C and 100 kPa. NTP stands for *normal* temperature and pressure, given at 20°C and 1 atm. pressure. Different countries use different standard temperature and pressure.

1.6   CONCLUDING REMARKS

The primary focus of the present book is the *big five* flow assurance solids: asphaltene, paraffin wax, gas hydrate, inorganic scale and naphthenate. Four of the solids are hydrocarbon-based and one of the solids is mineral-based. The hydrocarbon (organic) solids are relatively different in behaviour, some precipitating because of pressure reduction (asphaltene and naphthenate) and some due to cooling (paraffin wax and gas hydrate). The precipitation of inorganic scale is controlled mainly by the flashing of produced water. The different physical and chemical processes involved in the precipitation and deposition of the big five solids, makes the field of study a multidisciplinary endeavour.

The view of solids presented above, makes it clear that the use of the flow assurance concept in the present book is reasonable. The precipitation and deposition of solids, is an integral part of the *overall* flow assurance field of study. Synergetic effects are to be sought by a better appreciation and proper understanding of multiphase flow. The negative effects of emulsions and foams, as well as sand and corrosion studies, cannot be overlooked. What makes all the different effects special, is that they occur in deep wells and long flowlines. Depth and length mean that the flowing conditions of pressure, temperature and composition, are subject to considerable variations. These variations represent an important challenge.

The present book offers a description of each of the big five flow assurance solids. The descriptions are intended to lead of *learning and understanding*. The worst that can happen, is that something will be misunderstood. The main chapters are therefore supplemented by many appendices. Some of the appendices are *directly* relevant to the subject matter of the book, other appendices are more of *general* relevance. In some cases, the materials in the appendices may turn-out to be useful for other technical challenges and problems, faced by professional engineers and natural scientist in the E&P industry and supporting industries.

Chapter 2

Flow phenomena
The tools for study

The precipitation and deposition of solids are two different phenomena. The precipitation of flow assurance solids is chemistry-dominated, while the deposition is flowrate-dominated. The *chemistry* is manifested in chemical thermodynamics involving intensive properties (independent of size) such as temperature, pressure, concentration and chemical potential. The *flowrate* is manifested in the traditional transport phenomena, such as heat transfer, mass transfer and momentum transfer. The chemistry and flowrate phenomena share a commonality to the big-five flow assurance solids. The emphasis of the present chapter are the flowrate phenomena. The chemistry phenomena are more specific to each of the big-five solids, and are found in the following chapters. Extra materials on chemistry are presented in several appendices.

The literature on the big-five flow assurance solids, has been dominated by the various facets of solubility and chemical thermodynamics. The focus has been on the precipitation of solids from liquid solutions. The element of time (kinetics) does not enter into such considerations. In thermodynamics, the path from A to B is not important, just the initial and final states. In asphaltene precipitation, regular solution theory has met with success. In paraffin wax precipitation, chemical potential and fugacity relationships, have met with success. In natural gas hydrate precipitation, thermodynamic modelling of equilibrium lines has met with success. In inorganic scaling, the use of solubility product constants has met with success. In naphthenate precipitation, carboxylic acid chemistry has met with success.

Flowrate phenomena have *not* received enough attention in the literature. The bias against flowrate versus chemistry, needs to be balanced. Flowrate phenomena have only gradually emerged in software models to predict precipitation. The link between precipitation and deposition continues to be the weakness of software modelling. The deposition of solids is more complicated than the precipitation of solids. Laboratory flow-loop and field data are required to *couple* the phenomena of precipitation and deposition. Such data are expensive to gather and often difficult to interpret. Flowrate related phenomena are mentioned in each primary chapter on the big-five flow assurance solids.

The fluid mechanics of relevance in flow assurance are found in a multitude of text books and other publications, readily available for engineers and scientist in academia and industry. Referencing the fluid mechanics literature is therefore not necessary in the present text. Only the basic flow phenomena, most relevant for the deposition of the big-five flow assurance solids, will be presented below and in the several appendices.

2.1   BULK AND WALL TEMPERATURE

Reduction in temperature leads to the precipitation of *paraffin wax* from crude oil and the formation of natural *gas hydrate*. Pipeline temperature affects indirectly also the precipitation of inorganic scale and less directly asphaltenes and naphthenates. Pipeline and wellbore cooling lead also to the condensation of water vapour to liquid water. The reduction in temperature will be along a pipeline (bulk temperature) and across the wall of a pipeline. Both bulk temperature and wall temperature are important parameters in the precipitation and deposition of solids due to pipeline cooling. Temperature also affects the viscosity and density of fluids.

The *bulk temperature* in steady-state pipeline flow with constant ambient temperature is given by the equation

$$T_2 = T_o + (T_1 - T_o) \exp\left[\frac{-U\pi d}{mC_p}L\right]$$

where T_1, T_2 and T_o are the upstream, downstream and ambient temperature (o stands for outside), respectively. See derivation of the equation in *Appendix A – Temperature in Pipelines*. In oil and gas production, the upstream temperature will typically be the wellhead temperature feeding a flowline. The downstream temperature will be the temperature in the flowline at any distance L. The ambient outside temperature can be the average seabed temperature in offshore operations or the average outside air temperature in onshore operations. The equation can also be used for wellbores when adjusting for the geothermal gradient. The wellbore can be divided into segments, each with a constant surrounding formation temperature. The world-wide average geothermal gradient is about 25°C/km. However, it is common to use 30°C/km or greater for oil and gas provinces.

The above equation shows that the bulk temperature decreases exponentially with distance L. U stands for the overall heat transfer coefficient, which ranges typically from 15 to 25 W/m^2·K for bare steel pipes and 2 to 4 W/m^2·K for well insulated pipelines. Reported values for completed wellbores are in the range 4.9 to 5.3 W/m^2·K, according to Fossum (2007). It is possible to estimate the overall heat transfer coefficient from first principles by conduction and convection in series. However, pipelines on the seabed usually have a reinforced concrete coating, are partly buried and/or partly covered by rock and may have free spans, making it difficult to arrive at reliable values for the overall heat transfer coefficient.

Other parameters in the bulk temperature equation are pipeline internal diameter d m, mass flowrate m kg/s and heat capacity C_p kJ/kg·K. The equation can be used for single phase fluids and for mixtures (assuming homogeneous flow). For mixtures of oil, gas and water, the heat capacity can be estimated from the equation

$$C_{pM} = x_o C_{po} + x_g C_{pg} + x_w C_{pw}$$

where the x's stand for the respective *mass fractions*. The mass fractions are used because the heat capacities are commonly per unit mass. Heat capacities can also be given per mole C_p kJ/mol·K, in that case mole fractions should be used. The unit of heat capacity is the same as that of the gas constant R. Heat capacities change with temperature, but marginally with pressure. Polynomial expressions in temperature are

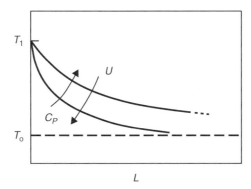

Figure 2.1 Bulk (downstream) temperature T_2 along a pipeline with inlet (upstream) temperature T_1 and surrounding (outside) temperature T_o. Rate of cooling decreases with increasing heat capacity C_p, and increases with increasing overall heat transfer coefficient U.

commonly used to correlate heat capacity data of gases and liquids. In cases where the temperature changes along a pipeline are large, the pipeline can be divided into length segments to improve the accuracy of calculations.

A schematic behaviour of the steady-state pipeline bulk temperature equation is shown in Figure 2.1. The bulk temperature along the pipeline approaches asymptotically the surrounding (ambient) temperature. With increasing heat capacity, the bulk temperature decreases more slowly with distance. Water having a relatively high heat capacity of about 4.2 kJ/kg · K, will cool down more slowly than an oil having a lower heat capacity of about 2.3 kJ/kg · K, assuming the same mass flowrate. One practical consequence being that as watercut increases with time, the temperature loss along a pipeline will be less. With increasing overall heat transfer coefficient, the bulk temperature decreases *more rapidly* with distance. One practical consequence being that thermal insulation of pipelines reduces the overall heat loss; the bulk temperature decreases *less rapidly* with distance.

Minimum cooling will occur for high velocity flow (high mass rate in small diameter pipe) in an insulated pipeline. However, high velocity flow gives high pressure drop in pipelines (hence potentially less oil production because of higher wellhead pressure) and thick insulation and/or heating gives higher investment costs. Each oilfield situation is different; a balance needs to be found for an optimum solution in a given situation. In situations where chemicals need to be injected before shut-in, a flowline system can be designed (for example, insulation) such that cooling during normal steady-state operations will not cause precipitation of natural gas hydrate.

Flowlines and pipelines are operated under three main conditions: start-up, steady-state and shut-down. The following equation expresses the temporal (pseudo-transient) temperature in a pipeline when it is shut-down

$$T_2 = T_o + (T_1 - T_o) \exp\left[\frac{-4U}{\rho C_p d} \Delta t\right]$$

The *shut-down temperature* equation has the same form as the *bulk temperature* equation; the bulk temperature at any location approaches asymptotically the surrounding (ambient outside temperature) temperature with time. In this case, T_1 is the pipeline inside temperature at some location L at the time of shut-down ($\Delta t = 0$). This temperature can/must first be estimated from the bulk temperature equation. And T_2 is correspondingly the pipeline inside temperature at the same location, after some given time ($\Delta t > 0$). As in the bulk temperature equation, T_o is the ambient temperature outside the pipeline. The other parameters are density, heat capacity and pipeline diameter. The above equation is also derived in *Appendix A – Temperature in Pipelines*.

For mixtures of oil, gas and water the mixture heat capacity equation (see above) and the following mixture density equation are to be used

$$\rho_M = \alpha_o \rho_o + \alpha_g \rho_g + \alpha_w \rho_w$$

where the α's are *volume fractions* of the respective phases. In terms of multiphase flow, the gas fraction α_g is the void fraction while the sum of the liquid fractions $\alpha_o + \alpha_w$ is the liquid hold-up. The volume fractions are used in the mixture density equation because the density is given per unit volume. It is not an uncommon mistake to use the mass fractions in the mixture density equation.

The shut-down temperature equation can be used to estimate the time it takes for a flowline (pipeline) to cool down to a temperature where solids will precipitate. It is the time available before a flowline must be started-up again to avoid solids precipitation. In scheduled shut-downs the operator will typically inject chemical into the flowline to protect against precipitation of solids. Injection of refined oil products can also be used, filling the pipeline with a non-precipitating hydrocarbon fluid, sometimes called hot oiling (also used to dissolve paraffin wax in pipelines). Alternatively, the operator can turn on heating during the shut-down period. Heating can be direct electrical heating or pipe-in-pipe heating (pipeline bundle). A pipeline bundle integrates the required flowlines and control systems necessary for subsea developments. The bundle internals are assembled within a steel carrier pipe.

Hydrocarbon fluids and mixtures entering a flowline will normally be at temperatures above the temperature at which solids precipitate. The fluids are undersaturated with respect to the solids that will precipitate due to cooling; for example, paraffin wax and natural gas hydrate. For some distance along a pipeline, the bulk temperature T_2 will be above the saturation temperature T_s (solubility limit, equilibrium line), illustrated by L_s in Figure 2.2. For distances greater than L_s, the bulk temperature will be below the saturation temperature. After some further distance, the bulk temperature approaches the ambient temperature T_o, after which no further cooling occurs and consequently no further precipitation. Precipitation will also occur at distances less than L_s because the *wall temperature* is lower than the bulk temperature. It should be noted that further cooling can occur in natural gas pipelines due to Joule-Thomson cooling.

Precipitation will occur between distances L_s and L_o because the bulk fluid is at temperatures below the saturation temperature. In the case where the fluid is saturated with one particular molecule (particular molecular weight) having one particular saturation temperature, the saturation temperature will be specific. In crude oil and condensate containing dissolved paraffin wax, the dissolved molecules have different molecular weights and hence different saturation temperatures.

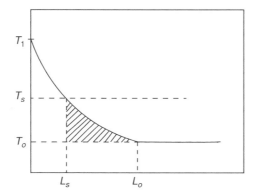

Figure 2.2   Bulk temperature T_2 along a pipeline illustrating cooling down from T_1 to saturation temperature T_s and surrounding temperature T_o. Bulk precipitation starts at L_s and extends to L_o, indicated by the hatched area.

The horizontal line for the bulk saturation temperature in Figure 2.2 represents the limit where precipitation starts. For paraffin waxes the limit is called the cloud point, also called the wax appearance temperature (WAT). With further cooling, subsequent waxes of lower molecular weights also precipitate. There comes a point where so much paraffin wax has precipitated that the oil in question no longer flows easily, compounded by low fluid viscosity due to the low temperature. This point is called the pour point; typically 5–15°C below the cloud point. The term gelling is also used for oils below the pour point.

The temperature difference between the cloud point and pour point is greater in crude oil and condensate than in model systems containing fewer n-paraffins. The cloud point and pour point are defined in standard tests. The cloud point of a given reservoir oil is lower than the cloud point of the same oil at stock-tank conditions, assuming the n-paraffins have not precipitated in the production and processing pipelines and equipment. It means that paraffin wax precipitates at a higher temperature from stock-tank oil than reservoir oil. In terms of degrees the difference could be 10–20°C.

Precipitation will also occur at distances less than L_s because the inside wall temperature will be below the saturation temperature. A temperature profile from a pipeline wall to bulk fluid is sketched in Figure 2.3. The pipe is cooled from the outside (for example, pipeline at seabed), resulting in an inside pipe wall temperature of T_w. The temperature is sketched with distance away from the wall. The temperature increases away from the wall temperature to the bulk temperature T_2 through the viscous sublayer, also called the boundary layer. At y_2 the temperature is approximately equal to the bulk temperature; in the sketch this occurs at a distance further away from the wall than y_s, which represents the distance where the temperature equals the T_s the saturation/solubility temperature.

Heat transfer in subsea pipelines is the subject of *Appendix B - Pipeline Wall Heat Transfer*. It turns out that most of the temperature drop from inside a subsea pipeline to the outside seawater, is outside the steel wall. The dominating outside temperature

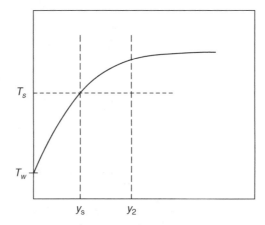

Figure 2.3 Temperature profile trough the viscous sub-layer away from wall T_w to bulk T_2. Saturation temperature T_s of precipitating solids is assumed to be reached inside the viscous sub-layer.

drop is due to the reinforced concrete anti-floating layer and/or the surrounding water-saturated mud and sediments. A steel-reinforced concrete layer 1–3 inches thick is used to add weight to a pipeline to make sure it lays stable at the sea bottom. Subsea pipelines are naturally buried in the mud and sediments on the sea bottom. In some cases, however, subsea pipelines are laid on artificial rock walls where free-span would otherwise exist.

When the radial temperature drop is primarily outside the steel wall of a subsea pipeline, the inside wall temperature is very close to the bulk temperature. Rough calculations for typical pipeline conditions show that the inside pipe wall temperature in subsea pipelines are only 1–2°C below the bulk temperature or even lower. The calculations were carried out using the temperature-of-the-wall equations given in *Appendix C – Boundary Layer Temperature Profile*. An important reason why the temperature difference between the bulk temperature and wall temperature is small, is the low overall heat transfer coefficient in subsea pipelines, typically 2–20 W/m^2 · K compared to 500–1000 W/m^2 · K in industrial heat exchangers. The overall heat transfer coefficient in oil and gas wells is typically 5 W/m^2 · K (between wellbore and surrounding formation).

The saturation/solubility temperature of a material that precipitates due to cooling is shown with a horizontal line in Figure 2.3 indicated by T_s. The saturation temperature was chosen to occur between the wall temperature and the local bulk temperature. The solid material will precipitate extremely close to the wall (within the viscous sub-layer) and not in the turbulent region (bulk temperature region).

Precipitated solids (in bulk and/or near wall) may or may not form a wall deposit in a pipeline. While precipitation is a prerequisite for deposition, a multitude of other known and unknown factors determine whether deposition actually occurs. And the critical factors determining/dominating deposition of different flow assurance solids are different.

2.2 PRESSURE DROP IN PIPELINES

Pressure drop occurs simultaneously with temperature drop in pipelines and wells. While temperature drop is the primary reason for solids precipitation and deposition, *simultaneous* pressure drop and associated fluid velocity are also important. In particular, pressure drop determines directly the degree of flashing of both hydrocarbons and water/brine. Many different flow situations are found in oil and gas production. The situations range from simple single-phase flow to complex multiphase flow. The literature is rich with articles and books on the fluid mechanics applicable to the oil and gas industry. It is beyond the scope of the present text to report and discuss this literature. Nevertheless, it is considered appropriate to include some basic pressure drop equations and relationships. An understanding of single-phase flow, is a prerequisite for understanding multiphase flow. The *flowrate aspects* of flow assurance complement the phenomena presented on the behaviour of big-five solids in oil and gas production operations, including the *chemical aspects*.

The total pressure drop in pipelines and wells consists of three main parts

$$\Delta p = \Delta p_g + \Delta p_a + \Delta p_f$$

where the subscripts g, a and f stand for gravitation, acceleration and friction, respectively. In single-phase liquid flow the three terms can be expressed as

$$\Delta p_g = \rho g \sin \alpha \Delta L$$

$$\Delta p_a = \rho u \Delta u$$

$$\Delta p_f = \frac{f}{2} \frac{1}{d} \rho u^2 \Delta L$$

The equations are based on the momentum equation for steady-state, isothermal one-dimensional flow in pipelines. The term $\sin \alpha$ is the sinus of the angle referenced to horizontal. The term ΔL is the actual length of a pipeline, not the horizontal route of the pipeline. The term Δu is the change in velocity from one cross-section to another. The other symbols have their usual meaning.

The last part of the total pressure drop equations above is the Darcy-Weisbach Equation, expressing pressure drop due to *wall friction* (see *Appendix D – Darcy-Weisbach Equation* for details). The symbol f is the friction factor. The friction factor is empirically determined from experiments and can be expressed in diagrams and equations. One well known diagram is the Moody-diagram, found in most books on fluid mechanics (McGovern, 2011). A simplified Moody-diagram is shown in Figure 2.4, illustrating log-log the friction factor versus the Reynolds number.

The Darcy-Weisbach equation can be used for both laminar flow and turbulent flow. In laminar flow (Reynolds number less than about 2000) the pressure drop does not depend on wall roughness. The friction factor in laminar flow can be calculated from the relationship

$$f = \frac{64}{\text{Re}}$$

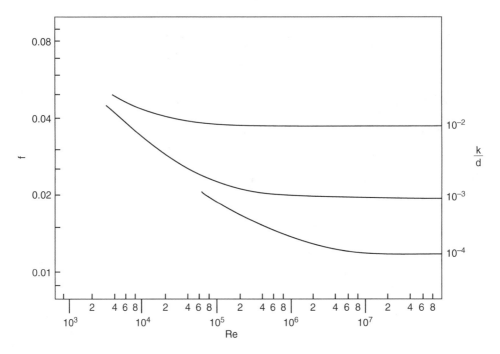

Figure 2.4   Log-log of friction factor versus Reynolds number. The ratio *k/d* is the relative roughness.

The structure of the flow is uncertain when the Reynolds number is in the range 2000–4000. In carefully conducted experiments, the laminar flow structure can be observed at Reynolds number above 2000. In industrial situations, the turbulent flow structure can be observed at Reynolds number below 4000. Note that the friction factor in the laminar-turbulent transition zone, is much lower (about one-half) in laminar flow compared to turbulent flow. One approach in carrying out calculations in the transition zone is to take the average of the laminar flow and turbulent flow friction factors.

In oil and gas pipeline operations the flow is almost always turbulent. In liquid flow the Reynolds number is typically in the range 10^5 to 10^6 such that the friction factor is typically in the range 0.01 to 0.02. In gas flow (large diameter, long distance pipelines) the Reynolds number is typically above 10^6 (primarily because the viscosity of gases is much lower than the viscosity of liquids) such that the friction factor is typically in the range 0.01 to 0.015.

In turbulent flow the pressure drop depends on the pipe wall roughness; the greater the roughness, the greater the friction factor. In general, the friction factor depends on both the Reynolds number and the relative roughness k/d. However, considering Figure 2.4, note that there is a transition from a Reynolds number and roughness depending region, to a roughness-only dependent region. Typical roughness values are shown in Table 2.1.

An explicit equation for the friction factor in turbulent pipe flow is that of Haaland (1983). The equation is recommended for use in the oil and gas industry. The general equation

$$\sqrt{\frac{1}{f}} = \frac{-1.8}{n} \log\left[\left(\frac{6.9}{Re}\right)^n + \left(\frac{k}{3.75d}\right)^{1.11n}\right]$$

uses $n = 1$ for liquids (oil and water) and $n = 3$ for gases (natural gas). The background for the Haaland n-factor is that in natural gas pipelines the transition from the hydraulically smooth to fully rough, is more abrupt in natural gas flow than in liquid flow (not as smooth as shown in Figure 2.4). The *Haaland Equation* is based on field data. The more abrupt transition may be due to moisture adsorbed between wall roughness elements that stay in place at low Reynolds number, but are carried with the flow at high Reynolds number. The data used date back to times when the control of moisture content of pipeline gas was perhaps not as stringent as today.

The ratio k/d is the relative roughness (see *Appendix F – Friction Factor of Structured Deposits*). The symbol k stands for sand-grain roughness. It is not the measured roughness of a pipe wall, but the roughness equivalent to the diameter of sand grains glued to a pipe wall, that gives the same pressure drop (Sletfjerding, 1999). See also Sletfjerding and Gudmundsson (2003), Abdolahi *et al.* (2007) and Langelandsvik (2008).

The *Haaland Equation* for turbulent flow of liquids gives the same friction factor as the *Colebrook-White Equation* (Colebrook and White 1937, Colebrook 1939). However, both are considered conservative in a good engineering tradition, giving a friction factor that is somewhat higher than measured in new pipelines, typically about 5% higher (own observation). The friction factor is called the Moody friction factor, the Darcy friction factor or the Darcy-Weisbach friction factor; they are all equal. They are not to be confused with the Fanning friction factor which is four times smaller

$$f_{Darcy} = f_{Moody} = 4f_{Fanning}$$

The wall roughness needs to be known to calculate the friction factor. Typical values of the equivalent sand grain roughness of pipes (based on oil-country tubular goods) is that of Farshad and Rieke (2005), are shown in Table 2.1. Bare carbon steel is common in oil and gas projects having an equivalent sand grain roughness of $k = 35.1\,\mu m$. Stainless steel (bare 13 Cr) is also commonly used, having an equivalent sand grain roughness of $k = 53.3\,\mu m$. Intuitively, the roughness of stainless steel would be assumed to be less than the roughness of carbon steel; but that is not the case. Long-distance, large-diameter natural gas pipelines are commonly lined with epoxy coating, here called plastic coated pipeline. The equivalent sand grain roughness quoted is $k = 5.1\,\mu m$. Even lower roughness values have been reported for epoxy coated natural gas pipelines (Sletfjerding, 1999).

The equivalent sand grain roughness values in Table 2.1 are recommended for engineering calculations. The values are perhaps slightly higher than measured under ideal conditions. Hence, the recommended values will result in a slightly higher calculated pressure drop than actual pressure drop.

Table 2.1 Equivalent sand grain roughness of pipe walls (Farshad and Rieke, 2005).

Wall material	Average absolute roughness inch	Average absolute roughness μm
Internally plastic coated pipeline	0.200×10^{-3}	5.1
Honed bare carbon steel	0.492×10^{-3}	12.5
Electropolished bare 13Cr	1.18×10^{-3}	30.0
Cement lining	1.30×10^{-3}	33.0
Bare carbon steel	1.38×10^{-3}	35.1
Fiberglass lining	1.50×10^{-3}	38.1
Bare 13Cr	2.10×10^{-3}	53.3

Calculations of pressure drop in liquid pipelines can be used to investigate the effect of deposits. The Darcy-Weisbach pressure drop equation can be expressed in terms of volumetric flowrate q m^3/s by the following

$$\Delta p_f = 8f \frac{L}{d} \rho \left(\frac{q}{\pi d^2} \right)^2 = \frac{fL\rho}{2\pi^2} \frac{q^2}{d^5}$$

Importantly, the pressure drop is inversely proportional to the pipe diameter raised to the *power of five*. Assuming the same volumetric flowrate, a small change in pipeline diameter will therefore have a significant effect on measured pressure drop. In fact, measured pressure drop is often used to estimate the thickness of deposits in test loops and operating pipelines (Botne 2012, Singh *et al.*, 2011). Further details are given in *Section 4.10 – Monitoring of Pipelines*.

Deposits in pipelines are commonly not evenly distributed from inlet to outlet. A short pipeline section with deposits (rest of pipeline without deposits) will exhibit a higher pressure drop than the same volume of deposits evenly distributed throughout the pipeline (Botne, 2012). Another complicating factor is the texture of deposits, as discussed in *Appendix F – Friction Factor of Structured Deposits*. Rippled deposits exhibit several times greater pressure drop than ordinary rough/smooth deposits.

Equations for pressure drop in natural gas pipelines and wells are derived in *Appendix G – Pressure Drop in Gas Pipelines and Wells*. In a *horizontal* pipeline the pressure drop is given by the equation

$$\frac{dA^2M}{fm^2zRT}(p_2^2 - p_1^2) - \frac{d}{f} \ln\left(\frac{p_2^2}{p_1^2} \right) + L = 0$$

The inlet pressure is p_1 and the outlet pressure p_2. The symbols have the usual meaning where A m^2 is the cross-sectional area of flow, M kg/kmol is the molecular weight of the natural gas and m kg/s is the mass flowrate. The compressibility factor z and temperature T K should be considered as the average values in the pipeline segment considered. The logarithmic term is relatively small and can be ignored in most pipeline cases.

Interestingly, recognizing that the density of natural gas is given by the equation

$$\rho = \frac{pM}{zRT}$$

and assuming average friction factor and density in a pipeline segment, it can be shown that the pressure drop in a horizontal natural gas pipeline equation can be simplified to the following *Darcy-Weisbach Equation*

$$p_1 - p_2 = \frac{\bar{f}}{2}\frac{L}{d}\bar{\rho}u^2$$

given that the natural logarithm term is ignored.

The frictional and hydrostatic pressure drop in *non-horizontal* natural gas pipelines and wells is given by the equation

$$p_2^2 = p_1^2 \exp(-2ag\sin\alpha L) - \frac{b}{a^2 g\sin\alpha}[1 - \exp(2ag\sin\alpha L)]$$

where

$$a = \frac{M}{zRT}$$

$$b = \frac{fm^2}{2A^2 d}$$

The derivation is given in *Appendix G – Pressure Drop in Gas Pipelines and Wells*. Note that the variable a includes gas properties while the variable b includes fluid flow related properties/values. The effect of acceleration on the overall pressure drop is not included in the above equation. It is usually quite small in pipe flow. Pressure drop due to acceleration may need considering in restrictions such as valves. The pressure loss coefficients of gate and sluice valves are presented in *Appendix X3 – Energy Dissipation and Bubble Diameter*.

In a static gas well, the mass flow rate is zero such that $m = 0$. It follows that the above equation can be simplified to

$$p_2 = p_1 \exp(-ag\sin\alpha L)$$

expressing the pressure increase with depth in a static natural gas well. The wellhead pressure is p_1 and the downhole pressure is p_2. The equation is solved by iteration, converging in relatively few steps. Average properties need to be used.

Monitoring the flow performance of pipelines can make it possible to discover whether solids are restricting the flow. In the case of liquid pipelines the *Darcy-Weisbach Equation* can be used to plot the total pressure drop divided by flowrate squared (volumetric or mass) against time. In the case of natural gas pipelines the pressure drop in a horizontal pipeline can be used to plot the square of the pressure drop divided by flowrate squared. In the absence of solid deposits the plotted line should be constant; in the case of solid deposits the line should increase with time (see *Section 4.10 – Monitoring of Pipelines*).

2.3 SURFACE ROUGHNESS

Sand-grain roughness assumes that the roughness elements are uniformly distributed on a surface. However, the roughness of solid deposits on pipe walls will not necessarily be uniformly distributed. Extreme examples of non-uniformly deposits are transversely rippled deposits as presented in *Appendix F – Friction Factor of Structured Deposits*. Experiments showed that the pressure drop in pipes (heat exchanger tubes) was 2–3 times greater for transverse rippled deposits than estimated for sand-grain roughness, having the same roughness height. Correlations developed for pressure drop in pipes with rectangular transverse roughness can be used to estimate the pressure drop (friction factor, roughness function) in pipes with rippled roughness. Such correlations are probably more applicable for thin deposits than thick deposits; say, when thickness $x < d/20$.

Increased surface roughness enhances *heat*, *mass* and *momentum* transfer at pipe surfaces. Increased deposit thickness will reduce heat transfer with time. The effect of increased surface roughness on overall heat transfer in pipelines and heat exchangers and friction factor (and thus pressure drop) with time are sketched in Figure 2.5. The pipe surface is assumed clean at $t = 0$ and the volumetric flow is assumed constant. The friction factor f and the overall heat transfer coefficient U have initial values, that change with time.

With time as deposition occurs and the wall surface becomes rougher, the overall heat transfer coefficient will increase. The friction factor will also increase. After a while the thermal insulation due to deposition decreases the overall heat transfer coefficient and from then on it decreases continuously (the form of the decrease depends on the type of solids deposition). The friction factor, however, increases gradually. The pressure drop will likely change more than the friction factor due to change in cross-sectional area; that is, reduced diameter. The sketch in Figure 2.5 is based on measurements in a pipe-in-pipe heat exchanger where transverse silica ripples formed (Gudmundsson, 1977; Figure 27.4 in Hewitt *et al.*, 1994). An *induction period* is observed in the data; the near-constant overall heat transfer coefficient U at the beginning.

The sketch in Figure 2.5 illustrates how difficult it can be to monitor the build-up of deposits in pipelines, heat exchangers and other equipment. Neither parameters reflect directly the thickness of the deposit, because the heat transfer performance and pressure drop depend on both the *thickness* and the surface *structure* of the deposit.

The height and width of transverse deposits and the spacing (also called pitch) determine the pressure drop (see *Appendix F – Friction Factor for Structured Deposits*). It turns out that the maximum theoretical pressure drop occurs when the spacing/height has a value around six (found in many natural systems). The pressure drop is less dependent on the height/width ratio. The spacing/height of the rippled silica deposits reported by Bott and Gudmundsson (1978) was about seven. That is, the naturally formed ripples had a geometry that maximised the pressure drop (by analogy, also heat and mass transfer).

Ripples are an intriguing phenomenon. In years past, it was suggested by Gudmundsson (1977) and reported by Bott and Gudmundsson (1978), that rippled deposits are formed due to flow *separation-reattachment*. The geometry of

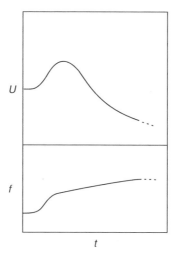

Figure 2.5 Overall heat transfer coefficient and friction factor with time during deposition of transverse silica ripples on heat exchanger pipe wall.

rippled silica deposits follows the laws of fluid mechanics. It was also suggested by the present author, that transverse rippled deposits are formed in flow situations where the deposition driving force is small and the deposited material strongly adherent. The rippled scale deposits are akin/similar to ripples on the seashore and in the desert. Because rippled surfaces are found in nature, their formation must be caused by a natural phenomenon (law of nature). The formation of ripples may be an *activator-inhibitor* system (Ball, 2015). The mathematician Alan Turing (the codebreaker), suggested that the formation of regular patterns/structures in nature, was controlled by the competition between an activating process and an inhibiting process (Turin, 1952). In the case of transverse ripples, the deposition of inorganic scale occurs because of simultaneous heat and mass transfer (the activation process). The deposition removes (dissolved) material from the flowstream locally, making less material available for deposition immediately downstream, thus lessening the deposition. Activator-inhibitor processes are diffusive, propagating in all directions. A third argument can be proposed to explain why transverse ripples form. Rippled surfaces enhance the transport of heat, mass and momentum. The transport processes are dispersive, as encapsulated in the concept of entropy, where an ordered system moves towards a disordered system. Transverse ripples lead to increased disorder in the transport processes. The deposition process (scale formation), occurs because of the reduction in chemical potential.

2.4 TWO-PHASE FLOW IN PIPELINES

Multiphase flow is inevitable in oil and gas production. The flowing phases are crude oil, associated gas, non-associated gas, gas condensate, water, brine and solids. The reductions in pressure and temperature from reservoir conditions to surface conditions, lead to continuous changes in phase distribution and chemical conditions. The natural gas can be associated gas (dissolved gas) and non-associated gas (gas reservoir). Gas condensate is light hydrocarbon liquid that condenses from high-pressure gas. Water is fresh water that condenses from natural gas. Brine is formation water with high salinity. Solid represents the range of solids co-produced with crude oil, including sand/fines from near-wellbore formations, corrosion products and the whole range of solids dealt with in following chapters: asphaltene, paraffin wax, natural gas hydrate, inorganic scale and naphthenate.

Computer software have been developed to simulate multiphase flow in pipelines and wells. Such software should be used in the analysis of flow in pipelines and wells. The dictum garbage-in, garbage-out is relevant for the use of multiphase flow software in the oil and gas industry. To help in avoiding *nonsensical* input and output data, the basics of two-phase flow will be presented. Going from multiphase flow to two-phase flow is commensurate with considering only gas-phase flow and liquid-phase flow. The gas-phase contains light hydrocarbons and water vapour. The liquid-phase contains crude oil and/or gas condensate mixed with condensed water and/or co-produced formation brine. Reliable methods are available to estimate the properties of natural gas mixtures. Methods to estimate the properties of liquid mixtures are less reliable and involve emulsion considerations.

Several modelling approaches have been used to analyse and calculate two-phase flow frictional pressure drop: (1) homogeneous flow (2) multiplier separated flow method (3) empirical separated flow and (4) mechanical separated models. In homogeneous flow the two phases have the same velocity (no slip) and average properties are used throughout (see *Appendix Y1 – Two-Phase Flow Variables and Equations*). The model gives reasonable results at high flow rates, but otherwise rather low pressure drop values. The multiplier method gives reasonable pressure drop values at low flow rates. Empirical separated flow models are perhaps the most widely used pressure drop models in industry. They are flow regime based and have criteria for transitions between the flow regimes and different formulation for each regime. The general experience is that they give too high pressure drops. Mechanistic models are the most recent addition to multiphase modelling. They are based on the physics of flow and empirical closure relationships. Extensive experimental data are used in all the modelling approaches. However, the data are gathered in relatively small-diameter flow-loops of limited length, necessitating scale-up to much larger diameters and actual flowline/pipeline lengths. Assuming that 10 parameters are involved in multiphase modelling, a three-point resolution requires 3^{10} data points (factorial experiment). A five-point resolution requires 5^{10} data points. At least three data points are needed to draw a curve; five data points makes it possible to draw a better curve.

The flow patterns in vertical pipes such as production tubing, are more evenly distributed circumferentially that in horizontal pipes. The radial distributions of the liquid-phase and the gas-phase in production tubing, are basically unaffected by gravitational forces. That is not the case in horizontal flowlines. The deposition processes of the big-five flow assurance solids, depend on the local turbulent flow structures, including the gas-liquid distribution in flowlines. Qualitative observations of flow patterns in horizontal pipes, have been quantitatively illustrated in flow pattern maps (flow regime maps).

The pressure drop in two-phase flow in *near-horizontal* pipelines (slight upward slope), can be illustrated by including the liquid inventory (liquid volume fraction).

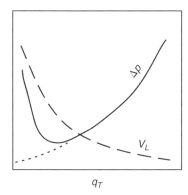

Figure 2.6 Steady-state pressure drop and holdup (liquid inventory) in a pipeline versus total volumetric flowrate.

Such an illustration is shown in Figure 2.6, where the pressure drop Δp and liquid inventory V_L are plotted versus the total (oil, gas, water) volumetric flowrate q_T. On the left-hand side, the pressure drop is *gravity-dominated*; on the right-hand side, the pressure drop is wall *friction-dominated*. The details of such diagrams/illustrations, depend on the particular flowline in question. The slugging nature of the flow will also depend on the particular situation. The minimum rate enabling a stable flow regime, is commonly called the turndown rate. The solid line is the pressure drop, the dashed line the liquid holdup (liquid inventory of the flowline). The dots in the lower left-hand corner, indicate the pressure drop performance in single phase flow, as illustrated in a figure *in Section 4.10 – Monitoring of Pipelines*. In the present context, the volume of the liquid inventory is not to be confused with liquid hold-up. The liquid inventory is a global value (whole pipeline) while the hold-up is a local value (pipeline segment).

Variables such as superficial velocity and mass flux of the flowing phases, have been used in constructing flow regime maps. The specific Froude number of the flowing phases have also been used, based on slightly different definitions. The Froude number expresses the ratio of inertial energy to gravitational energy in flowing systems. The main flow regimes in horizontal flow and a Froude number based flow regime diagram (map) are shown in *Appendix Y2 – Two-Phase Flow Regimes*. Only the basic flow regimes of bubble flow, stratified flow, slug flow and annular flow are illustrated. A hypothetical multiphase (oil, gas, water) flow situation was used to calculate the Froude number of the liquid-phase and gas-phase, illustrating that the particular flow situation was slug flow. It was pointed out that the flowline diameter is the only major parameter that can be changed in engineering design to control what flow pattern exists in a flowline. Installation of multiphase pumping at some later time, has the potential of ensuring steady-state flowrate.

Slug flow is the least desired flow pattern in oil and gas production. The specific Froude number is specific for two-phase flow, taking into consideration the densities of the gas-phase and the liquid-phase (combined oil and water/brine). In front-end-engineering (FEED), the flow pattern map in *Appendix Y2* can be used to *gauge the risk* of slugging flow in flowlines and other pipelines. The criterial for slug flow occurring are proposed to be the following

$$Fr_G < 1$$
$$0.1 < Fr_L < 10$$

These are approximate limits, only useful for a quick evaluation. Other flow regime maps may also be consulted. Given time and resources, full-scale simulation modelling will give more detailed information about the risk of slug flow.

The slug flow illustrated in the flow regime map in *Appendix Y2 – Two-Phase Flow Regimes*, refers to what is called *Hydrodynamic slugging*. The kind of slugging in horizontal flowlines and near-horizontal flowlines, where the hydrodynamics determine the flow regime. In oilfield situations, other slugging flow situations occur. *Terrain slugging* occurs when a flowline has a downward-to-upward profile; the two-phase flows down a hill and then up a hill. Liquid tends to accumulate at the bottom of the valley. *Operational slugging* occurs when sudden changes are made in the flowrate in a flowline; changes in up-stream and/or down-stream pressures. *Riser slugging* (also called severe slugging) occurs at the junction between a near-horizontal flowline and a platform riser (from seabed to dry deck). Riser base slugging is more controllable than the other slugging phenomena, for example through controlled choke valve adjustments or riser-base gas lift.

The multiplier model (see *Appendix Z – Multiplier Pressure Drop Method*) can be used to illustrate a well-proven two-phase flow modelling approach. The method has been used for a wide range of pressure drop situations in horizontal pipelines and channels. The method requires the calculation of pressure drop in *both* liquid-only and gas-only flow. The classical multiplier method *does not* include the effect of flow regime. The model has now been extended to include the effect of *interfacial effects* (interaction between gas-phase and liquid-phase), to make possible calculations in the different flow regimes (Muzychka and Awad, 2010). The authors provide a comprehensive review of the literature. The classical method was also the subject of an extensive review of 36 data sources having 7115 data points for adiabatic two-phase flow in channels (Kim and Mudawar, 2012). The authors listed 13 two-phase frictional pressure gradient correlations available in the literature. A multiplier-based correlation was developed, improving the match between model and measured data. It is heartening to observe that an old-times (seven decades) methodology continues to serve the multiphase flow community.

The experimental data of Lockhart and Martinelli (1949) were quantified by Chisholm (1967) in terms of two multiplier equations

$$\phi_L^2 = 1 + \frac{C}{X} + \frac{1}{X^2}$$
$$\phi_G^2 = 1 + CX + X^2$$

based on liquid-only and gas-only flow, respectively. Liquid-only and gas-only means that the superficial velocities are used. The Chisholm constant is given in Table 2.2. In flowlines in the oil and gas industry, both the gas-phase and liquid-phase are turbulent. The liquid-only multiplier ϕ_L against the Martinelli parameter X is shown

Table 2.2 Chisholm (1967) constant for Lockhart
and Martinelli (1949) pressure drop in
horizontal pipes.

Liquid-Gas	Constant
Turbulent-Turbulent	20
Laminar-Turbulent	12
Turbulent-Laminar	10
Laminar-Laminar	5

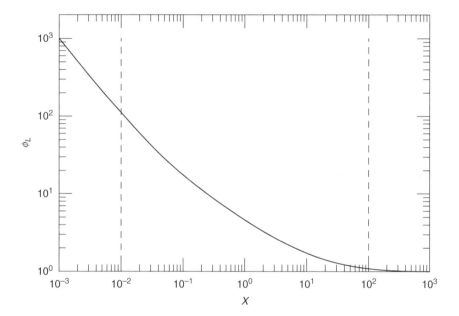

Figure 2.7 Liquid-only flow Lockhart-Martinelli multiplier ϕ_L against the parameter X (Muzychka and Awas, 2010).

on a log-log plot in Figure 2.7 for $C = 20$ only. The other turbulent/laminar configurations have a Chisholm constant $C < 20$, such that the two-phase curves are below the turbulent-turbulent curve. The two-phase flow region is in the Martinelli parameter range $0.01 < X < 100$. The left-hand dotted line marks the two-phase to gas-only parameter line; the right-hand dotted line marks the two-phase to liquid-only parameter line.

The multiplier method was used in *Appendix Z – Multiplier Pressure Drop Method*, to estimate the two-phase pressure drop was found to be about twice that assuming single-phase flow. The flowline data in *Appendix Y2 – Two-Phase Flow Regimes* can be used to estimate the two-phase pressure drop. The data given in the tables in the appendix, can be extended by including the viscosity, Reynolds number and friction factor. The viscosity of the gas was taken to be $0.015\,\mathrm{mPa \cdot s}$ and the viscosity of the liquid-phase was estimated to be $3.8\,\mathrm{mPa \cdot s}$, based on an oil-phase viscosity of 10 mPa · s and a water-phase viscosity of 2 mPa · s. The McAdams *et al.* (1942) method was used to estimate the effective viscosity, based on the mass fraction of each of the liquid phases.

The gas-phase pressure gradient, based on the gas superficial velocity, was calculated 3.13 Pa/m, using average properties in the *Darcy-Weisbach Equation*. The pressure gradient of the liquid-phase, consisting of equal amounts of oil and water, was calculated 34.4 Pa/m. The parameter X was therefore 3.3 and the liquid-phase multiplier $\phi_L = 2.7$. This result means that the two-phase pressure drop will be 2.7-times greater than the liquid-phase superficial velocity pressure drop. The multiplier two-phase flow model provides a quick and easy evaluation of the pressure drop in horizontal pipelines. Given time and resources, a full-scale simulation modelling should be employed.

The material presented above and in the associated appendices, has only addressed the frictional component in idealized horizontal flow, without upward and downward sections. A gentle slope of a few degrees, will change the nature of the flow, necessitating the inclusion of the gravitational (vertical) pressure drop component. In two-phase flows with a continuous fluid phase (liquid-dominated or gas-dominated), the gravitational component can be approximated. In segregated flows, this becomes more complicated, calling for the use of advanced simulation tools.

The material presented above, has not addressed vertical flow in wellbores (production tubing) and risers. In most oil and gas wells, the gravitational pressure drop component controls the flow. The frictional component may contribute a little bit to the overall pressure drop, but only in the uppermost part of wellbores (high-velocity gas-dominated flow). The flow in wellbores is more homogeneous than the flow in pipelines. The general statement can be made, that the pressure drop in flowlines is wall *friction-dominated* and that the pressure drop in wellbores is *gravity-dominated*. In both situations, the phase behaviour of the flowing fluids determines their relative volumes, adjusted for the slip-effect (ratio of gas velocity over liquid velocity).

2.5 CONVECTIVE MASS TRANSFER

In fully developed turbulent flow, the velocity profile at a pipe wall is given by the universal velocity profile. The corresponding temperature profile at a pipe wall is given by the temperature law-of-the-wall, which is based on the universal velocity profile. The concentration profile at a pipe wall is similarly related to the velocity profile, but only truly when the concentration is independent of temperature. In paraffin wax deposition, for example, the concentration profile depends on the temperature profile. Therefore, mass transfer of wax at the wall is not analogous to heat transfer at the wall. It follows that mass transfer correlations based on analogy with heat transfer correlations will not be directly applicable.

The temperature profile across the viscous sub-layer (and the buffer layer) is sketched in Figure 2.3 above; the temperature driving force is given by the difference $T_2 - T_w$. However, the concentration driving force, sketched in Figure 2.8, is based on the temperature difference $T_s - T_w$. Clearly, because $T_2 - T_w$ is greater than $T_s - T_w$, the heat transfer driving force is greater than the mass transfer driving force. It follows that a mass transfer correlation based on analogy with a heat transfer

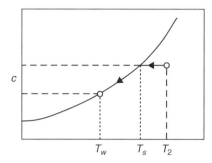

Figure 2.8 Solubility (concentration) of precipitating solid increases with temperature. Cooling from bulk temperature T_2 to solubility (saturation) limit T_s and further to wall temperature T_w.

correlation will over predict the rate of mass transfer. The degree of over prediction will depend on where, within the viscous sub-layer (and buffer layer), the saturation temperature is reached.

The challenge is that the concentration profile in the boundary layer (viscous sub-layer and buffer layer) depends on the temperature profile. Furthermore, that solids can precipitate within the boundary layer and not necessary at the wall. Solids formed within the boundary layer may or may not deposit. Indeed, a lifting force due to the velocity profile can move (transfer) solid particles away from the wall and into the turbulent core, due to the Saffman lift force (see *Section 4.3 – Viscosity of Oil-Wax Slurry* and *Section 5.8 – Prevention by Cold Flow*). The effect of shear forces on particles in fluids has been discussed by Siljuberg (2012).

The first problem (concentration driving force not independent of temperature driving force) has been addressed by Venkatesan and Fogler (2004). Extensive experimental work has shown that *heat transfer* in turbulent pipe flow can be correlated by semi-empirical equations such as

$$\mathrm{Nu} = 0.023 \mathrm{Re}^{0.8} \mathrm{Pr}^{0.33}$$

where Nu, Re and Pr are the well-known dimensionless numbers of Nusselt, Reynolds and Prandtl. See also *Appendix B – Pipeline Wall Heat Transfer*.

$$\mathrm{Nu} = \frac{hd}{k}$$

$$\mathrm{Re} = \frac{\rho u d}{\mu}$$

$$\mathrm{Pr} = \frac{C_p \mu}{k}$$

where the symbols stand for heat transfer coefficient h W/m$^2 \cdot$ K, pipe diameter d m, fluid thermal conductivity k W/m $\cdot$ K, fluid density ρ kg/m^3, fluid average velocity u m/s, fluid viscosity μ Pa $\cdot$ s and heat capacity C_p J/kg $\cdot$ K.

The semi-empirical equation for *mass transfer* in turbulent pipe flow is

$$\text{Sh} = 0.023\text{Re}^{0.8}\text{Sc}^{0.33}$$

where Sh and Sc are the dimensionless numbers of Sherwood and Schmidt

$$\text{Sh} = \frac{hd}{D}$$

$$\text{Sc} = \frac{\mu}{\rho D}$$

where D m^2/s is (molecular) diffusivity.

The central idea being that there exists an *analogy* between heat transfer and mass transfer. That is, given experimental results from heat transfer, the results can be used for mass transfer in the same flow situations, and vice-versa. The above heat and mass transfer equations are typical and well known. Other similar equations exist in the literature for the same and other flow situations.

For paraffin wax deposition, it was shown by Ventekatesan and Fogler (2004) that the ratio of the Sherwood to Nusselt number

$$\frac{\text{Sh}}{\text{Nu}} = \left(\frac{\text{Sc}}{\text{Pr}}\right)^{0.33}$$

should be replaced by the ratio

$$\frac{\text{Sh}}{\text{Nu}} = \left(\frac{dc}{dT}\right)_i \frac{\Delta T}{\Delta c}$$

The differential gradient is that at the solid-fluid interface (i = interface) while the overall gradient is from the solid-fluid interface to the bulk of the fluid. That is, the differential is taken locally at the interface while the overall gradient is taken across the whole boundary layer. Therefore, the solubility curve of the precipitating material is used to adjust for the *mismatch* between the concentration gradient and the temperature gradient.

The mass transfer coefficient obtained from the Sherwood number is traditionally used in the convective mass transfer equation

$$j = hA(c_2 - c_w)$$

The rate of mass transfer j mol/s is expressed in terms of the mass transfer coefficient h m/s, the wall surface area A m^2 and the concentration driving force between the bulk and wall concentrations. Note that j/A gives the mass flux of molecules from bulk to the wall, expressed by the symbol J below. The above equations apply to convective mass transfer in turbulent flow; that is, dissolved molecules, not particles.

The conservation of energy, mass and momentum are fundamental in engineering science and technology. The transfer of energy, mass and momentum can be described by fundamental laws on molecular scale in fluids and solids. However, flow assurance studies concern fluids in laminar and turbulent flow. The fundamental laws can be used for *laminar flow* but semi-empirical correlations are needed for *turbulent flow*. The fundamental transfer equations are presented in *Appendix E – Transfer Equations*.

The transfer of heat q J/s($=$ W) in turbulent flow is given by the expression

$$q = -(k + k_\varepsilon) A \frac{dT}{dx}$$

where k W/m·K and k_ε W/m·K are thermal conductivity and turbulent thermal conductivity, respectively.

The transfer of mass J mol/s·m^2 in turbulent flow is given by the expression

$$J = -(D + D_\varepsilon) \frac{dc}{dx}$$

where D m^2/s and D_ε m^2/s are molecular diffusivity and turbulent diffusivity, respectively.

The transfer of momentum is expressed by the wall shear stress τ N/m^2 in turbulent flow given by the expression

$$\tau = (\mu + \rho\varepsilon) \frac{du}{dx} = (\mu + \eta) \frac{du}{dx}$$

where ε m^2/s is eddy diffusivity, μ Pa·s dynamic viscosity and η Pa·s eddy viscosity. The fundamental properties k W/m·K, D m^2/s and μ Pa·s can be measured and are universal (can be used in all kinds of applications). However, the turbulent properties k_ε W/m·K, D_ε m^2/s and η Pa·s are system dependent. Turbulent flow in pipes is one such system.

Boundary layer theory has been developed to express momentum transfer (through the universal velocity profile and wall shear stress) and can be extended to heat transfer (through the law-of-the-wall presented in *Appendix C – Boundary Layer Temperature Profile*) and further to mass transfer (for example, through the transfer equations in *Appendix E – Transfer Equations*).

2.6 BOUNDARY LAYER THEORY

Boundary layer theory describes the velocity profile from a solid wall to the bulk of a flowing fluid. The velocity profile is an expression of momentum transfer and can be used to derive equations for the friction factor in pipe flow; see for example the thesis of Sletfjerding (1999) and textbooks on fluid mechanics. The velocity profile in turbulent pipe flow can be derived from what is called mixing length theory, presented in *Appendix H – Universal Velocity Profile*. Traditionally, the velocity profile is presented in dimensionless form by the equations

$$u^+ = y^+$$
$$u^+ = 2.5 \ln y^+ + 5.5$$
$$u^+ = 5.00 \ln y^+ - 3.05$$

Applicable for $(y^+ < 5)$, $(5 < y^+ < 30)$ and $(y^+ > 30)$, respectively. The above equations are from the early-days of fluid mechanics and illustrate the principles. Alternative equations based on later studies and experiments have the same form but slightly different constants. The dimensionless velocity and dimensionless distance are defined by

$$u^+ = \frac{u}{u^*}$$

$$y^+ = \frac{\rho u^* d}{\mu}$$

where u^* is the friction velocity, defined by

$$u^* = \sqrt{\frac{\tau}{\rho}}$$

in terms of wall shear stress τ N/m^2 and fluid density ρ kg/m^3. The friction velocity has typically a value of 10% of the mean flow velocity.

Based on boundary layer theory, expressions can be developed for the temperature profiles from a solid wall to the bulk of a flowing fluid. Typical expressions are presented in *Appendix C – Boundary Layer Temperature Profile*.

2.7 PARTICLE MASS TRANSFER

The mass transfer of solid particles in turbulent flow is different from that of molecules in solution. The diameter of individual particles in oil and gas operations will be in the 1–100 μm range. Sand and clay particles dislodged from reservoir formations are larger, perhaps 200–300 μm. Very small particles have a tendency to follow the movement of turbulent eddies in pipe flow, while the largest particles have a tendency to overshoot (keep moving in the same direction), when *turbulent eddies* change direction. In deposition studies, in addition to having a familiarity to turbulent eddies, the concept of *turbulent bursts* needs to be acknowledged. These are instantaneous, rapid and random fluid volumes that penetrate the boundary layer, impinging on the solid wall. Turbulent bursts are envisaged to enhance the break-up of deposits in deposition-release models (see *Appendix K – Deposition-Release Models, Section 3.12 – Deposit Buildup* and *Section 4.8 – Deposition Profiles*). Important concepts in fouling of heat exchangers (directly relevant to deposition of flow assurance solids), are presented in chapter 27 of Hewitt *et al.* (1993).

An early review of particulate fouling is that of Gudmundsson (1981). Published research on particle deposition in turbulent air flow was reviewed by Sippola and Nazaroff (2002). A more extensive treatise of particles in air including experimental work is that of Sippola (2002). A recent fluid mechanics review by Guha (2008) discusses the transport and deposition of particles in turbulent and laminar flow. Particle diffusivity in turbulent pipe flow has been presented by Gudmundsson and Bott (1977). Studies of particles in liquids are seldom compared to studies of particles in air. One example of particles in liquids is that of Chen (1993).

The concept of particle stopping distance, can be used to analyse the movement of particles in turbulent flow. See *Appendix I – Particle Stopping Distance* for derivations of particle stopping distance and particle relaxation time. The stopping distance λ is the distance a particle will travel through a stationary fluid, given by the relationship

$$\lambda = \tau u_i$$

The stopping distance depends on the initial velocity of the particle u_i and what is defined as the particle relaxation time

$$\tau = \frac{\rho_p d^2}{18\mu}$$

The density is that of the *particle* and the viscosity that of the *fluid*. For particle deposition in turbulent flow, the stationary fluid is conceived to be the boundary layer at the wall (viscous sub-layer and perhaps some of the buffer layer). The relaxation time is the time it takes a particle to reduce its velocity from the initial velocity (given by a turbulent eddy) to zero velocity.

In dimensionless form the relaxation time can be written

$$\tau^+ = \frac{\tau \rho_p (u^*)^2}{\mu}$$

where u^* is the friction velocity. A dimensionless mass transfer coefficient can be defined as

$$h^+ = \frac{h}{u^*}$$

In the case of particles depositing at a wall, the mass transfer coefficient is also called the deposition velocity (Guha, 2008) given by the expression

$$J = u_d c_p$$

The mass transfer rate to the wall J kg/s·m^2 depends on the deposition velocity u_d m/s and c_p kg/m^3 the mass concentration of particles. Note that the mass transfer is given in terms of mass, not mole, and the commonly used symbol for specific heat capacity is used for particle concentration. The dimensionless deposition velocity is expressed relative to the friction velocity

$$u_d^+ = \frac{u_d}{u^*}$$

Experimental work has shown that the mass transfer coefficient of particles in turbulent flow can be divided into three approximate regimes based on the dimensionless particle relaxation time as shown in Table 2.3. The mass transfer occurs at the wall in turbulent pipe flow; that is, from the turbulent core to the wall. In the diffusion regime, particles move due to molecular diffusion and Brownian diffusion. In the inertial regime, particles move according to the stopping distance concept (the stopping distance being of the same order of magnitude as the boundary layer thickness). In the impaction regime, particle mass and velocity (impacted by large turbulent eddies) are

Table 2.3 Particle deposition regimes and order of magnitude dimensionless relaxation time and dimensionless mass transfer coefficient (Gudmundsson, 1981).

Regime	τ_p^+	h^+ m/s
Diffusion	<0.1	$10^{-3} - 10^{-4}$
Inertia	0.1–10	$10^{-4} - 10^{-1}$
Impaction	>10	10^{-1}

so large that the stopping distance is an order of magnitude larger than the boundary layer thickness (all particles reach the wall).

The dimensionless mass transfer coefficient in the diffusion regime decreases from 10^{-3} to 10^{-4}. As the particles become larger, the Brownian diffusivity decreases. In the inertia regime, the dimensionless mass transfer coefficient increases by three orders of magnitude. As the particles become larger, the stopping distance also becomes larger and more particles reach the wall. In the impaction regime, the dimensionless mass transfer coefficient is nearly constant (decreases slightly with particle size; that is, with stopping distance) and all particles reach the wall. When the dimensionless relaxation time is greater than 100 the particles are so large that their trajectory is not affected by individual turbulent eddies.

The particle deposition regimes in turbulent flow are illustrated in Figure 2.9 (Guha, 2008). The figure is based on a large body of experimental data in the literature. Similar figures are found in Gudmundsson (1981) and Sippola (2002). The three deposition regimes have been given different names by different authors. Guha (2008) calls the three regimes turbulent diffusion, turbulent diffusion-eddy impaction and particle inertia moderated. Sippola (2002) calls the three regimes diffusion, diffusion-impaction and inertia moderated.

Assuming spherical particles 1, 10 and 100 μm in diameter, typical dimensionless relaxation times can be calculated, shown in Table 2.3. The numbers used in the calculations are for a 0.1 m diameter pipe flowing 700 kg/m² oil at 2 m/s. The Reynolds number is 280,000, the friction factor 0.0157 and the viscosity 0.5 mPa · s. Particle density was assumed in the range 930–970 kg/m³, typical for paraffin wax, gas hydrate and asphaltene. The 1 μm particle is in the diffusion regime, the 10 μm particle is in the inertia regime and the 100 μm particle in the impaction regime, approximately. Sand particles with density 2300 kg/m³ with the same diameters will have dimensionless relaxation times of 0.003, 0.3 and 30. Note that while the sand particles have a higher dimensionless relaxation time, they are within the same particle deposition regimes as the hydrocarbon particles.

Empirical equations have been proposed for the dimensionless particle deposition velocity (dimensionless mass transfer coefficient) in the diffusion, inertia and impaction regimes, respectively (Sippola and Nazaroff, 2002)

$$u_d^+ = k_1 Sc^{-2/3}$$
$$u_d^+ = k_2 (\tau_p^+)^2$$
$$u_d^+ = k_3$$

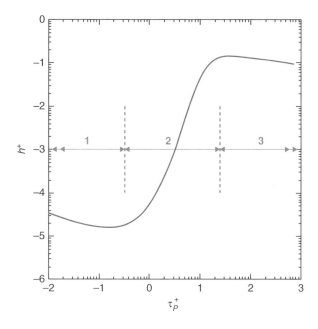

Figure 2.9 Dimensionless particle deposition velocity (mass transfer coefficient) against dimensionless particle relaxation time (Guha, 2008). Note logarithmic scales (for example, 10^{-3} is shown as -3). Regime 1 is diffusion, regime 2 is inertial and regime 3 is impaction.

Table 2.4 Particle diameter and dimensionless relaxation time.

$d \, [\mu m]$	τ_p^+
1	0.0012
10	0.12
100	12

where the Schmidt number is defined using Brownian diffusivity of a particle

$$\mathrm{Sc} = \frac{\mu}{\rho D_B}$$

The constants have approximate values of $k_1 = 0.06$, $k_2 = 0.5 \cdot 10^{-3}$ and $k_3 = 0.2$ for particles in air (Sippola and Nazaroff, 2002). It is unclear whether the same constants can be used for particles in liquids. Gudmundsson (1981) reports that the particle deposition velocity in the impaction regime is much lower in liquid systems than in gas systems (that is, lower in hydrosols compared to aerosols).

Brownian diffusivity of spherical particles in gases and liquids can be estimated from the Stokes-Einstein equation

$$D_B = \frac{k_B T}{3\pi \mu d}$$

where k_B is the Boltzmann constant and T absolute temperature (the other symbols have the usual meaning). The spherical particles are assumed much larger than the surrounding molecules. The Boltzmann constant has the same unit as entropy; namely energy per degree absolute temperature $k_B = 1.38 \cdot 10^{-23}$ J/K and equals the ratio of the gas constant R divided by Avogadro's number N_A.

Spherical particles 1, 10 and 100 μm in diameter in a 0.5 mPa·s viscosity liquid at 20°C have a Brownian diffusivity coefficient ranging from $0.858 \cdot 10^{-12}$ m²/s to $0.00858 \cdot 10^{-12}$ m²/s. For comparison, molecular diffusion coefficient of hydrocarbons ranges from $0.1 \cdot 10^{-9}$ m²/s to $10 \cdot 10^{-9}$ m²/s as presented in *Appendix J – Diffusion Coefficient*. The difference being four orders-of-magnitude different, illustrating clearly that the diffusivity of particles is substantially lower that the diffusivity of molecules.

Many other factors affect the transport of particles to the wall in turbulent pipe flow. Among these are gravitational settling $\tau_p g$ (see *Appendix I – Particle Stopping Distance*) and forces due to temperature (thermophoresis) and electrostatics. In thermophoresis, particles diffuse due to temperature gradient; away from hot regions. In electrostatics, particles diffuse due to electrical charges (on particles and surfaces). Guha (2008) and Sippola and Nazaroff (2002) offer reviews of these and other forces causing particle diffusion.

A distinction needs to be made between forces that *transport* particles in turbulent flow and forces that make particles *deposit* on a wall; that is, stick to a wall. Much of the literature on particle transport is based on 100% stickability and thus represents the maximum particle deposition velocity (maximum mass transfer). Deposition models based on experimental data, can be used to predict the build-up of deposits with time.

2.8 DEPOSITION MODELS

The deposition of solids on walls is time dependent. It is common to consider how deposit thickness increases with time. Linear and asymptotic increases with time are shown schematically in Figure 2.10. Both types of increase in deposit thickness are observed in pipes and equipment, depending on the precipitating solids involved. Linear increase in deposit thickness can be expressed by the simple relationship

$$\frac{dx}{dt} = k_1$$

where k_1 is an empirical constant or some deposition function; for example, convective mass transfer discussed in *Section 2.5 – Convective Mass Transfer*.

It can be reasoned that linear increase will occur when the deposition driving force is independent of heat transfer at the wall; for example, scaling due to mixing of fluids. Another example is the deposition of asphaltene due to changes in liquid density. It can

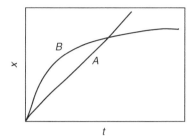

Figure 2.10 Linear increase A and exponential increase B in deposit thickness x with time t.

further be reasoned that exponential increase will occur when the deposition driving force depends on heat transfer at the wall; for example, paraffin wax and in some cases gas hydrate. Inorganic scaling can also be heat transfer controlled, when due to changes in solubility with temperature; for example, in a heat exchangers. It may be difficult to distinguish between an exponential and logarithmic deposit buildup with time. Logarithmic-like buildup may be the early behaviour of exponential buildup. Flow-loop testing in the laboratory may exhibit an exponential buildup, because the deposit strength may be less than in long-term field measurements, exhibiting either logarithmic or even linear buildup. This aspect of deposition of solids, is one of the conundrums researchers and field engineers are facing.

One model has been shown to work well for a range of deposition situations, called the deposition-release model (see *Appendix K – Deposition-Release Models*). Such models have been reviewed by Gudmundsson (1981), for example. In an exponential model the increase in deposit thickness with time is described by the equation

$$\frac{dx}{dt} = k_1 - k_2 x$$

where k_1 and k_2 are constants, coefficients or functions. The model implies that the initial rate of deposition is constant because $x \approx 0$. With increasing thickness, the net rate of deposition decreases. Integration of the above deposition-release model gives the thickness with time as

$$x = \frac{k_1}{k_2}[1 - \exp(-k_2 t)]$$

In other words, the deposit thickness increases exponentially with time to an asymptotic value at large times. Note that at large times the deposit thickness will be

$$x_\infty = \frac{k_1}{k_2}$$

For a particular deposition situation, experiments can be carried out to determine the values of k_1 and k_2. The constants (coefficients, functions) can be simple numerical values or semi-empirical correlations.

In an alternative expression/formulation of the exponential deposition-release model

$$\rho \frac{dx}{dt} = k_1 - k_2 x$$

the left-hand-side represents mass flux kg/s $\cdot$ m^2. The constants have the same subscript but will have different values compared to the equations above. Note that in steady-state mass transfer the mass flux is given by the expression

$$J = \frac{j}{A} = h \Delta c \quad .$$

where h is the mass transfer coefficient and Δc the concentration driving force. In some situations, the mass transfer flux will be equivalent to k_1 in the deposition-release model.

A more general form of the exponential deposition-release model

$$\rho \frac{dx}{dt} = \psi_D - \psi_R$$

expresses the mass flux in terms of a deposition function ψ_D and a release function ψ_R. Experimental results can be used to find the appropriate parameters in the two functions.

The deposition function describes the transport of solids to a wall and the release function the removal (shedding) of material from the wall. The overall deposition process is conceived to consist of continuous transfer of material to the wall and continuous removal of deposited solids. The continuous transfer of material to the wall is mass transfer controlled, while the removal of deposited solids is adhesion/cohesion controlled (Gudmundsson, 1977).

The wall shear stress in turbulent pipe flow fluctuates with time. It changes rapidly due to turbulent eddies and turbulent bursts that strike the viscous sub-layer and pipe wall. The behaviour of particles in funnel vortices at pipe walls has been reported by Kaftori *et al.* (1995). Eddies, bursts and vortices are among the shear stress phenomena that contribute to the removal of flow assurance solids from walls.

Various forms of the deposition-release functions can be proposed, including some form of temperature dependent stickability in the deposition function. For a particular situation, fluid velocity (alternatively, wall shear stress) is probably the most important variable in deposition of solids on walls. Mass transfer rates increase with fluid velocity and wall shear stress increases with fluid velocity. In other words, increasing fluid velocity transfers more mass to a surface and at the same time increases the wall shear stress causing removal of deposited material. Wall shear stress is a more fundamental variable than fluid velocity in modelling deposition and should be preferred when scaling-up

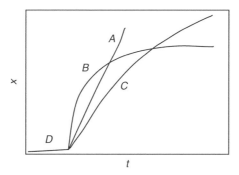

Figure 2.11   Increase in deposit thickness with time. A linear, B exponential (asymptotic), C logarithmic, all with D induction time.

flow loop data to real pipelines. Early in the present chapter, the labels chemistry-dominated was used for the precipitation from solution, and flowrate-dominated for the deposition process.

In heat transfer controlled situations, the deposit thickness will decrease the heat transfer (coefficient) with time. In long pipelines, this may result in the deposition zone gradually moving further downstream; for example, in paraffin wax deposition. An increase in deposit thickness, will simultaneously decrease the cross-sectional area available for flow, resulting in increased wall shear stress, assuming the same volumetric flowrate. However, note also that increased fluid velocity due to diameter reduction will also enhance mass transfer to the wall, making modelling the overall deposition process even more complicated. Hence the great value in performing experiments. Gudmundsson (1981) presents tables with deposition-release relationships obtained from experiments for various deposition and fouling processes/systems.

Increases in deposit thickness with time other than linear and asymptotic (shown in Figure 2.9) are found in the literature; for example, logarithmic increase with time (decreasing rate of deposition). An additional feature is that of an *induction time*; it takes some time for the deposit thickness to start forming. The length of the induction time varies in different systems. Asymptotic (exponential), logarithmic and linear increases in deposit thickness with induction time are illustrated in Figure 2.11 (shown with equal induction time for simplicity).

In the present work, a model for logarithmic increase in deposit thickness with time has been proposed

$$\frac{dx}{dt} = k_1 k_2^{-x}$$

where k_1 and k_2 are constants, coefficients or functions (see *Appendix K – Deposition-Release Models*). When $t \approx 0$ the rate of deposition is the initial rate of deposition $(dp/dx) = k_1$. Again, k_1 can be a constant determined from experiments or an appropriate mass transfer function. The same applies to k_2. Integration gives the deposit thickness with time as

$$x(t) = \frac{1}{\ln k_2} \ln[1 + (k_1 \ln k_2)t]$$

Alternatively, the model can be expressed by the empirical expression, where the *combined constants* need to be determined from experiments

$$x(t) = k_3 \ln(1 + k_4 t)$$

An overview of fouling of heat transfer equipment is given by Somerscales and Knudsen (1981); also by Bryers and Cole (1982). A more recent overview and treatise of deposition and fouling in heat exchangers is that of Bott (1995). Deposition-release models have been fitted to paraffin wax deposition experiments and an operating pipeline by Botne (2012). Modelling of paraffin wax deposition is reviewed and presented in *Chapter 4 – Paraffin Wax*.

2.9 CONCLUDING REMARKS

The flow phenomena presented and discussed above are based on traditional heat, mass and momentum transfer as found in numerous textbooks. The focus has been on phenomena considered relevant for flow assurance solids in oil and gas operations. An attempt has been made to present and discuss phenomena that are common to asphaltene, paraffin wax, natural gas hydrate and naphthenate. The basic phenomena are also relevant to the deposition of inorganic scale and other solids in general.

The field of flow assurance solids has increased in importance because of increased production of oil and gas offshore. In particular, subsea production and the transport of oil, gas and water from offshore fields to on-land receiving terminals. There is a lack of field data where flow assurance solids affect the transport of oil and gas in subsea pipelines. The present chapter deals primarily with single phase flow. An important complicating factor is the fact that most long-distance subsea pipelines carry multiphase fluids (oil, gas, water), along an uneven seabed with multiple upward and downward segments. The flow in such pipelines is complicated and difficult to model. The specialised multiphase flow literature must be consulted for details.

Appreciating and understanding the flow phenomena in production tubing and flowlines in the oilfield, will contribute to finding cost effective and suitable technical solutions. The merging together of heat, mass and momentum transfer processes with chemical precipitation processes, will contribute better big-five solids management in the oilfield. The fundamental transfer processes are available in the literature, as are the fundamental chemical thermodynamic processes. The deposition-release model approach, offers an opportunity to couple the fundamentals (chemistry-dominated and flowrate-dominated) to flow-loop and field observations. The goal being to provide the operators of oil and gas fields with tool for improved big-five flow assurance solids management.

Chapter 3

Asphaltene

The tar-like solids

Asphaltenes precipitate and form deposits in reservoirs, wellbores, flowlines, separators and other equipment. Precipitation and deposition occur due to changes in pressure, temperature and phase changes (composition) from reservoir conditions to surface conditions. Somewhere along the flowstream the conditions may favour precipitation and deposition. The problem is increasing in the oil and gas industry because more deep-water reservoirs are coming on-line. The implementation of enhanced oil recovery operations, based on miscible injection of gases, may lead to asphaltene precipitation and deposition due to lower solubility in lower density crude oils.

The asphaltenes content of reservoir fluids varies greatly. Precipitation and deposition of asphaltenes is most prevalent from light crude oils and least prevalent from heavy crude oils. In early times, this *counter intuitive* fact made it difficult to understand oilfield observations. Considerable amount of research, development and field work have been carried out in recent decades in the oil industry. The results achieved have dismissed the counter intuitive impression, and given way to useful models and guidelines for field developments and operations. Asphaltenes are no longer the enigma in crude oil characterization and production operations.

Asphaltenes are a poorly defined group of high molecular weight, highly aromatic, polar compounds found in crude oils. They are the most *polar* compounds in crude oil and are amorphous (non-crystalline) when precipitating from solution and depositing on substrates. The presence of asphaltenes stabilizes oil-water emulsions and oil-gas foams, making the separation of oil, gas and water difficult, affecting important sales and environmental specifications. The polar nature of asphaltenes contributes to the stability of emulsions and foams. The physical properties of crude oils are affected by the amount of asphaltenes present, including viscosity and surface tension.

Asphaltenes are both molecules and particles (colloid, micelle). Several fields of study are required to make progress in understanding asphaltenes: organic chemistry, colloidal/emulsion sciences, thermodynamics and phase behaviour. A number of doctoral theses have addressed the asphaltene problem, including: Yarraton (1997), Aske (2002), Ting (2003), Verdier (2006), Panuganti (2013), Balou (2014) and Teklebrhan (2014). The theses contain both general and specific knowledge about asphaltenes, useful in further studies. The theses were cursory used in the present chapter and are referenced where appropriate.

A review of the prediction, prevention and remediation of asphaltene deposition is that of Khaleel *et al.* (2015). Advances in asphaltene science were reviewed by Mullins *et al.* (2012). An industry based overview is that of Akbarzadeh *et al.* (2007). Screening for potential asphaltene problems was presented by Wang *et al.* (2006).

3.1 CLASSES OF CRUDE OIL

Crude oil (petroleum) is a continuum of several thousands of molecules; a complex mixture of hydrocarbons and trace amounts of heteroatoms and metals. Meaningful chemical analysis of crude oil is an impossible task. Instead, for practical reasons, crude oils are divided in four solubility classes: saturates, aromatics, resins and asphaltenes, commonly called SARA. The saturates, aromatics and resins are collectively called maltenes. The four classes apply to the liquid phase of crude oils at ambient conditions. Crude oil contains also the volatile compounds methane, ethane, propane and butane with boiling points $-162°C$, $-89°C$, $-42°C$ and $0°C$, respectively. The next alkane in the carbon series, pentane, has a boiling point of $36°C$ and remains liquid at atmospheric conditions.

Asphaltenes are insoluble in paraffinic (normal alkane) hydrocarbons, but soluble in aromatic hydrocarbons. Pentane n-C_5, hexane n-C_6 and heptane n-C_7 are paraffinic hydrocarbons. Benzene (C_6H_6) and toluene (C_7H_8) are aromatic hydrocarbons commonly used to dissolve asphaltenes. The asphaltene class is determined by mixing a crude oil sample with an excess volume of n-heptane, n-hexane or n-pentane in a volume ratio 40:1 (1 g crude oil and 40 mL precipitant), according to standard procedures. The filtered weight of solids precipitated from a particular crude oil are shown in Figure 3.1, using several paraffinic hydrocarbons (Speight and Moschopedis, 1981). Other crudes will show different values. The figure illustrates that if n-C_5 (heptane) is used, the weight percent is about 17.5%. If n-C_7 (pentane) is used, the weight percent is about 10.5%. The heptane value is 2/3 (66%) higher than the pentane value. Despite this large difference, the two paraffinic precipitates are commonly used to report the weight percent of asphaltenes in crude oils. The analysis procedures in different laboratories give slightly different results, making SARA analysis sometimes uncertain. A laboratory method based on normal hexane n-C_6 and high-pressure liquid chromatography (HPLC) was used by Aske (2002), instead of the more tedious ASTM-method.

A schematic illustration of the classification of crude oil is shown in Figure 3.2. Stock-tank crude oil (stabilized oil without volatiles) is mixed with either heptane or pentane in 40:1 volume excess (n-hexane is sometime used). After stirring and waiting (for example, 24 hours), the participated asphaltene is filtered off. The soluble part, called maltenes, is passed through liquid chromatographic columns using different absorbents and eluents to determine the weight percent of the saturates, aromatics and resins.

The organic chemistry classification of hydrocarbons is presented in *Appendix X1 – Crude Oil Composition* and will not be repeated here. However, the four fractions in SARA classification need to be explained. Descriptions of the classes/fractions are given by Aske (2002), Sjöblom *et al.* (2003) and Silset (2008). The following descriptions are based on these three references.

Saturates are non-polar hydrocarbons without double bonds, including straight-chain, branched alkanes and cycloalkanes (naphthenes). Cycloalkanes contain one

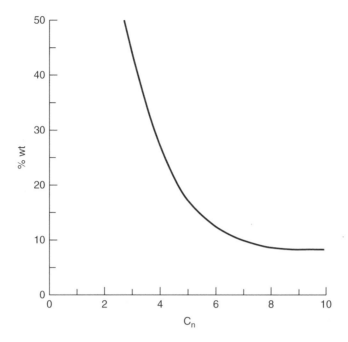

Figure 3.1 Weight percent asphaltene precipitating when mixed with excess volumes (40:1) of paraffinic (normal alkane) hydrocarbons (Speight and Moschopedis, 1981).

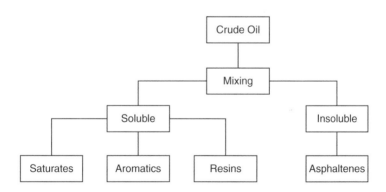

Figure 3.2 Schematic illustration of SARA classification of crude oil.

or more rings, which may have several alkyl side chains. An alkyl is an alkane with one hydrogen molecule removed. For example, methyl has the reduced methane formula $-CH_3$. The saturates are the lightest fraction of petroleum, their proportion (% wt.) decreasing with increasing molecular weight and density of crude oil.

Aromatics are hexagonal hydrocarbon structure with alternating double covalent bonds, like benzene and its derivative structures. Aromatics are common in all crude

Table 3.1   SARA analysis of 11 North Sea stock-tank crude oils at standard conditions (Aske, 2002). Reproduced with permission.

Density kg/m³	Saturates	Aromatics	Resins	Asphaltenes
0.796	79.8	16.5	3.6	0.1
0.796	65.0	30.7	4.3	0.0
0.839	82.7	13.4	3.9	0.0
0.840	55.4	28.3	12.9	3.4
0.844	62.7	23.6	12.2	1.5
0.857	60.6	30.0	9.2	0.2
0.885	50.9	34.6	14.0	0.5
0.898	50.3	31.4	17.5	0.7
0.914	41.8	38.8	18.7	0.6
0.916	48.0	37.5	14.2	0.3
0.945	35.3	36.8	24.5	3.5

oils. The majority of aromatics contain alkyl chains and cycloalkane rings, along with additional aromatic rings. Aromatics are often classified according to the number of rings. High-molecular weight, polar aromatics may overlap in weight with the resins or asphaltenes fractions.

Resins consist of large polar molecules, often containing heteroatoms such as nitrogen, oxygen and sulphur. Because resins are defined as a solubility class, an overlap with the aromatic and asphaltic fractions is expected. Resins have not been as much studied as asphaltenes, perhaps because the latter represent a serious deposition problem. Resins affect the physical properties of crude oils and reportedly play an important role in the stability of asphaltenes. Alkanes are more abundant in resins than in asphaltenes. Resins are structurally similar to asphaltenes but have a lower molecular weight. Naphthenic acids (see *Chapter 7*) are commonly regarded as a part of the resin fraction.

Asphaltenes contain the largest amounts of heteroatoms (nitrogen, oxygen and sulphur) and metallic constituents (Ni, C and Fe), compared to the other fractions of crude oil. The structure of asphaltenes consists of polycyclic aromatic clusters, substituted with varying alkyl side chains. The molecular weight of asphaltenes is uncertain because of self-aggregation. Molecular weights in the region 500–2000 kg/kmol are believed to be reasonable. An average molecular weight of 750 kg/kmol is frequently assumed. The diameter of monomer molecules (non-aggregates) of asphaltenes are believed to be in the range 12–24 Å (1.2–2.4 nm).

3.2   CRUDE OIL SARA VALUES

Consistent SARA values were presented by Aske (2002) for 11 North Sea crudes six West Africa crudes. Normal hexane (n-C_6) was used to precipitate the asphaltenes fraction from the maltenes fraction. The results are shown in Table 3.1, where the data have been organized with increasing stock-tank oil density. The SARA method used by Aske (2002), was also described in Sjöblom *et al.* (2003).

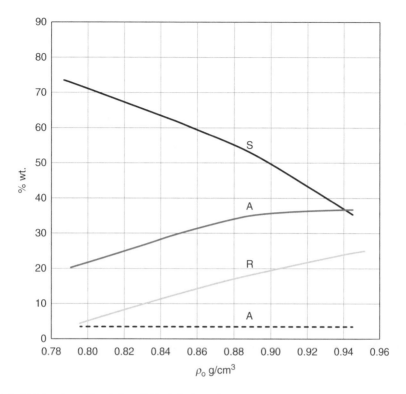

Figure 3.3 SARA-values (% wt.) from Table 3.1, plotted against stock-tank crude oil density at standard conditions. Normal hexane was used in the mixing/precipitation.

The SARA data in Table 3.1 are plotted in Figure 3.3. The figure shows the weight percent of the four fractions against the crude oil density at standard conditions. Straight lines can be drawn through the data for saturates, aromatics and resins. The data from the third line in Table 3.1 are not included in Figure 3.3 because the values are deemed out-of-bounds. The gradient of the saturates data is linearly downwards and the gradients of the aromatics and resins in parallel linearly upwards. The asphaltenes values were relatively low, ranging from 0% wt. to 3.5% wt. Unfortunately, whether asphaltene problems are associated with the crudes, was not reported.

The SARA-values from West Africa reported by Aske (2002), if plotted in Figure 3.3, would coincide with the North Sea data. However, the stock-tank oil density was much narrower, in the range 0.873–0.921 g/cm^3, making it unreliable to evaluate the trends in the SARA fractions. The asphaltenes-values were in the range 1–6.6% wt., higher than the North Sea values.

An example of SARA analysis is that of Fan *et al.* (2002). Six medium-gravity, stock-tank crude oils were used in the study (see Table 3.2) where n-heptane (n-C$_7$) was used. They evaluated the accuracy of the ASTM D2007-93 standard method and own HPLC (high pressure liquid chromatography) method. They found that the asphaltene % wt. was the same in the two methods when using hexane as the precipitate. For the

Table 3.2   Crude oil sample properties used in SARA analysis (Fan *et al.*, 2002). Density and RI at 20°C. Column n-C$_7$ is asphaltenes.

Crude oil	°API	ρ kg/m^3	M kg/kmol	n-C$_7$ % wt.	RI*
A	25.2	896	236	8.7	1.5128
B	22.6	916	268	2.8	1.5137
C	31.3	867	235	1.9	1.4851
D	28.8	880	240	5.8	1.4976
E	37.2	841	213	1.3	1.4769
F	31.1	869	270	4.1	1.4906

*Oil sample before addition of heptane.

six stock-tank crude oil samples, they found that the % wt. asphaltenes varied depending on whether hexane or heptane was used. No particular trend could be observed. The reason may be that the stock-tank crudes used were from different oil provinces, having different properties. The same observation was made in the present work when plotting the SARA value against the stock-tank oil density.

The refractive index RI of the Fan *et al.* (2002) stock-tank samples is shown in Table 3.2. Not shown is the RI measured when asphaltene first appeared upon gradual addition of the precipitant heptane. The reported measurements showed that the RI had a consistent average value of 1.44 when asphaltenes began to precipitate. It was concluded that RI could be used to determine the onset of asphaltenes precipitation from crude oils. The authors (Fan *et al.*, 2002) used 67 crude oil samples to derive the empirical correlation for the refractive index of stock-tank crude oils

$$\text{RI} = \frac{1}{100}[1.4452 \cdot S + 1.4982 \cdot A + 1.6624 \cdot (R + As)]$$

The symbols *S*, *A*, *R* and *As* stand for the four SARA fractions. The purpose of such a correlation is to enable prediction of asphaltenes precipitation for crude oil and crude oil fractions. Samples would be tested for asphaltene precipitation and the refractive index measured (RI is the same as *n* in *Appendix L1 – Dipole Moment of Hydrocarbons*). Fan *et al.* (2002) pointed out that the use of various ratios of the components of SARA analysis have failed in predicting when precipitation occurs. The central unresolved problem is finding ways to use SARA analysis to predict the precipitation of asphaltenes.

A range of other scientific methods are used to detect and identify the onset of asphaltenes precipitation: focused-beam laser reflectance (Marugán *et al.*, 2009), high-pressure near-infrared spectroscopy (Aske *et al.*, 2002), nuclear magnetic resonance spectroscopy (Calles *et al.*, 2008), ultra-violet spectroscopy (Edmonds *et al.*, 1999) and optical microscopy (Hirschberg *et al.*, 1984; Karan *et al.*, 2003; Angel *et al.*, 2006; Maqbool *et al.*, 2009).

The Fan *et al.* (2002) correlation was used to calculate the RI of the consistent SARA values in Table 3.1 (and Figure 3.3). The RI values of the ten North Sea stock-tank crudes are shown in Figure 3.4. The index *increases* linearly with crude oil density,

Table 3.3 SARA fractions of crude oils in Table 3.3 based on the ATMS methodology (Fan *et al.*, 2002). Note different alkane precipitate in the two tables. Column n-C₆ is asphaltenes.

Crude oil	Volatiles* wt.% wt.	Saturates wt.% wt.	Aromatics wt.% wt.	Resins % wt.	n-C₆ % wt.
A	6.7	46.2	19.7	18.6	8.8
B	10.3	38.8	23.6	23.9	3.4
C	2.4	68.7	17.4	9.9	1.6
D	16.3	47.0	19.4	14.7	2.6
E	6.3	45.6	27.8	14.2	6.1
F	3.2	59.0	22.9	11.7	3.2

*Lost during ASTM chromatographic analysis.

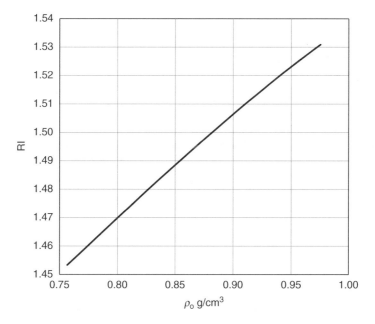

Figure 3.4 Refractive index (RI) with density of ten North Sea stock-tank crudes. Data from Table 3.2.

measured at standard conditions. The linearity applies also for mixtures of crude oils and hydrocarbon solvents. This property is useful when measuring the RI of crude oils that are dark and do not transmit much light. Dilution with a light hydrocarbon with known RI makes it possible to extrapolate to zero diluent and 100% crude oil.

The refractive index of crude oils *decreases* linearly with temperature (George and Singh, 2015). The following two correlations were reported for a light crude (API °46.2) and a heavy crude (API °16.3), respectively.

$$n = RI = 1.46118 - 3.83967 \times 10^{-4} \, T$$

$$n = RI = 1.68029 - 6.736 \times 10^{-4} \, T$$

where the temperature is °C. For temperature 15°C, 20°C and 25°C the RI for the light crude ranged from 1.455 to 1.452 and for the heavy crude 1.670 to 1.663, being consistent with the values in Table 3.2 above.

3.3 ANALYSIS OF ELEMENTS

One method to probe the composition of asphaltene deposits, is to measure the concentration of the constituent elements. Asphaltenes are considered to be composed of polyaromatic condensed rings and aliphatic chains, containing both heteroatoms such as nitrogen, oxygen and sulphur, plus metals. The same can be stated for petroleum, being a mixture of compounds consisting of hydrocarbons, heteroatoms and metals. The elemental ratio H/C is commonly used to characterize petroleum, petroleum fractions and hydrocarbon molecules. Asphaltenes were reported to have a H/C ratio in the range 0.9–1.2 and resins in the range 1.2–1.7 (Aske, 2002). The exact ratios for each fraction depend on the crude oil source: region, basin/province, field/reservoir.

The metals found in asphaltene deposits are commonly vanadium, iron and nickel. They are transition metals having atomic numbers 23, 26 and 28, respectively. The metals can have several oxidation states, 2+ and 3+ being most common. The molecular weights are 50.9, 55.8 and 58.7 kg/kmol, respectively.

The composition of hydrocarbons in crude oil consists of paraffins, naphthenes, aromatics and asphaltics (*Table XI.2 – Composition by weight of hydrocarbons in crude oil*). The paraffins have saturated chains and $H/C \simeq 2$. The naphthenes (cycloparaffins) have saturated rings and $H/C \simeq 2$ also. The aromatics have unsaturated rings and $H/C \leq 2$, meaning that they have less hydrogen than paraffins. Asphaltics have $H/C \simeq 1$. Ratio values in the range 1–2 are representative for the lower and upper bounds expected for crude oil samples.

Elemental compositions of asphaltene fractions (precipitated by n-pentane and n-heptane) of stock-tank crude oils from four countries of the world, are shown in Table 3.4. The table numbers/values are from Speight and Moschopedis (1981), also

Table 3.4 Elemental compositions of asphaltene from four countries obtained by n-pentane and n-heptane precipitation (Speight and Moschopedis, 1981).

Source	C	H	N	O	S	H/C	N/C	O/C	S/C
Canada									
n-C_5	79.50	8.00	1.20	3.80	7.50	1.21	0.013	0.036	0.035
n-C_7	78.40	7.60	1.40	4.60	8.00	1.16	0.015	0.044	0.038
Iran									
n-C_5	83.80	7.50	1.40	2.30	5.00	1.07	0.014	0.021	0.022
n-C_7	84.20	7.00	1.60	1.40	5.80	1.00	0.016	0.012	0.026
Iraq									
n-C_5	81.70	7.90	0.80	1.10	8.50	1.16	0.008	0.010	0.039
n-C_7	80.70	7.10	0.90	1.50	9.80	1.06	0.010	0.014	0.046
Kuwait									
n-C_5	82.40	7.90	0.90	1.40	7.40	1.14	0.009	0.014	0.034
n-C_7	82.00	7.30	1.00	1.90	7.80	1.07	0.010	0.017	0.036

Table 3.5 Elemental analysis of three n-heptane asphaltene samples (Balou, 2014). Elements in % wt. and metals in ppm (mg/kg).

Sample	C	H	N	O	S	H/C	Ni	V
A1	81.50	7.79	1.05	2.21	7.20	1.15	154	479
B1	77.32	7.54	0.93	2.41	8.99	1.17	197	574
K1	84.25	6.36	1.29	1.91	4.50	0.91	185	571

given by Thou *et al.* (2002). The table shows the % wt. of the molecules C, H, N, O and S, and the atomic ratio of H/C, N/C, O/C and S/C (the amounts of heteroatoms relative to carbon). The atomic ratios are based on the % wt. values and the atomic weight of the molecules.

To find the H/C atomic ratio, the atomic weight of carbon is about 12 kg/kmol and that of hydrogen about 1 kg/kmol. The % wt. values in Table 3.4 were divided by the respective atomic weights. For example, dividing 79.5 by 12 and dividing 8.0 by 1, gives the H/C atomic ratio as 1.21. In the table the amounts of carbon and hydrogen vary over a very small range: $82 \pm 3\%$ and for carbon $8.1 \pm 0.7\%$ for hydrogen. This indicates that the H/C ratio of asphaltenes is fairly constant at $1.15 \pm 0.05\%$. An almost constant H/C ratio suggests that the asphaltenes have a definite composition. The following high-to-low weight percentages can be deduced from the table: $C \gg H \simeq S > O > N$.

The total amount of asphaltene precipitated for the samples in Table 3.4 was not reported. An examination of the table, however, shows that the general trend from n-C_7 to n-C_5 precipitant, is that the C and H % wt. increase and the heteroatoms N, O and S decrease. The average H/C was 1.15 for the n-C_5 asphaltene and 1.07 for the n-C_7 asphaltene. Clearly, the elemental analysis depends on the precipitant used. This is one of the challenges in studying asphaltenes precipitation and deposition. Asphaltenes represent a continuum of molecules, just as petroleum consists of a continuum of hydrocarbon molecules from the lightest (methane) to the heaviest (commonly asphaltenes). Molecular weight and density are not the same property. Heavy alkanes may have a higher molecular weight that asphaltenes; the density of asphaltenes may be higher than that of long-chain alkanes.

Asphaltenes were n-heptane (n-C_7) precipitated from three crude oil samples by Balou (2014). The details of the crude oils were not given/identified. The crude oil to heptane ratio used was 1:25 and the mixture was left for 24 hours before filtration. The elemental analyses of the three samples are shown in Table 3.5. The average of the three H/C values was 1.08, almost the same as the average value above. However, the average was dominated by the lowest H/C value, the two other values were similar to the n-C_5 values above. The metal ions Ni and V were also reported, usefully.

In connection with molecular dynamic simulation (MDS) studies, Rogel and Carbognani (2003) carried out measurements on 12 asphaltene samples. The samples were labelled stable, unstable and deposit. The stable and unstable asphaltene samples were precipitated from stock-tank oils. The deposit samples were asphaltene solids of an unreported origin. The density of the three kinds of asphaltenes are shown in Table 3.6. The tendency to precipitate increased with decreasing H/C ratio (increasing aromaticity) giving and increase in asphaltene density.

Table 3.6 Measured H/C ratio and pycnometer density of n-heptane precipitate asphaltene (Rogel and Carbognani, 2003).

Asphaltene	Samples	H/C	ρ_a (g/cm^3)
Stable	6	1.11–1.23	1.16–1.19
Unstable	4	0.96–1.05	1.20–1.26
Deposit	2	0.98–1.00	1.25–1.28

The SARA classification scheme has been widely applied to stock-tank crude oils. An important challenge is to find methods, based on SARA measurements, to predict which crudes will suffer from asphaltene precipitation and deposition. Attempts have been made to plot asphaltene % wt. against API gravity (alternatively density), with some success, but the demarcation between problem crudes and non-problem crudes are unclear. Attempts have been made to plot the ratio of saturates over aromatics against asphaltenes over resins. Again, some success has been achieved, but the demarcation has been unclear. Attempts have been made to plot the refractive index (RI) against stock-tank crude oil density, as shown in Figure 3.4. Although RI has been used to localize the start of asphaltene precipitation (upper bound of the asphaltene phase envelope) during pressure reduction of known problem crudes, the same has not been shown for crude oils in general. The intrinsic problem is that SARA and elemental analyses are not consistent for samples from outside a particular production region, sedimentary basin and subsurface reservoir. The number of consistent samples is usually too low to formulate guidelines to differentiate between problem and non-problem crudes.

3.4 MOLECULES AND PARTICLES

Asphaltenes are both molecules and particles. The molecules are the largest in petroleum, consisting of a collection of hydrocarbon molecules. The size of asphaltenes molecules is comparable to that of colloidal particles. This fact has given rise to the uncertainty of whether asphaltenes are molecules or colloidal particles. For the sake of comparison, colloidal particles are found in sizes ranging from nanometres to micrometres. Formation fines produced with crude oil are clays and silts, less than 60 micrometres in diameter. Sand particles in oil reservoirs can be classified as *fine* (0.006–0.2 mm), *medium* (0.2–0.6 mm) and *coarse* (0.6–2 mm). In terms of micrometres, the sand particles classified as fine are 6–200 micrometre in diameter. It means that the smallest fine sands are of the same order-of-magnitude in size as o the largest colloidal particles.

In the literature, there is seldom a distinction made between a molecule and a particle. For example, particle sizes of asphaltenes in a crude oil at reservoir conditions were reported by Akbarzadeh *et al.* (2007) to be in the range 1–10 μm. When the pressure decreased below the asphaltene onset pressure, the particles sizes range increased to 1–20 μm. On further pressure decrease, the particle size range increased to 1–50 μm. A stock tank-oil (API 20°) from Iran was studied by Ashoori (2005). Mixing with n-heptane gave an asphaltene concentration of 8% wt. The oil mixture

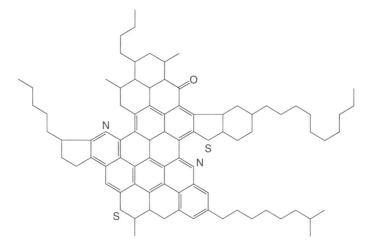

Figure 3.5 A hypothetical structure of an asphaltene molecule proposed by Angle (2001). Figure from Silset (2008).

was centrifuged and the size of the solid particles measured by SEM (Scanning Electron Microscope). The particles were large, in the size range 3–500 μm. Without centrifugation, the particles in the crude oil n-heptane mixture were measured using an optical microscope. The particles were observed to be plate-like in the size range 1–4 μm. The molecular weight of the asphaltene precipitate was measured 750 kg/kmol and the density 1.1 g/cm³ (Ashoori, 2005).

A hypothetical structure of an asphaltene molecule is shown in Figure 3.5 (Angle, 2001). The molecule consists of 15 hexagonal rings, seven with double bonds and eight without double bonds. Also shown are two pentagonal rings and several alkyl chains. The heteroatoms N, O and S are also shown. Not shown are OH and NH that are present in both resins and asphaltenes (Firoozabadi, 2016). The heteroatoms give different functionalities; for example, N and NH act differently. The structure in Figure 3.5 is hypothetical; similar structures of an asphaltene molecule are found in the literature.

Taking the average molecular weight of an asphaltene molecule as 750 kg/kmol and knowing that of benzene is 78 kg/kmol, a simplistic estimation can be made. Assuming that a asphaltene molecule consists *only* of benzene molecules, which of course is not the case, it will take 9–10 benzene molecules to make up the molecular weight of the asphaltene. The hypothetical structure in Figure 3.5 contains 17 cycloalkane rings and several alkyl chains. The H/C of benzene is one (=6/6), which is lower than the assumed 1.15 average for asphaltene. The H/C for straight-chain paraffins is 2. This simplistic consideration/estimation means that the molecular structure of asphaltene is more aromatic than paraffinic.

Visualizations of the progression of asphaltene molecules to particles have been presented by Spiecker *et al.* (2003), Mullins *et al.* (2012), Zuo *et al.* (2012), Zuo *et al.* (2013) and Khaleel *et al.* (2015). The original references may be different. Schematic illustrations of asphaltene monomers and their aggregates, in the absence and presence

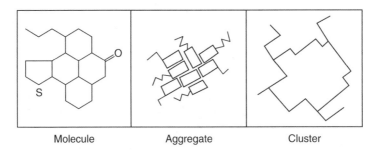

Molecule   Aggregate   Cluster

Figure 3.6 Visualization of asphaltenes, ranging from dissolved molecules ($p = p_R$) to aggregated colloids ($p \geq p_b$) and further to clustered colloids ($p \leq p_b$).

of resins, were presented by Spiecker *et al.* (2003). Mullins *et al.* (2012) added the size of an asphaltene molecule as ≈ 1.5 nm, an asphaltene nanoaggregates as ≈ 2 nm and a clusters of nanoaggregates as ≈ 5 nm. Keeping in mind that colloidal particles are nm–μm in size, the Mullins *et al.* (2012) sizes fall in the *lower* colloidal range. Ashoori (2005) reported particle sizes in the *upper* colloidal range, as did the sizes reported by Akbarzadeh *et al.* (2007). In the visualization presented by Khaleel *et al.* (2015), dissolved (nano-aggregate) asphaltene was 3 nm in size, precipitated primary particles were <1 μm and aggregated (micro-aggregate) was >1 μm. The dissolved and primary asphaltene particles are within the colloidal range while the micro-aggregates are larger. The reported sizes of asphaltenes are relatively small and will deposit due to molecular and Brownian diffusions (see *Section 2.7 – Particle Mass Transfer*).

An alternate illustration/visualization of the progression from asphaltene molecules to aggregates, and further to clusters, is shown in Figure 3.6. The molecule-fraction illustrated is only a small part of the molecules shown in the hypothetical structure in Figure 3.5. Real asphaltenes molecules contain the heteroatoms N, O and S and the transition metals Ni, V and Fe. The aggregates are an intermediate stage before the formation of clusters. The aggregates and clusters have a plate-like structure. Instead of using uncertain molecule and particle sizes to demarcate the transitions from molecules to aggregates to clusters, the *in-situ* pressure conditions can be used. The visualization scheme shown in Figure 3.6 is based on the suggestion that *molecules* existing at reservoir pressure ($p = p_R$) and the slightly lower well-flowing pressure ($p_{wf} < p_R$). *Aggregates* are envisaged to exist at pressures slightly above the bubble point pressure ($p \geq p_b$) but below the pressure of the *upper bound* of the asphaltene phase envelope. Furthermore, that *clusters* exist at pressure below the bubble point ($p \leq p_b$), down to the *lower bound* of the asphaltene phase envelope. Asphaltene aggregates and clusters are supposedly non-uniform in size and are commonly called polydispersed.

The asphaltene fraction and the resin fraction of crude oils are both polar in nature, the asphaltenes being more polar than the resins (Goual and Firoozabadi, 2002). The saturates and aromatics fractions are non-polar in comparison. Upon aggregation, asphaltenes may be subjected to induced polarization, making the formation of clusters easier. Resins are generally considered to act as solvation agents for asphaltenes, resins keep asphaltenes in solution, giving *increased solubility*. The presence of resins reduces the aggregation of asphaltenes through interfering with bonding between the monomers (Spiecker *et al.*, 2003).

Contradictory to Spiecker *et al.* (2003), it was demonstrated by Goual and Firoozabadi (2004) that the presence of resins led to increased precipitation of asphaltenes, meaning *decreased solubility*. The laboratory tests showed that the strength of the polarity of each fraction (strong-strong, strong-weak, weak-weak, weak-strong) could give an opposite effect (increased precipitation).

Resins are less aromatic than asphaltenes. The H/C ratios of resins are in the range 1.3 and 1.5, based on SARA analysis of four crude oils studied by Spiecker *et al.* (2003). The asphaltene concentration of the crude oils was 6.7 to 14.8% wt. Because of their low aromaticity compared to asphaltenes, resins have a higher solubility than asphaltenes. Resins have the ability to dissociate intermolecular asphaltene bonds, resulting in reduced aggregate size. The colloidal perspective emphasizes that crude oil can be divided into polar and non-polar fraction. The saturates and aromatics make up the non-polar fraction while resins and asphaltenes make up the polar fraction.

Asphaltenes are stabilized in crude oil by the presence of resins, according to Edmonds *et al.* (1999). When the oil is diluted by light hydrocarbons the concentration of resins decreases. At some dilution, the asphaltenes are no longer stable and therefore flocculate to form a solid phase. Other words used to describe the formation of a solid asphaltene phase are aggregate and precipitate. Because the stabilising action of resins works through polar interaction, the effect of resins becomes weaker with increasing temperature, causing flocculation. An opposite effect is that asphaltenes become more soluble in crude oil with increasing temperature.

3.5 POLARITY AND DIPOLE MOMENT

Light hydrocarbons are non-polar, while the high molecular weight asphaltenes are polar. Large hydrocarbon molecules in size between light hydrocarbons and asphaltenes can be slightly polar (Nalwaya *et al.*, 1999). The polarity of the four SARA fractions are very different, increasing with molecular weight (from left-to-right). Asphaltenes are more polar than resins, while the polarity of saturates and aromatics is miniscule. The dipole moment μ (not to be confused with viscosity) of asphaltenes of eight crudes (see Table 3.7) was reported in the range 3–7 D, that of resins in the range 2–3 D and that of the remaining stock-tank oil (after separation of asphaltenes and resins), less than 1 D (Goual and Firoozabadi, 2002). The *Onsager Equation* (formulation) was used in the calculations/measurements. The unit D (=Debye) is equivalent to 3.34×10^{-30} C·m (=Coulombe·metre). The *Onsager Equation* and the dipole moment of three hydrocarbons (pentane, benzene, toluene) are presented in *Appendix L1 – Dipole Moment of Hydrocarbons*, to show what physical parameters are involved.

The polarity of a molecule arises when one end of the molecule is positively charged and the opposite end is negatively charged. In non-polar molecules, the electrons are distributed more symmetrically and cancel each other out. A polar molecule does not mix with a non-polar molecule. For example, water does not mix with petroleum which consisting primarily of non-polar molecules (saturates and aromatics). However, water and alcohol can mix because both are polar molecules. Dipole moments occur when

Table 3.7 Properties at 20°C of eight stock-tank crude oils used to measure dipole moment (Goual and Firoozabadi, 2002).

Crude	Density g/cm^3	API	RI	Description
A	1.0089	9	–	Extra heavy
B	0.9990	10	–	Extra heavy
C	0.8588	33	1.487	Medium
D	0.8372	37	1.479	Medium
E	0.8520	35	1.488	Medium
F	0.8627	32	1.472	Medium
G	0.8550	34	1.491	Medium
H	0.7554	56	1.431	Light

there is a separation of charges. The larger the difference in the charges at each end of a molecule, the larger the dipole moment (the definition is charge × distance). The dipole moment is a measure of the polarity of molecules. In the case of asphaltenes and resins, the polarity depends largely on the heteroatom (N, O, S) content.

The properties of eight stock-tank oils from different regions of the world are shown in Table 3.7 (Goual and Firoozabadi, 2002). The RI of the two *extra heavy* crudes was not measurable. The descriptions of the crudes are from Table X1.1, being slightly different from the descriptions used by the authors. Three eluents used were: n-C$_5$, n-C$_7$ and n-C$_{10}$. The weight of solids precipitated *decreased* with increasing alkane number, as widely reported in the literature and illustrate in Figure 3.1. Based on n-C$_5$, the approximate average molecular weight of the eight crude oil samples, was 900 kg/kmol for asphaltenes, that of resins 700 kg/kmol and that of the remaining oil 300 kg/kmol. The average molecular weight of the *five medium density* crude oils (the remaining fraction, maltenes) was close to 290 kg/kmol. The Goual and Firoozabadi (2002) paper contains a wealth of additional data.

The measured dipole moment of the SARA fractions (asphaltenes, resins and saturates+aromatics) by Goual and Firoozabadi (2002) are shown in Figure 3.7. The values are from *only* the medium crude oils in Table 3.7 were used. The average densities of the asphaltenes, resins and remaining oil (saturates and aromatics) were reported 900 kg/kmol, 700 kg/kmol and almost 300 kg/kmol, respectively. The *average* of the dipole moments reported were 5 D for asphaltenes, 2.5 D for resins and 1 D for the remaining saturates plus aromatics. The figure shown that the dipole moment increases linearly from almost zero, the intercept starting at a finite molecular weight of less than 100 kg/kmol (see *Appendix L1 – Dipole Moment of Hydrocarbons*), to the resins value and then more steeply up to the asphaltenes value. It needs to be emphasized that the values are only from five stock-tank crude samples in one particular study.

Asphaltenes are defined as the heaviest components (the molecular weight of heavy alkanes may be similar) of petroleum fluids, insoluble in light n-alkanes such as n-pentane (n-C$_5$) and n-heptane (n-C$_7$) but soluble in aromatics such as toluene (Goual, 2012). Based on the principle that polar molecules are mutually soluble (like-dissolves-like), polar hydrocarbons should be soluble in polar water. The dipole moment (and solubility parameter) of selected hydrocarbons and water are shown

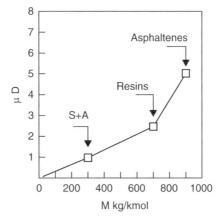

Figure 3.7 Diagram of measured dipole moment against molecular weight of asphaltenes, resins and saturates+aromatics fractions of five medium stock-tank crude oil samples.

Table 3.8 Measured properties of selected pure hydrocarbons and water from different literature sources.

Component	Formula	ρ g/mL	μ D	δ (MPa)$^{1/2}$
Pentane	C_5H_{12}	0.626	0.00	14.3
Hexane	C_6H_{14}	0.655	0.00	14.9
Heptane	C_7H_{16}	0.680	0.00	15.3
Benzene	C_6H_6	0.879	0.00	18.8
Toluene	C_7H_8	0.867	0.36	18.3
Methanol	CH_3OH	0.791	1.70	29.6
Ethanol	C_2H_5OH	0.789	1.69	26.0
Water	H_2O	1.000	1.85	47.9

in Table 3.8. The shown values are from different sources at standard conditions (atmospheric pressure) and either 20°C or 25°C.

The first group in Table 3.8 are three n-alkanes. Their solubility in water is reported in the mg/L range. The second group are two aromatic hydrocarbons. Their solubility in water is reported in the g/L range. The third group are two common alcohols. They are mutually soluble (miscible) in water. The measured dipole moment of benzene and toluene can be compared with the theoretically calculated values of 0.14 D and 0.37 D in *Appendix L1 – Dipole Moment of Hydrocarbons*. The two toluene values are practically equal, but not the benzene values. The difference in the dipole moment of the alcohols and water is about 0.15 D (mutually soluble). The difference of the aromatics and water is about 0.50 D (sparingly soluble). Whether these are demarcating differences in general, is not known.

The saturates + aromatics in Figure 3.7 have a measured dipole moment ≤1 D. The resins have a dipole moment in the range 2–3 D. Both of these fractions have a difference in dipole momentum numerically larger than 0.15 D and are therefore supposedly insoluble (not miscible) in water. That asphaltenes with a dipole moment in the range 3–7 D are soluble in toluene ($\mu = 0.36$), does not agree with the like-dissolves-like statement, based on dipole moment. Similar intermolecular interactions and solubility parameters may be necessary for dissolution to occur.

The solubility parameter has a clear physical meaning (Firoozabadi, 2016). The difference in the solubility parameters of two substances, ranges from small to large, resulting in appreciable solubility to negligible solubility. The solubility parameter of asphaltene, toluene and heptane at 20°C, were reported 19.4 MPa$^{1/2}$, 18.2 MPa$^{1/2}$ and 15.3 MPa$^{1/2}$, respectively. The small difference asphaltene-toluene of 1.2, correlates with asphaltene being soluble in toluene. The larger difference asphaltene-heptane of 4.1, correlates with asphaltene being insoluble in heptane (at least, very low solubility).

3.6 RESERVOIR TO SURFACE CONDITIONS

The solubility of asphaltene in crude oil depends on pressure, temperature and composition. The unusual feature of asphaltene precipitation is that it occurs at high pressure, unlike the other pure hydrocarbon solids, paraffin wax and naphthenes. Paraffin wax precipitates due to reduction in temperature and naphthene precipitates due to the interaction with above neutral pH formation brine. To provide a background to the physical and compositional parameters affecting asphaltene precipitation, their changes from reservoir to surface conditions need to be known. Crude oil density has been shown to be the dominating parameter in asphaltene precipitation.

The density of crude oil ρ_o at reservoir pressure above the bubble point, relates to the compressibility κ_o and formation volume factor B_o through

$$\kappa_o = -\frac{1}{V}\left(\frac{\partial V}{\partial p}\right)_T = -\frac{1}{B_o}\left(\frac{\partial B_o}{\partial p}\right)_T = \frac{1}{\rho_o}\left(\frac{\partial \rho_o}{\partial p}\right)_T$$

all at constant temperature (isothermal compressibility). The formation volume factor is the volume of oil plus dissolved gas at reservoir pressure, divided by the volume of oil at stock-tank conditions. Reservoir crude oil can be saturated with dissolved gas and the crude oil can be undersaturated. The same kind of consideration needs to be considered with respect to dissolved asphaltene; reservoir crude oil can be either saturated or undersaturated with asphaltenes.

The general features of the formation volume factor, crude oil (liquid) density and solution gas ratio R_s are shown in Figure 3.8. The solution gas ratio is the volume of gas at standard conditions divided by the stock-tank volume (degassed crude oil), the unit being m^3/m^3. The formation volume factor decreases with pressure from any pressure reservoir pressure to the bubble point pressure p_b. A typical formation volume factor would be 1.4–1.5 at reservoir conditions, decreases therefrom to the bubble point pressure; the reservoir crude oil volume expands on reduced pressure. Not shown in Figure 3.8 is the crude oil viscosity; it exhibits a similar qualitative behaviour as the crude oil density. The solution gas ratio remains constant above the bubble point pressure, thereafter decreasing to its value at standard conditions.

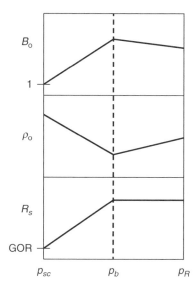

Figure 3.8 General features of crude oil properties/phases from reservoir pressure to bubble point pressure and thereafter to surface pressure at standard conditions (stock-tank).

Table 3.9 Typical division of reservoir crude oil in terms of gravity and gas-oil-ratio (Zuo *et al.*, 2013).

Crude Oil	°API	GOR (m³/m³)
Volatile	40–50	338–570
Black	25–40	20–267
Heavy	10–25	0–20

One reason for presenting the generalized properties of crude oil in Figure 3.8, is the relationship between density kg/m³ and specific volume m³/kg and therefore the molar volume m³/kmol. As an example, for a crude oil density of 880 kg/m³ and molecular weight of 245 kg/kmol (average values based on Table 3.2), the molar volume will be 245/880 = 0.278 m³/kmol. The molar volume is used in the *Flory-Huggins Solution Model* of asphaltenes.

As the name implies, gas bubbles begin to evolve from a crude oil when the bubble point pressure is reached. Below the bubble point pressure, the liquid-phase crude oil density increases as more and more lighter components are degassed to the gas phase. When the *in-situ* pressure has reached stock-tank standard conditions, the oil formation volume factor will have been reduced to one (=1). The crude oil density will have reached a value equivalent to its quoted API gravity. At standard conditions, the solution gas ration will have reached the quoted standard GOR value. For reference, it was stated by Zuo *et al.* (2013) the crude oils can be divided into volatile, black and heavy oils, having the API gravity and GOR ranges shown in Table 3.9. The divisions are different from the classification shown in Table X1.1.

Table 3.10 Comparison of properties of crude oils with and without asphaltene problems. Five North Sea crudes *without* problems and four North Sea and one Kuwait crude *with* severe problems (de Boer et al., 1995).

	Few/No Problems	Severe Problems
Crude Composition		
C1–C3	<27% mole	>37% mole
C7+	>59% mole	<46% mole
Asphaltenes	>3% wt.	<0.5% wt.
Stock-Tank Oil		
Saturates	≤62 % wt.	≥75 % wt.
Aromatics	≥26 % wt.	≤22 % wt.
Heavy ends	>11% wt.	<4% wt.
Asphaltenes	>3% wt.*	≤1 % wt.
Properties		
Bubblepoint	<6.2 MPa	>10 MPa
Reservoir pressure	<35 MPa	>40 MPa
Compressibility (R)	$<1.6 \times 10^{-9}$ Pa^{-1}	$>2.3 \times 10^{-9}$ Pa^{-1}
Compressibility (b)	$<1.0 \times 10^{-9}$ Pa^{-1}	$>1.2 \times 10^{-9}$ Pa^{-1}

*One North Sea crude with 0.3% wt. asphaltenes.

An interesting comparison of the properties of crude oils with and without asphaltene problems was presented by de Boer *et al.* (1995), shown in Table 3.10. The most striking feature is that the crude oils with *low* asphaltenes contents has severe deposition problems. The crude oils with problems were lighter and more compressible, both at reservoir (=R) and bubblepoint (=b) pressures.

3.7 SOLUTION MODEL

Quantifying the solubility of asphaltenes in crude oil has been a difficult task. The precipitation occurs at high pressure and the molecules tend to agglomerate and form clusters (see Figure 3.6). A break-through was made by Hirschberg *et al.* (1984) and subsequently by de Boer *et al.* (1995). The solubility of asphaltenes was formulated by using the *Flory-Huggins Theory* (also called Flory-Huggins Regular Solution Theory), widely used in polymer chemistry. In the petroleum engineering application, two phases are envisaged: an asphaltenes phase and a maltenes phase (the crude oil without asphaltenes). The theory is based on the mixing of the two phases/components. The dissolution of asphaltenes is the reverse process to precipitation.

Dissolution is expected to occur when the *Gibbs Free Energy* of mixing

$$\Delta G = \Delta H - T \Delta S$$

is negative (see *Appendix L2 – Regular Solution Model*), signifying an exothermic (gives energy) process. On the other hand, a positive ΔG signifies an endothermic (takes energy) process where precipitation occurs. In both cases, the change in *Gibbs Free Energy* becomes zero at equilibrium. For a particular reversible process, the *numerical* value of a negative ΔG value will equal to the positive value. The dissolution/solubility of asphaltene molecules is a *reversible* process (Hirschberg *et al.*, 1984) while the precipitation of asphaltene aggregates and clusters (particles) is *non-reversible* (Leontaritis and Mansoori, 1988; Ashoori, 2005). The vaporisation of asphaltene molecules may be an issue in some situations.

A seminal paper on asphaltenes solubility and precipitation is that of de Boer *et al.* (1995). The solubility/precipitation of asphaltenes c_a m^3/m^3 in crude was shown to depend primarily on pressure through crude oil density and therefore the molar volumes. The expression presented was

$$c_a = k_1 \exp\left[v_a \left(\frac{1}{v_o} - \frac{(\delta_a - \delta_o)^2}{RT} \right) - 1 \right]$$

The symbol v_a m^3/kmol stands for the molar volume of asphaltene as solute and v_a m^3/kmol the molar volume of the solvent oil. More accurately, the crude oil was assumed to consist of two components, the pure asphaltenes and the rest-of-the-oil, namely the maltenes. The deltas δ_a Pa$^{1/2}$ and δ_o Pa$^{1/2}$ are the *Hildebrand Solubility Parameters*. The unit used by de Boar *et al.* (1995) was Pa$^{1/2}$ while MPa$^{1/2}$ is more common (see Table 3.8). The RT term has the usual meaning, the gas constant times absolute temperature. The constant k_1 was called a correction terms and was assumed unity ($=1$). The work of de Boar *et al.* (1995) was unequivocally based on the ground-breaking work of Hirschberg *et al.* (1984), in particular the approximate *Flory-Huggins Theory*. The theory was independently derived by Paul J. Flory (1910–1985) and Maurice L. Huggins (1897–1981) in 1941 and published separately in 1942.

The *Hildebrand Solubility Parameter* (Hildebrand and Scott, 1962) is defined by the expression

$$\delta \equiv \left(\frac{\Delta H_{LV} - RT}{v} \right)^{1/2}$$

where ΔH_{LV} is the heat of vaporisation (from liquid L to vapour V). The solubility parameter is the square-root of the *cohesive energy density*. Checking the units

$$\frac{\Delta H}{v} \left(\frac{kJ}{kmol} \right) \left(\frac{kmol}{m^3} \right)$$

$$\frac{RT}{v} \left(\frac{kJ \cdot K}{kmol \cdot K} \right) \left(\frac{kmol}{m^3} \right)$$

gives kJ/m^3. The cohesive energy is the amount of energy needed to completely remove a unit volume of molecules from neighbouring molecules to infinite separation (an ideal gas). It is equivalent to the measurable heat of vaporization of a compound, divided by the molar volume of the condensed (liquid) phase.

In order for a material to dissolve, molecules must be separated from each other and surrounding solvent. The molecular interactions for dissolution are the same as must be overcome for vaporization. The idea behind the *Hildebrand Solubility Parameter* is that

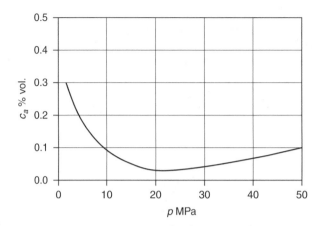

Figure 3.9 Calculated maximum solubility of asphaltene in light crude oil (40° API). Volume % of asphaltenes in oil against pressure MPa. Bubble point pressure at about 26 MPa.

the cohesive energy density kJ/m^3 is a numerical value representing solvency (ability to dissolve another compound). It has also been called the free energy parameter as mentioned in *Appendix L2 – Regular Solution Model*).

To calculate the solubility of asphaltene, using the simplified *Flory-Huggins Theory*, values are required for the molar volumes and the solubility parameters. The molar volume of crude oil can be found using a cubic EoS (Equation of State), similarly the heat of vaporization. The molar volume of asphaltenes must be given. Because of imperfections in the theory, the molar volume of asphaltene used by de Boer *et al.* (1995) was $1 \, m^3/kmol$, instead of the actual value (not specified). The solubility parameters were determined experimentally, based on measurements carried out at reservoir pressure p_R and at standard conditions p_{sc} for the same crude oil. The maximum solubility of asphaltene in a light crude oil ($40°API$, density $750 \, kg/m^3$) was calculated at pressures from reservoir pressure to pressures below the bubble point, shown in Figure 3.9. The reservoir temperature was constant and the below the bubble point the temperature, most likely, followed an adiabatic process (no heat added or removed). The concentration c_a decreased from reservoir pressure to bubble point pressure. It thereafter increased at pressures below the bubble point because the composition of the crude oil changed due to flashing, causing the liquid density to increase. The curve shown in Figure 3.9 has similar shape as the oil density ρ_o curve in Figure 3.8.

Although the *Flory-Huggins (Solution) Theory* has met with great success in the asphaltenes applications presented by Hirschberg *et al.* (1984) and de Boer *et al.* (1995), there is always room for improvements. Wang *et al.* (2004) presented an integrated approach to predict the precipitation of asphaltenes. The work was followed up by Wang *et al.* (2006), who used 27 crude oil samples to compared their improved method to that of de Boer *et al.* (1995). An important feature of the work of Wang *et al.* (2006) was the discovery of linear trends in solubility parameter values, that could be extrapolated from stock-tank oil values to reservoir values. One of the conclusions was that the original de Boer *et al.* (1995) method was too conservative, meaning that it predicted problems where the improved method did not, what the authors called false positives.

The *Flory-Huggins Theory* was applied to heavy oils and bitumens by Akbarzadeh *et al.* (2005). A monodispersed model was presented by Mohammadi and Richon (2007), the model being more general that previous models. The development of a specific equations of state for asphaltenes has been important in studies published by Zuo *et al.* (2012 and 2013). One of the results being the estimation of asphaltenes concentration with depth in reservoirs containing volatile, black and heavy oil. Asphaltenes gradients are corollary of other compositional gradients in oil reservoirs.

3.8 SATURATION LIMITS

The precipitation of asphaltenes can be directly correlated to the *in-situ* density of the crude oil. This fact/perspective culminated in the publication of de Boer *et al.* (1995). A simplified plot illustrating the precipitation limits of crude oils is shown in Figure 3.10 (adapted from Wang *et al.*, 2006). The plot (also called the *de Boer plot*) shows the pressure reduction from reservoir pressure p_R to bubble point pressure p_b expressed in MPa, against *in-situ* oil density ρ_o kg/m^3. The plot is based on the solution/solubility model presented above, assuming the crude oil was saturated with asphaltenes at reservoir conditions. Three asphaltene precipitation regions were identified in the plot: severe problems, slight problems and no problems (see Table 3.10 for details).

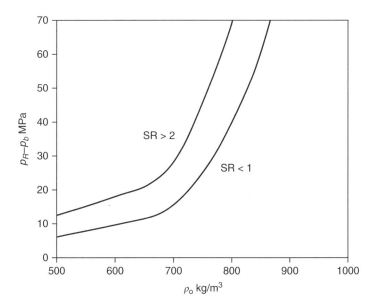

Figure 3.10 Pressure difference MPa (reservoir pressure minus bubble point pressure) against *in-situ* crude oil density kg/m^3.

Table 3.11 Asphaltene saturation ratio SR
at three solubility conditions.

Condition	SR
Undersaturated	<1
Equilibrium	=1
Supersaturated	>1

The precipitation limits between the three regions were based on the asphaltenes saturation ratio

$$SR = \frac{c}{c_b}$$

where the concentration (volume ratio m^3/m^3) at any pressure is referenced to the solubility at the bubblepoint pressure. The relevant SR's are shown in Table 3.11. The lower limit, between no problems and slight problems, was drawn for $SR = 1$, meaning at equilibrium (solubility limit). The upper limit, between slight problems and severe problems, was drawn for $SR = 2$, meaning that the asphaltene did not precipitate until its concentration was twice (double) the solubility limit. The intermediate band of precipitation limits drawn, is perhaps akin to the metastable zone illustrated in *Appendix V – Solubility and Nucleation*, well knowing that asphaltenes are non-crystalline (amorphous). The reservoir temperature used to construct the plot in Figure 3.10 was apparently 80°C (Wang *et al.*, 2006).

Measurements and prediction of asphaltene precipitation from live oils were reported by Edmonds *et al.* (1999). Based on eight crude oil samples, containing 0.1 to 8.5% wt. asphaltenes, it was concluded that when the saturation ratio $SR > 1.1$, precipitation/flocculation may occur in practice. The extend of the metastable zone illustrated in Figure 3.10, will likely depend on several factors. The factors are perhaps similar to the factors that affect normal crystallization, in the case of asphaltenes, precipitation. Turbulence, wall shear, high temperature and the presence of fines and other particles, may narrow the band of slight asphaltene problems.

In-situ density is markedly lower than stock-tank density (crude oil at reservoir conditions contains dissolved gas and the temperature is higher). A cubic Equation-of-State (EoS) can be used to find the *in-situ* density based on the composition of the stock-tank oil and the liberated gas. Of course, the whole gamut of semi-empirical relationship in the petroleum literature can also be used. Such use is beyond the intention of the present text. However, an examination of stock-tank oil densities may be useful. Stock-tank values in the range 796–945 kg/m³ were reported by Aske (2002) and shown in Table 3.1. Values in the range 841–916 kg/m³ were reported by Fan *et al.* (2002) and shown in Table 3.2. Stock-tank density values in the range 852–863 kg/m³ were reported by Goual and Firoozabadi (2002).

Light crudes contain more of light component than heavy crudes, illustrated in the SARA-plot Figure 3.3. The formation volume factor B_o gives the ratio of actual volumes to stock-tank volumes (see Figure 3.8). The factor is not known for the crude oils in Tables 3.1, 3.2 and 3.7. Using the *guestimate* that light crudes have an *in-situ*

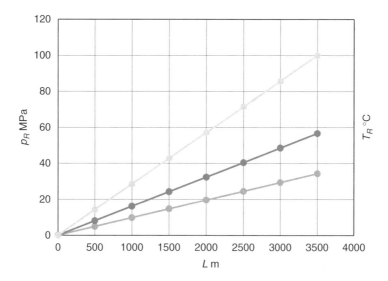

Figure 3.11 Idealized subsurface (reservoir) temperature (topmost), lithostatic (middle) and hydrostatic (bottom) pressures with depth.

density 20% less than stock-tank density and heavy crudes have an *in-situ* density 10% less than stock-tank density, the density ranges quoted above become 663–710 kg/m^3 for the light crudes and 785–859 kg/m^3 for the heavy crudes. Comparing these guestimate values to the limits in Figure 3.10, is appears that the light crudes may have slight-to-severe problem, depending on the reservoir pressure. The heavy crudes will not, unless pressure difference (p_R–p_b) is greater than about 50 MPa (500 bara), which is an unlikely scenario for conventional reservoirs. High-pressure, high-temperature (HPHT) reservoirs have temperature in the range 150–200°C and pressure in the range 100–150 MPa (see *Section 6.2 – Produced Water*).

Reservoir pressure and temperature increase with depth. Average pressure with depth follows roughly the basic equation

$$p_R = \bar{\rho} g D$$

where $\bar{\rho}$ is the average density, g the gravitational constant and D depth. The average density with depth can be that of liquid (combination of water/brine and hydrocarbons) or the overburden. Two pressure boundaries can be calculated, based on the hydrostatic gradient and the lithostatic gradient. Assuming an average density of 1000 kg/m^3 for the former and 1650 kg/m^3 for the latter, the density with depth were calculated and are shown in Figure 3.11. The middle curve is the lithostatic gradient and the lowest curve the hydrostatic gradient. Also shown is the temperature profile (the upper-most curve), assuming a geothermal gradient of 30°C/km. For a reservoir depth of 3000 m, the pressures are 30 MPa and 50 MPa, hydrostatic and lithostatic, respectively. Real reservoir pressures are somewhere between these two values. The reservoir temperature estimate was 90°C. On-land and off-shore pressures and temperatures need to be adjusted to the solid surface (sea bottom) conditions. The reservoir pressure will be higher in a sub-sea well than an equally deep on-land well. The starting temperature at the surface will be different for an on-land well and a sub-sea well.

3.9 PRECIPITATION ENVELOPE

The use of phase envelopes (pressure versus temperature) for hydrocarbon/petroleum fluid mixtures is well known in the oil and gas industry. Schematic phase diagrams are shown in Figure X1.3 for black oil, volatile oil, gas condensate and gas. Equations of State (EoS) are used to construct phase envelopes based on the mole fractions of a hydrocarbon fluid, composition examples being shown in Table X1.4. Inside a phase envelope, two phases exist, liquid and gas. The upper-most-side of a phase envelope is called a bubblepoint curve and the right-hand-side is called the dewpoint curve. It is beyond the scope of the present text to discuss the details of the construction of phase envelopes and the nature of the fluids involved.

Familiarity with hydrocarbon phase envelopes is advantageous when considering the precipitation of asphaltenes in crude oil. The advantage was used by Edmonds *et al.* (1999), who were perhaps the first to present a asphaltene phase envelope. The phase envelope was schematic, but it was based on detailed studies of eight crude oils using an in-house cubic EoS and field-based data and correlations. The crude oils studied contained from 0.1 to 8.5% wt. asphaltenes. A schematic representation of a asphaltene precipitation envelope is shown in Figure 3.12, along with a part of the traditional hydrocarbon phase envelope (bubblepoint curve).

Shown in Figure 3.12 is an isothermal line from above the precipitation envelope to below the envelope. An isothermal line is used because there is no change in crude oil temperature in reservoir liquid flow. However, at pressures below the bubble point curve, flashing occurs to release the lightest alkanes (gas), resulting in cooling (heating may occur in some hydrocarbon mixtures). The vertical isothermal line starts in the liquid phase A, at some reservoir pressure above the precipitation envelope. The line crosses the precipitation envelope at some pressure above the bubblepoint curve of the hydrocarbon phase envelope; this fact is one of the *most striking feature* of asphaltene precipitation. Further, the isothermal line crosses the bubblepoint curve (saturation curve) at B and leaves the asphaltene phase envelope and decreases further into the two-phase region to point C. The point to emphasize here is that asphaltene precipitation *occurs only* within the asphaltene phase envelope; it starts at the upper bound and stops at the lower bound. The asphaltene precipitation envelope is governed by pressure, temperature and composition. These three extensive properties, in turn, affect the molar volume and solubility parameter of the asphaltenes phase and the maltenes phase (crude oil minus asphaltenes). The square in Figure 3.12 mark the upper bound and the lower bound of the precipitation envelope.

Shown in Figure 3.13 is the concentration (solubility) of asphaltene in crude oil from reservoir pressure A to and through breadth the phase envelope E, down to the bubble point and up to the two-phase region C. The squares in the figure indicate the asphaltene precipitation envelope. How quickly the precipitation occurs and how much asphaltenes precipitate, is uncertain. Precipitation may depend on the rate of pressure change. Because the precipitation process involves large molecules that aggregate and form clusters (see Figure 3.6), the precipitation process is unlikely to be rapid.

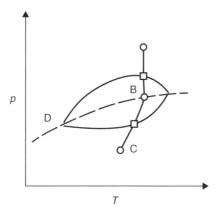

Figure 3.12 Asphaltene precipitation envelope and hydrocarbon bubblepoint line (D). Pressure shown to decrease isothermally from reservoir pressure A to bubblepoint pressure B and a lower pressure below the envelope, into the two-phase fluid region. Modified and redrawn after Edmonds *et al.* (1999).

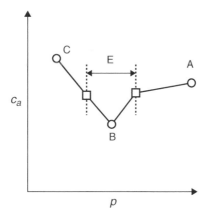

Figure 3.13 Asphaltene concentration/solubility in crude oil from reservoir pressure at A, to the precipitation envelope E (upper and lower bounds). The bubblepoint is shown as B and the flashed oil as C. Modified and redrawn from Edmonds *et al.* (1999).

The precipitation process is endothermic (takes energy), as shown in *Appendix L2 – Regular Solution Model*. An endothermic process taking place in parallel with crude oil flashing (below the bubblepoint) is unlikely to favour precipitation. Almost all of the experimental work on asphaltene precipitation has been carried out at quiescent laboratory conditions without gauging the energy balance.

Near-infrared (NIR) spectroscopy was use by Aske (2002) to measure the onset of asphaltene precipitation in a recombined (stock-tank oil and separated gas) crude oil. The crude oil contained 0.8% wt. asphaltenes. The recombined crude oil was pressurized to 30 MPa and thereinafter gradually depressurized. The isothermal depressurization is shown in Figure 3.15 (also in Sjöblom *et al.*, 2003). The temperature was 125°C and the starting pressure 30 MPa. The upper bound of the asphaltenes

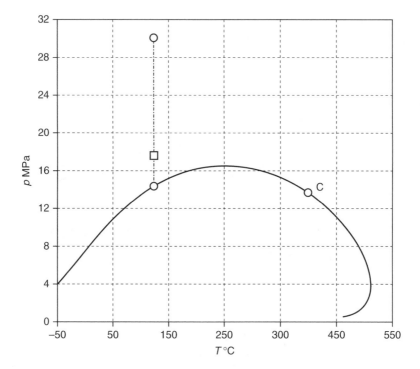

Figure 3.14 Phase envelope (pressure versus temperature) of recombined crude oil and isothermal depressurization from simulated reservoir pressure down to bubble point pressure (Aske, 2002; Sjöblom *et al.*, 2003).

precipitation envelope was reached at 18 MPa and the bubblepoint at 15.5 MPa. The phase envelope of the recombined hydrocarbon crude oil mixture is also shown in Figure 3.14, where C indicates the critical point. The molar composition was not made available/listed. The Aske (2002) example illustrates the same behaviour as illustrated in Figure 3.12 in pressure, from reservoir pressure down to bubblepoint pressure. Near-infrared spectroscopy was used by Khaleel *et al.* (2015) to identify the upper bound of the asphaltene precipitation envelope and the bubble point. It was pointed out that the method may not detect the first small sized asphaltene particles that form.

3.10   INSTABILITY LINE

The refractive index (RI) of hydrocarbons, including crude oil, can be used to determine a range of chemical and physical parameters. Asphaltene precipitation and solvent properties of crude oils were studied by Buckley *et al.* (1998). The behaviour and properties of asphaltenes in crude oils were studied by Panuganti (2013). The refractive index was essential in the two studies

$$\mathrm{RI} = n = \frac{c}{u}$$

The symbol c (not to be confused with concentration) stands for the speed of light in vacuum and u the speed of light in a particular medium (air, water, oil). The value of $n > 1$ because light travels faster in vacuum than in a particular medium. The refractive index determines how much a light beam is bent (refracted) when entering a material, as expressed by *Snell's Law*. The degree of refraction depends on the molecules of a medium, hence the ability of RI to detect asphaltenes and measure various crude oil properties. The wavelength of light depends on the refractive index through

$$\lambda = \frac{1}{n}\lambda_o$$

where the subscript o indicates the wavelength in vacuum. The question of wavelength (actual versus infinite) is important because it is used in the *Onsager Equation* (see *Appendix L1*) and various other relationships. The properties are referenced to infinite wavelength, at zero frequency.

It was convincingly demonstrated by Buckley *et al.* (1998) that a linear relationship existed between known variables involving the refractive index and frequency of light. The variables could be extrapolated to zero frequency. Furthermore, that the Lorentz-Lorenz refractive index variable

$$\frac{n^2 - 1}{n^2 + 1}$$

correlated linearly with n-alkane (n-heptane) and crude oil volume fractions over the whole range (0–1). The linear correlation makes possible the additivity of RI values of the components of a hydrocarbon mixture. The linearity arises because hydrocarbon molecules are non-polar (with the exception of the small fractions of asphaltenes, and perhaps resins). The value of the gradient (Lorentz-Lorenz index variable versus volume fraction) of the lines dependent on the crude oil used.

Eleven stock-tank crude oil samples were used by Buckley *et al.* (1998). The asphaltenes concentration in nine of the samples were reported. The RI of all the samples were reported. The weight percentage of asphaltene (n-heptane) in nine of the samples are plotted in Figure 3.15 against API gravity. The samples came from various oil fields in different sedimentary basins, so the exact relationship between the asphaltene % wt. versus API gravity is tenuous. The two parallel dashed lines indicate the lower and upper bounds of the plotted values. Buckley *et al.* (1998) did not find a correlation between the asphaltenes content and the API gravity.

Contrary to the tenuous correlation between % wt. asphaltene versus API gravity, the same was not true for the RI. The refractive index of the nine stock-tank crude oils are shown in Figure 3.16. The values are shown to fall on a curve not unlike a logarithmic behaviour, suggesting $\ln RI \propto API$ (proportionality). The outlier, well above the correlation line, did not have an associated % wt. asphaltene value reported. Based on five crude oil samples in another study, Bolou (2014) found that the % wt. asphaltenes correlated poorly with crude oil density while correlating reasonably well with the RI.

A methodology developed by Buckley *et al.* (1998) made it possible to calculate the refractive index of live oils. The method is based on knowing the RI of the stock-tank crude oil, the molar refraction of the separated gas, the formation volume factor B_o

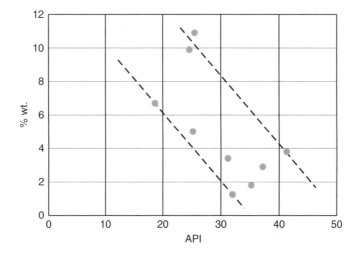

Figure 3.15 Asphaltenes % wt. (n-heptane) against API gravity in nine stock-tank crude oil samples from different fields/reservoirs.

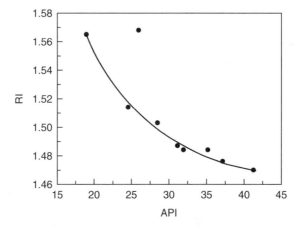

Figure 3.16 Refractive index versus API gravity of nine stock-tank crude oils (Buckley *et al.*, 1998).

and the solution gas/oil ratio R_s. The last two parameters are illustrated in Figure 3.8. The additivity of specific refractive indices was used in arriving at the RI of live crude oils. For details, consult Buckley *et al.* (1998).

The refractive index was measured for ten stock-tank crude oils, one more than shown in Figure 3.16. It was found that while the RI varied for the different crude oil samples, the RI at which asphaltenes precipitation occurred was remarkably constant at RI $\simeq$ 1.44. This value demarcates the horizontal *Instability Line*, shown in Figure 3.17. The measurements and calculations of Buckley *et al.* (1998) show that the RI decreases from reservoir pressure to the first appearance of precipitation (upper bound of asphaltene phase envelope) at approximately 53 MPa. The decrease in RI accelerates

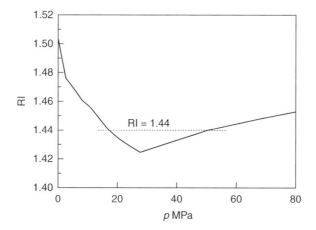

Figure 3.17 Typical refractive index upon lowering of crude oil pressure from reservoir to wellbore and surface conditions. Asphaltene precipitation envelope below the instability line. Modified and redrawn from Buckley *et al.* (1998).

towards the bubblepoint pressure of about 28 MPa. With lowering pressure, the RI increases as the density of the crude oil increases upon flashing (loss of gas). The lower bound of the asphaltene precipitation envelope occurs at approximately 18 MPa. At pressures below the instability line implies that the crude oil is supersaturated with asphaltenes. It was assumed by Buckley *et al.* (1998) that dispersive forces dominated the flocculation and aggregation of asphaltenes. As mentioned in *Appendix L2*, these (van der Waals) forces are the weakest intermolecular forces.

3.11 PROPERTIES IN MODEL

The *Flory-Huggins Theory* was derived for the dissolution of long-chain molecules (polymers) in solvents. In the case of asphaltenes in crude oil, the long-chain molecules have become large solute molecules dissolved in smaller solvent molecules. Apart from changes in entropy, there are two enthalpy terms involved in the dissolution/precipitation process: (1) enthalpy of mixing and (2) enthalpy of vaporization. See *Appendix L2 – Regular Solution Model* for details. The enthalpy of mixing is expressed by

$$\Delta H_M = RT \chi_1 n_1 \phi_2$$

where χ_1 is the *Flory-Huggins Interaction Parameter* (dimensionless). The other symbols represent the number of moles of the solvent (crude oil) and the volume fraction of the solute (asphaltenes). The interaction parameter is given by the expression

$$\chi_1 = \frac{v_1(\delta_2 - \delta_1)^2}{RT} = \chi = \frac{v_o(\delta_a - \delta_o)^2}{RT}$$

such that the solvent becomes the crude oil and the solute becomes asphaltenes. The molar volume of oil

$$v_o = \frac{M_o}{\rho_o}$$

refers strictly to the oil without the asphaltenes (molar volume of maltenes). Practically, it can be assumed equal to the molar volume of the whole oil, especially in low asphaltenes crude oils. The molar volume of asphaltenes, used in the de Boer *et al.* (1995) dissolution/precipitation equation, was given by the same kind of fundamental expression

$$v_a = \frac{M_a}{\rho_a}$$

The deltas are the *Hildebrand Solubility Parameters* of asphaltenes δ_a and the maltenes phase δ_o (Hildebrand and Scott, 1962). The Hildebrand parameter for any compound is defined by

$$\delta \equiv \sqrt{\frac{\Delta H_{LV} - RT}{v}}$$

where ΔH_{LV} is the heat of vaporization, from liquid to vapour (L to V). The solubility parameter unit kJ/m^3, shown above, is the same as kPa ($= kN/m^2$). A practical and common unit for solubility parameters is MPa.

The solubility parameters of pure hydrocarbons are well known and widely reported, some of which are shown in Table 3.8. The solubility parameter of crude oil can be estimated using a cubic EoS (Equation of State). The solubility parameter of asphaltenes is not directly measurable. Therefore, it is quite often used as an adjustable parameter, to match observed dissolution/precipitation to a particular *Flory-Huggins* model. The molar volume of asphaltenes was chosen 1000 kg/m^3 to match model to experimental data by de Boer *et al.* (1995). The solubility parameter of crude oil can be measured at standard conditions (stock-tank oil at standard conditions). Procedures/methodologies have been developed and tested to extrapolate ambient measurements to *in-situ* reservoir conditions (see for example, Wang *et al.*, 2006).

The solubility parameter of stock-tank crude oils can be evaluated from the following correlation (Buckley *et al.*, 1998; Wang *et al.*, 2006)

$$\delta_{sc} = 20.426 - 0.078 \times \text{API}$$

The solubility parameter of crude oil at the bubble point, based on pVT data and compositional data, was reported

$$\delta_b = \frac{1}{B_{ob}}\delta_{sc} + 2.904\left(1 - \frac{1}{B_{ob}}\right) + 3.91 \times 10^{-4}\frac{GOR_b}{B_{ob}}RI_m$$

where the formation volume factor (FVF) and gas oil ration (GOR) are the values at the bubblepoint pressure. The RI_m term is the *molar refractive index*, equivalent to

$$RI_m = RI \frac{M}{\rho}$$

determined from the *light mole fractions* of the hydrocarbon composition of a crude oil, from the correlation

$$RI_m = 23.0 - 15.9 \frac{x_1 + x_2 + x_3}{x_1 + \cdots + x_6}$$

To make possible a comparison of solubility parameters at stock-tank (standard conditions) and bubble point pressure conditions, an API gravity of 40, a formation volume factor of 1.5 and a GOR of 338 m³/m³ (from Table 3.9) were assumed. The light-end molar refraction index was approximated based on the molar composition of the black oil shown in Table X1.4. The numerator was taken as 0.5 and the denominator (sum of mole fractions from C_1 to C_6) was taken as 0.6, such that $RI_m = 9.8$ for use in the equation for the solubility parameter at bubble point pressure. The solubility parameters at stock-tank pressure (atmospheric) and bubble point pressure are shown in Figure 3.18. The value of the solubility parameter of the *in-situ* 40° API was 17.3 MPa$^{1/2}$. The figure shows that the solubility parameter at stock-tank conditions is higher than at *in-situ* (bubble point pressure) conditions. This agrees with the fact that asphaltenes precipitate, are less soluble, when mixed with normal alkanes. The figure shown that the solubility parameters degrease with increasing API gravity and hence decreasing crude oil density. The GOR in the correlation was specified at the bubble point pressure. This may be a misunderstanding because the GOR is by definition referenced to standard conditions. What the specification may mean is that the crude oil should be assumed saturated (not undersaturated nor without free gas).

The approximate version of the *Flory-Huggins Regular Solution* model presented by Hirschberg *et al.* (1984) and subsequently by de Boer *et al.* (1995) was

$$\ln c_a = \frac{v_a}{v_o} - \frac{v_a(\delta_a - \delta_o)^2}{RT} - 1$$

The concentration of asphaltene has the unit m³/m³, meaning volume ratio of asphaltene dissolved in crude oil (maltenes).

To compare the contribution of the two right-hand-terms in the above equation to solubility, Table 3.12 was constructed. The molar mass (molecular weight) of crude oil was based on the average in Table 3.2, despite being lower than the 300 kg/kmol used in Figure 3.7. The molar mass of asphaltene was the value most often quoted in the literature. The oil density was arbitrarily chosen, equivalent to about 35 degrees API gravity. The asphaltene density was chosen based on a value within the ranges quoted in the literature. The molar volumes were calculated from the values above in the table. The solubility parameters were chosen based on average values found in the literature.

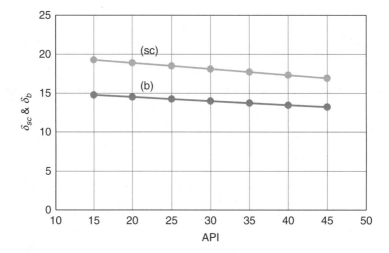

Figure 3.18 Solubility parameter against API gravity for stock-tank crude oil at standard conditions (upper line) and recombined oil at bubblepoint pressure (lower line).

Table 3.12 Selected average properties of solvent (crude oil) and solute (asphaltene).

Property	Solvent	Solute	Unit
M	240	750	kg/kmol
ρ	850	1200	kg/m^3
v	0.28	0.63	m^3/kmol
δ	14	20	MPa$^{1/2}$

The temperature chosen was $80°C = 353°K$ based on a reservoir depth 2500–3000 m in Figure 3.11. The term $RT = 2.935\,MJ/kmol$ such that $v_a/RT = (0.63/2.935) = 0.215\,(MPa)^{-1}$. The delta-difference square-term $(20-14)^2 = 36$. Overall, based on the numbers selected

$$\ln c_a = 2.25 - 0.215 \times 36 - 1 = 2.25 - 7.74 - 1 = -6.49$$

$$c_a = 1.5 \times 10^{-3}\frac{m^3}{m^3} = 0.15\%\,vol.$$

Based on the properties assumed, the crude oil at bubble point conditions contains 0.15% vol. asphaltenes, at chemical equilibrium assuming the reservoir crude oil was saturated with asphaltenes (SR = 1). Compared with the de Boer *et al.* (1995) graph in Figure 3.9, this value compares reasonably well. Considering the *numerical values*, the molar volume ratio contributes approximately 22.5%, the solubility parameter term 77.4% and the not-to-be-forgotten negative minus one (=−1) term, approximately 10%. Of course, these contributions are only illustrative for the assumed mix of parameters.

The dissolution/precipitation of asphaltenes is sensitive to the solubility parameter of crude oil. In the sample calculation above, the parameter was assumed to be $14\,MPa^{1/2}$. For illustration purposes, changing only the crude oil solubility parameter to $15\,MPa^{1/2}$ and to $13\,MPa^{1/2}$ the concentration of asphaltenes increases to 1.6% vol. and decreases to 0.01% vol., respectively. Therefore, the higher the crude oil (solvent) solubility parameter, the higher the equilibrium concentration of asphaltenes.

3.12 DEPOSIT BUILDUP

The asphaltene flow assurance challenge has two major parts: (1) estimating the equilibrium concentration (dissolution/precipitation) of asphaltenes in crude oils from reservoir to surface conditions and (2) estimating how much of the precipitated asphaltenes actually deposit and where, in reservoir formations, production tubing, wellhead equipment, flowlines and processing facilities. The first challenge (the concentration challenge) has been the focus of most studies in the literature. The second challenge (the deposition challenge) has hardly been addressed in the open literature. What is found in the literature are anecdotal comments based on field observations, lacking quantitative foundation. One such comment might be that not all precipitated asphaltene forms solid deposits on surfaces (sand grains in reservoirs and/or metal surfaces in production flowstreams).

Two groups of researchers have tackled the deposition challenge. Panuganti (2013) studied the deposition tendency of asphaltene on a quartz crystal microbalance. Asphaltene precipitated from crude oil using primarily n-pentane (and n-heptane), was mixed with toluene and flowed over the microbalance at very low flowrate. The flow was laminar; the Reynolds Number being of-the-order-of one (=1). The following effects were studied: bulk temperature, flowrate, n-pentane versus n-heptane and metal surface/substrate (gold and steel). The effect of bulk temperature is shown in Figure 3.19. The duration of the experiments was 15,000 seconds = 250 minutes, well over four hours. The deposit build-up increases with bulk temperature. The diffusion coefficient (diffusivity) D m^2/s obeys the following proportionality

$$D \propto \frac{T}{\mu}$$

where T is temperature and μ viscosity $mPa \cdot s$ (see *Appendix J – Diffusion Coefficient*). The deposition process, therefore, conforms with the nature of diffusivity, suggesting that advection effects are not important (convection is advection plus diffusions). A noteworthy feature of Figure 3.19 is that the initial rate of deposition is constant (same initial gradient). The shape of the curves suggests that the build-up can be described as being logarithmic as illustrated in Figure 2.10.

Measurements by Panuganti (2013) suggest that the initial rate of deposition increased with flowrate (may not hold in other situations). The shape of the deposit build-up curve was asymptotic (exponential) in nature, reaching an apparent constant

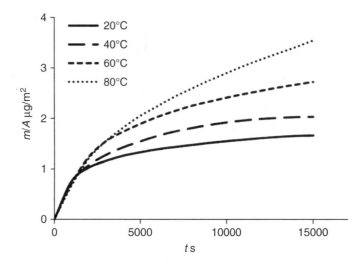

Figure 3.19 Mass of deposit per area with time and at four bulk temperatures. Redrawn from Panuganti (2013).

value at all high-rates. Therefore, the temporal effects of temperature and flowrate were different. The deposition rate of n-pentane asphaltenes was greater than the build-up rate of n-heptane asphaltenes. Deposition on carbon steel was greater than that on gold. Steel being magnetic and gold non-magnetic, may have contributed to the difference.

In a series of papers (Aslan and Firoozabadi, 2014a, 2014b; Hashmi *et al.*, 2015; Hashmi and Firoozabadi, 2016) asphaltenes deposition was studied in a small laboratory flow loop. Capillary tube 30 cm in length and 1 mm ID (internal diameter), was used to study the deposition of n-heptane asphaltenes. Controlled amounts of n-heptane were injected into a stock-tank oil sample to induce asphaltenes precipitation. The flow was laminar, having a Reynolds number of-the-order-of five. The experimental results were correlated using the Péclet number for mass transfer

$$\mathrm{Pe} = \frac{Lu}{D}$$

expressing the ratio of the rate of advection to the rate of diffusion. L stands for distance (here, the capillary tube) and u for flow velocity. In mass transfer, $\mathrm{Pe} = \mathrm{Re} \times \mathrm{Sc}$ where Sc is the Schmidt number (see *Section 2.5 – Convective Mass Transfer*). The authors listed above, modelled the deposition of asphaltenes as diffusion controlled. Equations were developed to quantify the deposition process, as was done by Panuganti (2013). The experiments and modelling are useful to explore how the various parameters affect asphaltene deposition. However, because all the work applies to laminar flow at quite low Reynolds numbers, the results cannot be applied to turbulent flow conditions. The results are perhaps best applied to Darcy-type flow in porous media.

The deposition-release modelling approach has been found suitable for buildup of deposits in turbulent pipe flow. The initial rate of deposition on a clean surface is greater than the subsequent buildup of solids on alike substrates. In a deposition-release model the build-up of solids is governed by continuous deposition and release mechanisms. If the deposit material is hard and shear-resistant, deposition will continue as long as there exists a driving force, internal or external. An internal driving force would be due to changes in bulk temperature, pressure and composition, as is the case in asphaltene deposition. An external driving force would be due to outside wall cooling of the flowstream, as is the case in paraffin wax deposition.

The build-up of deposits can follow several types of curves, presented in *Section 2.8 Deposition Models* and *Appendix K – Deposition-Release Models*. Three types of models have been presented: linear, exponential and logarithmic. The rate of deposition (deposit thickness) in these models can be expressed by the following relationships, respectively

$$\frac{dx}{dt} = k_1$$

$$\frac{dx}{dt} = k_1 - k_2 x$$

$$\frac{dx}{dt} = k_1 k_2^{-x}$$

The k's can represent a *parameter*, a *coefficient* or a *function*, depending of the deposits in question. Perhaps an s-shaped sigmoid curve should be added to the above list (needs to be derived), to follow the deposition shapes found by Hashmi and Firoozabadi (2016). The above deposition-release models can be used to correlate experimental data.

Asphaltenes precipitate due to changes in pressure, temperature and composition. The carrying crude oil becomes supersaturated ($1 < SR < 2$) and asphaltenes precipitate (Figure 3.10). The sizes of asphaltenes are commonly considered 1.5 nm for molecules, 2 nm for aggregates and 5 nm for clusters (Mullins *et al.*, 2012). Aggregates of asphaltenes depositing were reported 1–2 μm by Aslan and Firoozabadi (2014). For the present purposes, the nano-sized asphaltenes will be considered to behave as molecules, not as particles. The micro-sized asphaltenes, however, will be considered to behave as particles.

The effect of molecule/particle size on deposition from bulk to wall is presented in *Section 2.7 – Particle Mass Transfer*. Very *small particles* are transported by diffusion as molecules while *medium particles* are transported by own inertia and have a given stopping distance (see *Appendix I – Particle Stopping Distance*). And, *large particles* are transported independent of diffusion and own inertia and impact a wall at practically all flowing conditions. Whether particles stick to a wall (stickability) may be affected by the polar nature of the material in question. Because asphaltenes are strongly polar they may have a greater stickability than paraffin waxes that are much less polar in nature, for example.

An illustration of asphaltenes precipitation and deposition is shown in Figure 3.20. On-the-left, asphaltene particles are precipitated in the bulk fluid (crude oil). The

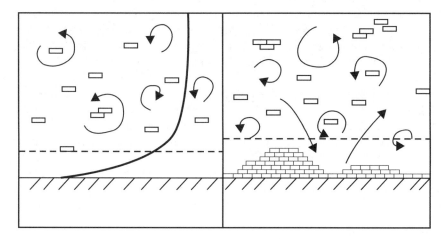

Figure 3.20 Illustration of precipitation (on-the-left) and deposition-release (on-the-right) of asphaltene aggregates and clusters. Shown are metal wall, viscous sublayer, velocity profile (on-the-left) and turbulent eddies. On-the-right, incoming and outcoming particles are shown.

circular arrows represent turbulent eddies and the sloping-to-vertical line is the velocity profile. The velocity profile increasing sharply within the viscous sublayer and marginally at greater distances from the wall. The dashed line just above the metal wall is the viscous sublayer (*Appendix H – Universal Velocity Profile*). The sublayer has the thickness $y^+ < 5$, making it possible to estimate its thickness based on the same properties as used in the Reynolds number. Assuming and oil density of $820 \, kg/m^3$, an oil viscosity of $5 \, mPa \cdot s$, the viscous sublayer thickness becomes approximately $8 \, \mu m$ and $16 \, \mu m$ for average turbulent flow velocities of $4 \, m/s$ and $2 \, m/s$, respectively. The sublayer thickness is several-times larger than the size of agglomerated asphaltenes particles, but of the same order-of-magnitude.

The illustration on-the-right in Figure 3.20 shows an asphaltenes deposit. Several circular turbulent eddies as shown to emphasize that the flow is turbulent. The thickness of the deposit is large enough to disturb the viscous sublayer such that a roughness element protrudes into the near-wall turbulent flow (see *Appendix F – Friction Factor of Structured Deposits*). Also shown in Figure 3.20, are the incoming and outcoming cluster/particles. The incoming is deposition and the outcoming is release. In the deposition-release model there is continuous deposition and release, analogous to chemical equilibria. When the particles have good stickability, rapid deposition is favourable; when the particles have poor stickability, the build-up with time will be less than favourable. The structure of the deposit may be porous and easily sheared off. The overall balance of deposition and release determines the deposit buildup with time. Competing transport forces include increased mass transfer (deposition) and increased wall shear stress (release) on increased flow velocity.

Asphaltenes are the most polar molecules found in petroleum. The aggregates and clusters of asphaltenes must likewise be polar. The polar nature of asphaltenes makes the molecules and particles subject to electrical double-layer phenomena, such as attraction and repulsion. The use of dispersants to prevent the agglomeration of asphaltenes is based on molecule/particle repulsion. Formation brine is produced simultaneously with crude oil, resulting in oil-continuous emulsion or water-continuous emulsion. Aspects of oil-water interfaces are discussed in *Section 7.6 – Interface Processes* and near the end of *Section 6.9 Carbonate Scale*. The deposit buildup of asphaltenes in the flowstream from reservoir to processing facilities is likely to depend on electrical double layer phenomena of oil-brine mixtures and flow channel surfaces (porous media, production tubing, pressure control (choke valves), flowlines and processing equipment).

The effect of water on the deposition and aggregate size of asphaltenes was reported by Aslan and Firoozabadi (2014). They found that the presence of water *delayed* the agglomeration and deposition of asphaltenes, but did not alter the size of the aggregates/clusters. The effect of water was related to the wetting and sticking behaviour of asphaltenes.

The flow of electrolytes (strong to weak) through tight spaces, such as porous media and capillary tubes, is known to induce a voltage difference over a distance, so-called *streaming potential*. In the shear plane between the liquid flowing and the solid surface, electrical changes can be segregated and set up an electrical potential perpendicular to the flow direction. Molecules, aggregates and clusters having a negative charge, will be attracted to a positively charged surface, and the reverse (positive charge particles attracted to negative charged surface). The hydroxylation of molecular/particulate surfaces may contribute to electrical potential effects. The electrical properties of oil-in-water and water-in-oil emulsions are commonly used in multiphase metering (Falcone *et al.*, 2010). The electrical conductivity and the electrical permittivity are dependent on the individual properties of the phases and their volume fractions in a particular emulsion. Therefore, both water-in-oil and oil-in-water emulsions have the potential to establish electrical charges that affect the deposition of asphaltenes solids in porous media and metal pipes and equipment.

The streaming potential be represented by the equation (Glover *et al.*, 2012)

$$\Delta E = \frac{\varepsilon \zeta}{\mu k} \Delta p$$

The potential has the unit V and the pressure drop the unit Pa. The coupling coefficient parameters are the electrical permittivity ε F/m of the flowing fluid, the zeta-potential ζ V is at the shear plane at the wall. The fluid viscosity μ Pa·s and electrical conductivity k S/m are properties of the fluid. In some texts the above equation is called the *Helmholtz-Smoluchowski Equation*.

Finally, consider what happens in porous media, the productive formation. The reservoir pressure p_R decreases with cumulative production, meaning with time. The greatest *in-situ* pressure drop is in the near-wellbore region, where the pressure decreases to p_{wf}, the well flowing pressure. It means that the greatest potential for asphaltenes deposition (bridging) in a productive formation (reservoir) is in the near-wellbore region. Considering asphaltenes deposition in flow channels in porous media, the principle the-best-go-first applies. Where crude oil flow preferably because of low pressure drop (most open channel), the deposition of asphaltenes will be the greatest.

The most open channels will become less open and eventually be blocked, the permeability will be impaired. Asphaltenes deposits in porous media may also reverse the wettability, from water-wet to oil-wet, for example (Kaminsky and Radke, 1997).

3.13 CONCLUDING REMARKS

Field developers want to know *whether* asphaltenes will form and *where* in the flowstream. Operators want to know how to deal with asphaltenes deposition in near-wellbore formations, production tubing and surface flowlines and equipment. Field developers and operators alike want to assess the *risk* and extend of asphaltenes flow assurance challenges. Engineers and scientist have worked hard to understand the chemical and thermodynamic factors/parameters that can lead to asphaltenes precipitation. However, what has not yet been elucidated in the literature, is the relationship between precipitation and deposition in operational environments. Experiments have been carried out at quiescent laboratory conditions, but long-term field testing has been absent/missing.

Dispersant chemicals can be applied/injected to halt the agglomeration of precipitated asphaltenes. The injection can be downhole and/or at the surface. Chemical and service companies offer proprietary compounds for this purpose. Testing reported in the open literature has been limited. Traditional and formulated hydrocarbon solvent have been used to dissolve asphaltenes deposits *in-situ*. Strong solvents are expensive. Acidic formulations attacking hydrogen bonding associated with heteroatoms and metals ions, may provide for a more practical approach to asphaltenes dissolution/removal. The prevention and removal of asphaltenes deposits continues as a major challenge for field developers and operators in the oil and gas industry.

Crude oil (petroleum) consists of a continuum of hydrocarbons from the lightest to the heaviest. Asphaltenes are also a continuum of the heavies and most polar fraction of crude oil. This detail is illustrated by how the different n-alkanes precipitate different amounts (% wt.) of asphaltenes. The corollary being that the heavies asphaltenes precipitate first when the pressure, temperature and composition change. Asphaltenes are described as molecules, agglomerates and clusters, each having a size distribution. The effect of time on the transition from a molecule to particles has not been widely addressed. The temporal aspect of precipitation and deposition relates directly to the mechanism of transport from bulk fluid to conduit wall (clean metal and/or solid deposit). The three transport regimes identified are *diffusion* (molecular and/or Brownian), *inertia* and *impact*, largely dependent on particle size and turbulent flow conditions.

Deposition-release modelling has proved useful in matching experimental data to semi-empirical relationships in a wide-range of industrial situations. Establishing the function form of deposit buildup in the laboratory, facilitates the monitoring and understanding of deposits found in field operations. Similarly, field observations can be matched with the appropriate deposition model, to make possible adjustments in operational conditions and to optimize remedial actions.

The classification of crude oils using SARA analysis has been a tradition in the oil industry. Notwithstanding variable results depending on laboratory procedures and different n-alkanes used, SARA analysis gives a semi-qualitative assessment of the risk of asphaltenes precipitation. However, numerous efforts to *quantify* the risk of asphaltenes precipitation have been disappointing, including the plotting of the different classification ratios. For the future, it would be useful to establish alternative repeatable standardized procedures, for example that include the size distribution of asphaltenes and resins. Other properties than size may also be relevant.

The effect of dissolved water (very small concentration) and emulsion/free water (from small to dominating) may be important in asphaltenes aggregation and clustering, and in deposit buildup. Formation water is always present in produced crude oil. The volume fraction of formation water/brine increases with time. In the early phase of field development, the produced crude oil will be an oil-continuous emulsion. In later phases of field development, the produced crude oil will be a water-continuous emulsion. Water will unlikely affect the solubility of asphaltenes in crude oils, but it will likely affect the behaviour of small and large asphaltene particles in both laminar and turbulent flow.

The major break-through in asphaltenes precipitation studies was the introduction of the approximate *Flory-Huggins Solution Theory*. Asphaltenes are considered the solute in a crude oil (maltenes) solvent. Apart from conventional parameters availed by cubic equations of state (EoS), the specific parameters of importance are the *solubility parameters* of the solute and solvent. In general, the solubility parameter of the solvent (crude oil) can be estimated from an EoS (based on composition of crude oil). The solubility parameter of the solute (asphaltenes) needs to be determined experimentally. In fact, the asphaltenes solubility parameter is used as a variable to match observed onset of precipitation in a particular crude oil (maltenes) at *in-situ* temperature and pressure.

Great strides have been made in improving the solute-solvent modelling of asphaltenes in crude oil. Extrapolations have been developed to estimate *in-situ* solubility parameters based on stock-tank measurements. Parameters such as crude oil refractive index have been added to the tool-box of asphaltenes (flow assurance) engineers and scientists. A new EoS has been developed specifically for asphaltenes. But the conundrum of molecule versus particle lives on in the analysis of asphaltenes precipitation.

The precipitation models are directly related to the saturation ratio of asphaltenes: undersaturated, equilibrium, supersaturated. When a solute-solvent model predicts precipitation, an asphaltene molecule transfers from solution to solid. The rate of the precipitation process is not known. The rate of deposition of the precipitated asphaltenes is neither known. Therefore, the solute-solvent models predict the *maximum* asphaltenes deposition potential. The situation is similar to what is the case for the other hydrocarbon deposits (paraffin wax and naphthenates) and inorganic scale. Flow assurance engineers and scientist may benefit from considering prospective parallels in the different flow assurance challenges. The whole field of study is multidisciplinary.

Chapter 4

Paraffin wax

The candle-like solids

Paraffin wax deposition is a serious problem of long standing in the oil and gas industry. When crude oil and condensate are cooled in a pipeline, paraffin wax precipitates when the temperature decreases to 30–40°C and below. Long-distance on-land pipelines can be insulated to prevent excessive cooling. The cooling of subsea pipelines is more severe than that of on-land pipelines. Heating of pipelines is practiced in some locations, but is expensive. Long-distance (50–150 km) subsea pipelines are usually bare (not insulated). Paraffin wax is usually allowed (!) to precipitate and deposit. The use of specialty chemicals is not common. The deposited wax is removed by mechanical scraping at appropriate time intervals. The frequency of scraping depends on the fluids and pipeline capacity. The acronym PIG (Pipeline Integrity Gauge) has been created, to avoid the use of the animal's name in some parts of the world. Chemical additives are used in some situations to prevent the agglomeration of precipitated wax. Paraffin wax particles are known to stabilize oil-water emulsions, making gas-oil-water separation in processing facilities difficult.

Early reviews of paraffin wax deposition are those of Reistle (1932), Shock *et al*. (1955) and Gudmundsson (1975). More recently, detailed reviews are those of Southgate (2004), Lee (2008), Handal (2008) and Kasumu (2014). The following theses on paraffin wax are directly relevant to the present chapter: Rosvold (2008), Botne (2012), Siljuberg (2012), Stubsjøen (2013) and Galta (2014).

The compositional thermodynamics of paraffin wax, the hydrodynamics (turbulent flow) with simultaneous heat and mass transfer, as well as deposit aging, were presented by Zhang *et al*. (2014). Laboratory and field data were used to adjust the fundamental and empirical models used in a proprietary software. The fundamentals presented in the paper resonate with many of the materials covered in *Chapter 2 – Basic Phenomena* and the associated appendices. What Zhang *et al*. (2014) call the shearing effect, corresponds to the deposition-release modelling approach presented in the mentioned chapter and appendices. Detailed experiments in a flow-loop using a real gas condensate, including measurements of entrapped oil in wax deposits, were reported by Hoffmann and Amundsen (2010). The basic heat and mass transfer relationships were evaluated. Both of the papers are useful in understanding the precipitation and deposition of paraffin wax in pipelines.

4.1 WAX IN OIL AND CONDENSATE

Paraffin wax precipitates when crude oil is cooled below 30–40°C. The much lighter gas condensates, will precipitate paraffin wax at considerably lower temperatures.

The temperature at which paraffin wax starts to precipitate, is called the wax appearance temperature (WAT). The term cloud point (temperature) is also used, based on a standardized measurement procedure of long standing. As crude oil and gas condensate are cooled further below the cloud point, more and more paraffin wax is precipitated. At some temperature, the hydrocarbon fluid paraffin wax mixture becomes a soft solid (called gel) and stops flowing. The precipitation and deposition of wax from gas condensate is more prevalent in situations where two liquid condensates of different compositions/origins are mixed.

The temperature at which gelling happen is called the pour point temperature (PPT), based on a standardized measurement procedure of long standing. Stationary (non-flowing) solids can increase in stiffness (shear strength) with time, making start-up of pipelines difficult if not impossible. Further cooling below the pour point will increase the gel stiffness.

It may seem contradictory that paraffin wax precipitates not only from crude oil, but also from gas condensate, consisting of relatively light hydrocarbons. Paraffin waxes are normal alkanes (straight chain) with the general formula C_nH_{2n+2} where the carbon number is typically in the range 20–40. Different authors report different ranges, often to higher carbon numbers. Pure (relatively) paraffin waxes are solids at ambient conditions. The paraffin wax that precipitates from crude oil and gas condensate, consists of a *distribution* of normal alkanes such that the heaviest molecules deposit first, thereafter lighter and lighter molecules. The exact composition of paraffin wax deposits depends not only on the crude oil composition, but also the cooling rate applied and the fluid flow conditions. Wax deposits are often found comingled with sand, fines, corrosion products and deposits of the other big-five solids.

Paraffin wax is the major constituent of solid deposits from crude oils. Two types of waxes are commonly encountered in crude oils: (1) macro-crystalline paraffin wax composed of mainly straight-chain n-alkanes ranging from C_{20} to C_{50} and (2) micro-crystalline waxes containing higher proportions of iso-paraffins (branched chain alkanes) and naphthenes (cyclic alkanes) with somewhat higher carbon numbers ranging from C_{30} to C_{60} (Tiwary and Mehrotra, 2004). Depending on the source of crude oil, the wax composition may range from predominantly n-alkanes of relatively low molar mass (molecular weight) to high proportion of heavier iso-alkanes and cyclic-alkanes. Typically, waxy deposits from crude oils consist of about 40–60% wt. paraffin waxes and less than 10% wt. microcrystalline waxes. Fresh wax deposits contain also a significant oil fraction. With time, the oil fraction decreases (due to diffusion) and the wax becomes harder (wax aging).

Measurements on *mixtures* of paraffin wax in *refined oil* were reported by Abdel-Waly (1997). The reported measurements serve as an example of paraffin wax in liquid hydrocarbons. The molecular weight of the refined oil was reported 447 kg/kmol and the API gravity of the oil 28 degrees. Compared to values reported in *Appendix M1 – Viscosity and Activation Energy*, there seems to be a mismatch between the gravity and molecular weight. In fact, the molecular weight estimated based on API gravity derived below, gives a molecular weight of 247 kg/kmol. Therefore, there is a misprint in the Abdel-Waly 1997) article; 447 should be 247. Gas-liquid chromatography analysis was used to determine the normal paraffin distribution of the wax, shown in Figure 4.1. The properties of the paraffin wax used (dissolved in the refined oil) were not reported beyond that shown in the figure. This in contrast to the work of Tiwary and Mehrotra

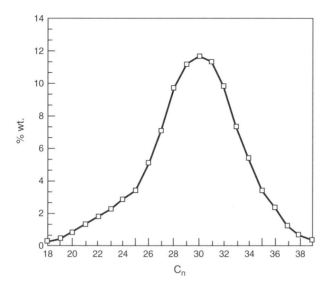

Figure 4.1 Distribution of paraffin wax dissolved in a refined oil (Abdel-Waly, 1997).

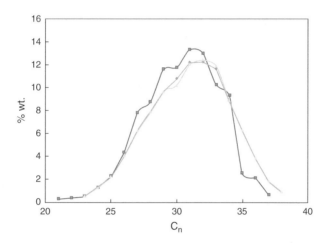

Figure 4.2 Weight of paraffin wax versus carbon number in three crude oil samples (Taiwo *et al.*, 2012).

(2004), who presented the compositional distribution of a 59–62°C melting point paraffin wax used in slurry viscosity measurements.

The weight distributions of paraffin wax in three crude oils (Niger Delta) were presented by Taiwo *et al.* (2012), shown in Figure 4.2. The specification of the three crude oils was not given, but it was stated that the sample densities ranged from 847–869 kg/M^2 and the API gravities were in the range 24.4–36.5 degrees. Based on Table 7.1, therefore, the crude oils ranged from medium to light quality. The mass

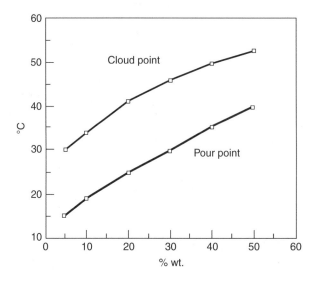

Figure 4.3 Cloud point (upper curve) and pour point (lower curve) temperatures versus % wt. of paraffin wax dissolved in refined oil (Abdel-Waly, 1997).

distribution of the paraffin waxes was remarkably similar, and within the carbon number range reported by others in the literature.

The *cloud point* and *pour point* of the wax-oil mixture were measured by Abdel-Waly (1997) at temperatures ranging from 5°C to 50°C, shown in Figure 4.3. The average pour point temperature was measured about 15°C *below* the average cloud point temperature. Standard methods were used to measure the cloud and pour points of the mixtures. It was reported by Taiwo *et al.* (2012), that the pour point temperature of Niger Delta crude oils was typically 10–20°C below the cloud point temperature. It is not uncommon to used mixtures of a hydrocarbon liquid and solid paraffin wax (dissolved in the liquid) to study deposition on cooled surfaces. One such example is that of Bott and Gudmundsson (1976) that reported heat exchanger deposition measurements of wax dissolved in kerosene.

Paraffin wax deposition flow-loop testing for gas-oil slug flow in a horizontal flow-loop, was reported by Rittirong *et al.* (2017) and Rittirong (2014). The oil phase was a API 42° gas condensate and the gas phase was commercial pipeline natural gas. The API gravity values was obtained from Sarica and Volk (2004), who gave the WAT values as 34°C, couple of degrees lower than given in Table 4.7. The experimental conditions are shown in Table 4.6 in *Section 4.7 – Flow Loop Observations*. The experimental work was detailed and included analysis of the normal alkanes in the liquid gas condensate phase and in the solid paraffin wax deposited. The % wt. of the alkanes from C_{13} to C_{54} are shown in Figure 4.4 (log-linear scale), for the gas condensate and for the paraffin wax deposits. The wax deposits contained normal alkanes from C_{30} with a maximum at C_{42}, decreasing to the maximum measured. The liquid gas condensate in question has been widely used in flow-loop experiments and analysis, including that reported by Zhang *et al.* (2014).

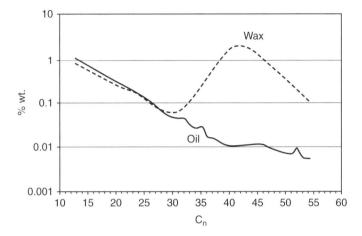

Figure 4.4 Log-linear weight percentage (% wt.) versus normal alkanes in gas condensate (oil line) and paraffin wax deposits (Rittirong *et al.*, 2017).

The amount of paraffin wax that precipitates from crude oils and gas condensates depends on the degree of cooling below the wax appearance temperature. The amount of paraffin wax that actually deposits will be less than that precipitated, in most situations. An equation for the calculation of bulk temperature in pipelines subject to cooling is presented in *Appendix A – Temperature in Pipelines*. Based on cooling from typical wellhead temperatures of 60–80°C to ambient temperature, the amount of wax precipitating (reduced concentration in flowing oil) will typically be in the range 5–15% wt. for different crude oils and gas condensates. These % wt. values are based on waxy crude oils that exhibit operational problems in pipelines. Not all crude oils and gas condensates exhibit paraffin wax problems.

4.2   PRECIPITATION CURVES

Two typical *precipitation curves* are sketched in Figure 4.5, showing the mass percentage precipitated (inverse of solubility curve) against temperature, starting at the WAT. The temperatures and % wt. shown as precipitate, are based on field values presented by Taiwo *et al.* (2012). The properties of the two crude oils were not available, only that they were typical Niger Delta waxy crudes. The WAT range from 30–40°C. The PPT is commonly 10–20°C lower than the WAT, measured 15°C lower in Figure 4.3, for example.

The exact shape of a particular precipitation curve depends on the chemical composition/make-up of the wax and crude oil. The curves can be straight or curved (concave or convex). Precipitation curves can be measured and/or estimated by modelling. Different authors have used different measurement methods and different modelling approaches. The precipitation curve close to the cloud point temperature is often relatively flat, making it difficult to determine the exact WAT. The different

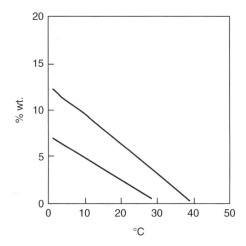

Figure 4.5 Schematic precipitation curves for two waxy crude oils. Wax precipitates <40°C in the heavier crude and <30°C in the lighter crude.

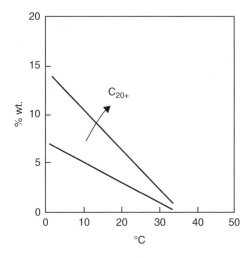

Figure 4.6 Schematic precipitation curves for crude oils with different molecular weight distributions. The upper curve contained significantly more C_{20+} than the lower curve.

measurement methods can give different results, commonly affected by the cooling rate and stirring (in pipes, turbulence), for example.

Eight crude oils were used by Lira-Galeana *et al.* (1996) to model precipitation curves. Two of the crude oils had much higher C_{20+} content and exhibited different curves compared to the other six crude oils. The difference is schematically shown in Figure 4.6. The WAT of most of the crude oils was about 35°C, with a range of about ±5°C. The difference in the two schematic curves exhibits how different molecular distributions can affect the precipitation curves (crude oils from different fields). The upper curve contained significantly more C_{20+} than the lower curve.

Table 4.1   Chemical formula, molar mass, melting point and density of three pure
paraffin waxes (Griesbaum et al., 2012).

Paraffin	Formula	M kg/kmol	T_m °C	ρ kg/m³
Icosane	$C_{20}H_{42}$	282	36.4	789
Triacontane	$C_{30}H_{62}$	422	65.8	810
Tetracontane	$C_{40}H_{82}$	562	81.5	820

Table 4.2   Solubility parameters MPa$^{1/2}$ of three normal alkanes versus temperature.

Temperature	n-Pentane	n-Hexane	n-Heptane
0°C	14.95	15.46	15.80
25°C	14.37	14.88	15.20
45°C	13.91	14.41	14.74
60°C		14.07	14.39
Formula	C_5H_{12}	C_6H_{14}	C_7H_{16}
Boiling point	36°C	69°C	98°C
Weight	72 g/mol	86 g/mol	100 g/mol

Upon cooling, the heaviest paraffin waxes precipitate first, because they are the least soluble in crude oils and condensates. On successive cooling, the medium and thereafter lighter paraffin waxes precipitate. Phrased differently, long-chained paraffins precipitate at higher temperature than medium-chain paraffins and thereafter short-chain paraffins. A relevant implication of the heavy-medium-light effect, is that the paraffin wax density changes with distance in subsea pipelines, for example. The physical properties of three pure paraffin waxes are shown in Table 4.1. The waxes span the carbon numbers common for deposits from crude oils, from C_{20} to C_{40}. The melting temperature and the pure wax density increase with carbon number.

The solubility of normal paraffins in crude oils can be modelled using the *Flory-Huggins Theory*, including the *Hildebrand Solubility Parameter*, as presented in *Appendix L – Regular Solution Model*. The solubility parameter of alkanes increases with the carbon number and decreases with temperature. Heavy paraffin wax molecules have a larger solubility parameter than light paraffin wax molecules. The solubility parameters for pure n-alkanes (pentane, hexane, heptane) are shown in Table 4.2. The parameters were obtained from the definition

$$\delta = \left(\frac{\Delta H_{vap} - RT}{v} \right)^{1/2}$$

where the enthalpy term is the heat of vaporisation, the RT term is the gas constant and absolute temperature, the whole divided by the molar volume. Because the solubility parameter is used as a difference, it needs to be accurate within ± 0.05 MPa$^{1/2}$ (Akbarzadeh et al., 2005). The details are shown in *Appendix L – Regular Solution*

Model. The solubility parameter of a mixture can be estimated by a summation of the individual values times their volume fraction

$$\delta_M = \sum \delta_i \phi_i$$

The *Regular Solution Model* can be expressed by the *simplified* expression

$$\ln(x) = \frac{-M_w}{\rho_w RT}(\delta_w - \delta_o)^2$$

where the activity coefficient of the solid has been assumed unit (=1), and the number of solvent (oil) volumes also assumed unity. The subscripts stand for wax and oil and the *x* is the mole fraction of wax (Schabron *et al.*, 1999; Schabron *et al.*, 2010). The simplified expression is presented here to illustrate the use of solubility parameters in estimating the solubility of paraffin wax in crude oil. The expression is directly related to the model presented in *Section 3.7 Solution Model*, for asphaltene solubility (Hirschberg *et al.*, 1984; de Boar *et al.*, 1995).

A great number of papers in the literature deal with the solubility of paraffin wax in crude oil and condensate, including Coto *et al.* (2008a), Coto *et al.* (2008b), Martos *et al.* (2008), Coto *et al.* (2010). Furthermore, Roenningsen *et al.* (1991), Pedersen *et al.* (1991), Hansen *et al.* (1991) and Pedersen *et al.* (1991), have presented review papers on wax precipitation from North Sea crude oils. Edmonds *et al.* (2008) and Coutinho *et al.* (2006), have presented modelling of paraffin wax deposition. Additional papers are those of Hoffmann and Amundsen (2010) and Han *et al.* (2010). The gelling (gelation) of paraffin wax has been reported by Paso *et al.* (2005, 2009).

4.3 VISCOSITY OF OIL-WAX SLURRY

Paraffin wax causes two types of operational problems in pipelines. *The first*, is the increase in bulk viscosity due to wax precipitation. *The second*, is the wall deposition of solid wax, reducing pipeline internal diameter. The various physical (fluid flow) effects involved are presented in *Chapter 2 – Basic Phenomena* and in several appendices. Considerations of the viscosity of liquids are presented in *Appendix M1 – Viscosity and Activation Energy*. The pressure drop in pipelines depends on the inverse of the internal diameter to the power five (!), exerting a significant detrimental effect. Separating the *viscosity effect* and the *deposition effect* in waxy crude pipeline flow is not a trivial task.

An increase in bulk viscosity lowers the Reynolds number and increases the friction factor. The flowing temperature in pipelines decreases exponentially with distance, due to external cooling, especially in subsea pipelines. The viscosity of crude oils increases with increasing cooling (lower bulk temperature). The bulk viscosity of paraffinic crudes increases markedly when cooled below the wax appearance temperature (WAT), when measured in a conventional viscometer (laminar flow shear rate). Experience with hydrate particles in water and diesel oil at *turbulent flow* conditions (tube viscometer), shows that the pressure drop in slurry flow is the same as in particle-free flow (Andersson, 1999; Andersson and Gudmundsson, 1999). Some of the details are presented in *Section 5.8 – Prevention by Cold Flow*. Whether the same effect applies to wax particles in crude oils is not known. Nevertheless, it seems plausible that the same fluid flow conditions may exist, especially for very small paraffin wax crystals/particles.

The *Eyring Equation* is commonly used to correlate liquid viscosity data, as presented in *Appendix M1 – Viscosity and Activation Energy*. The natural logarithm form of the equation

$$\ln(\mu) = \ln(A) + \left(\frac{E_a}{R}\right)\frac{1}{T}$$

expresses the dynamic viscosity μ in terms of a pre-exponential constant A, an activation energy of viscous flow E_a, the gas constant R and the absolute temperature T. The *Eyring Equation* difference from the *Arrhenius Equation* in that the exponential term is positive, not negative.

The *viscosity* of n-paraffin in refined oil mixture was measured from 70°C and down to 5°C (Abdel-Waly, 1997). The n-paraffin concentrations shown in Figure 4.3 were used to find/obtain the following empirical correlation

$$\ln(\mu) = 1.4280 - 8.7446\left(\frac{1+w}{T}\right) + 2269.3\left(\frac{1+w}{T}\right)^2$$

The units express dynamic mixture viscosity μ mPa·s, wax mass fraction w and temperature T °C. The parameter w is not the % wt. wax content in Figures 4.1 and 4.3. The correlation can be used to estimate the effect of n-paraffin wax on the viscosity of refined oil (and condensate and light crude oil) in pipelines. The correlation has the same form as the *Eyring Equation*, but with an added inverse T-squared term, making it a typical polynomial (non-linear regression). The literature shows that the viscosity of paraffinic crudes depends on the rate of shear applied, expressed by $\gamma = u/y$ where the u is velocity and y a distance (gives the velocity gradient). It follows that the unit of shear rate is 1/s. The shear stress established in a liquid in motion, is commonly expressed by $\tau = F/A$, where F is the force applied to an area A. For a Newtonian liquid, dynamic viscosity is given by $\mu = \tau/\gamma$ such that the unit becomes Pa·s. In Newtonian liquids, the dynamic viscosity increases linearly with the shear stress over shear rate. Liquid exhibiting non-Newtonian characteristics exhibit a non-linear behaviour.

The above viscosity correlation (Abdel-Waly, 1997), was based on a shear rate of 300 1/s. The measurements were made in a rotating cylinder viscometer. For a weight fraction of 0.15 (=15% wt.) at 25°C, the mixture viscosity is calculated 30 mPa·s (=cp). For a weight fraction of 0.0 (no wax dissolved) at 25°C, the correlation gives a viscosity of 110 mPa·s. This oil-value is higher than can be read from a diagram in the Abdel-Waly (1997) paper. For a pure oil at 25°C the viscosity should be closer to about 10 mPa·s, as borne out by Figure M1.1. It was concluded that the empirical viscosity correlation cannot be used for pure hydrocarbon liquids, just for liquids containing wax particles. The mixture viscosity estimated from the correlation, seems reasonable, as borne out by Figure M1.1.

The frictional *pressure gradient* was calculated for the pure oil and the oil-wax mixture. Average velocity in pipelines is commonly in the range 2–4 m/s. A value of 3 m/s was selected. The temperature selected was 25°C and the pipeline internal

Table 4.3 Comparison of calculated pressure gradient Pa/m in a pipeline carrying crude oil and crude oil/wax mixture.

Wax % wt.	μ mPa·s	Re	f	$\Delta p/\Delta L$
0	11	3056	0.0425	1340
15	340	1000	0.0641	2021

diameter (ID) 0.127 m (5 inch). The calculation results are shown in Table 4.3. The Reynolds number in the pipeline carrying crude oil was 3056 (due to low viscosity) and turbulent, compared to a Reynolds number of 940 and laminar, in the oil-wax mixture pipeline. The friction factors were estimated from the *Blasius Equation* for turbulent flow (assuming smooth pipe flow) and the *Poiseuille Equation* for laminar flow. The density of paraffin wax is not very different for that of crude oil. An API 28 gravity oil will have a density of about 890 kg/m^3. This density was used in the oil-wax calculations.

The pressure gradient for laminar flow (oil-wax mixture) is about twice that for turbulent flow (pure oil). Although the numbers calculated/estimated are rather approximate, they suggest that the pressure drop in pipelines carrying oil-wax mixtures will be much higher than in pipelines carrying pure oil. A caveat to this result may be that it is not correct. The reason being, that experiments in a tube-viscometer with hydrate solids in water and diesel fuel, have shown that the pressure drop in turbulent flow is not affected by the presence of small-size solids. The particles are envisaged to move away from the wall by the *Saffman Lift Force* (Philip G. Saffman, 1931–2008). The viscous sub-layer controlling pressure drop, is therefore practically empty for particles; the pure fluid viscosity controlling the pressure drop. See details in *Section 5.8 – Prevention by Cold Flow*.

The *Saffman Lift Force* is due to the pressure distribution developed on a particle, because of the rotation induced by the velocity gradient in a viscous sublayer. The higher velocity at the upper (top) side of a particle in a sublayer, and the lower velocity at the lower (bottom) side, give rise to the lift force (Saffman, 1965). Particle rotation will also contribute to the lift through the Magnus Force (Heinrich G. Magnus, 1802–1870), also called the *Magnus Effect*. The upper side faces the boundary between the viscous sublayer and the turbulent core, and the lower side faces the solid pipe wall (metal when clean surface, wax deposit with time). The *Saffman Lift Force* (Zheng and Silber-Li, 2009) can be expressed as

$$F_{lift} = 1.615 \mu \Delta u r^2 \left(\frac{\gamma}{\eta}\right)^{0.5}$$

where the variables are viscosity μ, particle radius r, shear rate γ and kinematic viscosity η. The lift force is greater in liquids with large viscosity, hence greater in crude oil than water. The variable Δu is the relative velocity of the particle, in the longitudinal direction. The assumption being that the viscous sublayer velocity is greater than the particle velocity. The details have not been considered in the present text, but the slip velocity can perhaps be quantified by *Stokes Law*, where the drag force on a particle can be expressed by

$$F_{drag} = 6\pi\mu r \Delta u$$

The symbols have the same meaning as above. In the original Saffman (1965) paper, there is a misprint that has propagated into some of the subsequent literature. In the original paper, the constant had the value 81.2 whereas it should have been 1.615 (Siljuberg, 2012). Even the paper by Zheng and Silber-Li (2009), has the wrong value of the constant (perhaps the value depends on the units used). Nevertheless, they studied 0.2 μm diameter polystyrene particles in water, measuring the particle distribution at 0.25–2.0 μm away from a wall in a microchannel. Zheng and Silber-Li (2009) found that the lift force was dominant in the range $2 < z^+ < 6$. The dimensionless $z^+ = z/d$ where z is the distance away from the wall and d the particle diameter. The particle sizes are similar to sizes found in paraffin wax and in gas hydrate. The thickness of the viscous sublayer is approximately 100 μm at turbulent flow conditions (see *Appendix C – Boundary Layer Temperature Profile*).

The Abdel-Waly (1997) slurry viscosity correlation, is but one of many published for paraffin wax in oil/crude oil. The phase transformation and rheological behaviour of highly-paraffinic waxy mixtures was studied by Tiwary and Mehrotra (2004). Two model oils, n-dodecane C_{12} and n-hexadecane C_{16} were used as solvents. The solute paraffin wax was specified with a melting temperature in the range 59–63°C and an average molecular weight of C_{30}. Viscometer tests were carried out for three compositions, 2 cooling rates and three shear rates (5–30 1/s). The literature on shear rate, composition and time-dependent rheological response was reviewed. The effect of shear history and cooling were reported as significant. Wax concentrations from about 5% wt. to about 30% wt. were tested. The viscometer testing resulted in the following first-order polynomial

$$\ln(\mu) = A + B\frac{1}{T} + \gamma\frac{1}{C^{0.5}} + Dw$$

A regression analysis gave the apparent viscosity of oil-wax mixtures in terms four model parameter, A, B, C and D. The variables are absolute temperature T, the shear rate γ and mass fraction (not % wt.) of wax w.

More oil-wax viscosity correlations are available in the literature. A correlation involving four model parameters, one exponent and three variables, was reported by Al-Zahranig and Al-Fariss (1998). They used an API 28° crude oil in the experiments.

4.4 THERMODYNAMICS OF PRECIPITATION

Vapour-Liquid-Solid (VLS) equilibria/thermodynamics are used to estimate the amount of paraffin wax dissolved in hydrocarbon liquids. Precipitation is the opposite of dissolution. VLS equilibria involve the various concepts used in chemical thermodynamics and physical properties. Commercial VLS softwares are based on an equation of state

(EOS), and a range of correlations to estimated PVT (pressure-volume-temperature) properties. The volume is the molar volume. It is beyond the scope of the current text to present the details of an/any EOS and associated correlations for real petroleum fluids. However, it may be useful to present some of the concepts in the literature, to smooth the transition from the current chapter/text, to a better understand and use of the available tools and software.

The thermodynamics of wax precipitation in petroleum mixtures was presented by Lira-Galeana *et al.* (1996). The framework presented is an example of the fundamentals required to model the equilibria between the vapour, liquid and solid phases in hydrocarbon mixtures. The formulations used by Lira-Galeana *et al.* (1996), will serve as scaffolding for the following text.

At equilibrium (see *Appendix S – Gibbs Free Energy* and *Appendix T – Chemical Potential*), the chemical potential of component i will be equal in the three phases

$$\mu_i^V = \mu_i^L = \mu_i^S$$

where the superscripts stand for vapour, liquid and solid. As an example, the chemical potential of an ideal gas is given by the relationship

$$\mu = \mu^o + RT \ln\left(\frac{p}{p^o}\right)$$

where the superscript o signifies a reference state. As the system pressure approaches the reference pressure, the natural logarithmic terms will be smaller and smaller. When $p = p_o$, the logarithmic term becomes zero. At even lower pressures, the natural logarithmic term becomes negative, and at very low pressure, its value approaches negative infinity. To avoid difficulties in equilibrium calculations for real fluids, the concept of *fugacity* was proposed by Lewis (1901). For real solutions

$$\mu = \mu^o + RT \ln\left(\frac{f}{f^o}\right)$$

The advantage of using fugacity, is that the reference state can be chosen fairly arbitrarily. In practice, convenient values can be chosen, depending on the equilibrium situation (vapour, liquid, solid). Although the unit of fugacity is pressure, it is not actually a pressure, but a useful thermodynamic variable.

Analogous to chemical potential, the fugacity of a component i in a vapour, liquid, solid system, will exhibit equilibrium when

$$f_i^V = f_i^L = f_i^S$$

Fugacity is a measure of the tendency of a component to leave a phase. A noteworthy advantage of using fugacity (a concept defined for partial pressure in gases) is that is can also for liquids and solids. A molecule/component in the gas-phase, can also be found in the liquid-phase and even the solid-phase. In principle, when temperature and pressure are changed, the component phase with the lowest fugacity, will be the most favoured (lowest *Gibbs Free Energy*). A schematic figure of the vapour-liquid-solid phases found in hydrocarbon fluid systems is shown in Figure 4.7.

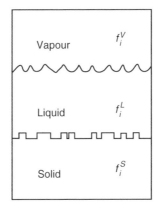

Figure 4.7 Schematic figure of vapour-liquid-solid phases and corresponding fugacity of a molecule/component in hydrocarbon equilibrium systems.

At this juncture, it may be useful to state two coefficients used in chemical equilibrium relationships: (1) activity coefficient and (2) fugacity coefficient (see *Appendix Q – Chemical Equilibrium*). The *activity coefficient* gives the effective concentration of a chemical component in a mixture (correction for molecular interactions), because it is affected by the surrounding molecules. The activity coefficient is expressed as

$$\gamma_i = \frac{a_i}{m_i}$$

The molal concentration of the component is m_i while a_i is the chemical activity of the component/molecule. The concentration can be expressed using other units. The numerical value of the activity coefficient will be different depending on what concentration measure is used. The *fugacity coefficient* gives the effective partial pressure of a chemical component in a gas mixture, expressed as

$$\phi_i = \frac{f_i}{p_i}$$

The fugacity expresses the deviation from ideal gas behaviour. Equations of state (EOS), with all their constants, coefficients and factors, are used to calculate/estimate the activity coefficient and fugacity coefficient in real hydrocarbon systems. Also required are the critical temperature and pressure of the constituent components, as well as their molar volumes.

The fugacity of a component in the *vapour phase* can be expressed by

$$f_i^V = \phi_i^V y_i p$$

where y_i is the mole fraction of component i. The system pressure is given by p such that $y_i p$ is the partial pressure. In the literature, the fugacity of a component in a mixture, is sometimes signified with the symbol circumflex, for example $\hat{f}$. The symbol is commonly called a *hat*, the same as the human head-wear. It will not be used in the present text, to avoid cluttered symbols.

The fugacity of a component in the *liquid phase* can be expressed by

$$f_i^L = \phi_i^L x_i^L p = \gamma_i^L x_i^L f_{i,pure}^L$$

and in the *solid phase*

$$f_i^S = \gamma_i^S x_i^S f_{i,pure}^S$$

following Lira-Galeana *et al.* (1996). The same kind of expressions are found throughout the chemical equilibrium literature. The subscript *pure* indicates the fugacity of the pure component at the system temperature and pressure. The subscript *pure* indicates *Raoult's Law*, where the partial pressure is given by the product of mole fraction and the saturation pressure of the pure component (see *Appendix Q1 – Solubility of Gases in Water*). The saturation pressure of a pure component can also be written as p^* in the literature. For pure components, both x_i and y_i are both equal to one.

Ideal models can be used for dilute fluids and when molecular interactions are negligible; when molecules of similar size and character are mixed together. When dealing with systems involving materials such as water, organic acids, amines and alcohols (hydrate inhibition, naphthenate formation), there will likely be deviations from ideality. An equation of state (EOS) model provides the PVT relations used to predict thermodynamic properties. There are several types of EOS used in the oil and gas industry.

In hydrocarbon systems, the liquid-phase activity coefficient γ_i^L is generally greater than one. Therefore, the fugacity of a component in a mixture will be higher than that of the pure component. It means that the same component will have a higher tendency to vaporize from a mixture that from its pure state. The reason being that there will be increased repulsion between molecules in mixtures.

An EOS can be used to describe the *vapour* phase. The *liquid* phase can either be described by an activity coefficient model or by an EOS. The *solid* solution is often described by an activity coefficient model. For vapour-liquid equilibria, it is common practice to use K factors, defined by

$$K_i = \frac{y_i}{x_i^L}$$

where y_i is the mole fraction in the vapour phase and x_i^L the mole fraction in the liquid phase. K factors are also called partition coefficients. It can be shown (Lira-Galeana *et al.*, 1996) that

$$K_i = \frac{\phi_i^L}{\phi_i^V}$$

The fugacity coefficients can be found from an EOS. For solid-liquid equilibria, there is the analogous K factor

$$K_i = \frac{x_i^S}{x_i^L}$$

Similarly, it can be shown that

$$K_i = \frac{\gamma_i^L}{\gamma_i^S} \left(\frac{f^L}{f^S} \right)_{i,pure}$$

At any temperature and any pressure, the fugacity ratio for pure component i, can be calculated/estimated from melting temperature, the melting enthalpy, heat capacities and the densities of pure liquid i and pure solid i. The effect of pressure is usually negligible. Details of the modelling of paraffin wax precipitation using an EOS and various property correlations, are found in Lira-Galeana et al. (1996). The details are too many to present in the present text. Papers by Chen and Zhao (2006) and Liu et al. (2015) provide similar modelling approaches and property correlations.

> The fugacity of the vapour phase is referenced to an ideal gas and the fugacity of the liquid phase is referenced to an ideal solution. In an ideal gas, the molecules themselves occupy no volume and have no intermolecular interactions. Gases at low pressure, approximate this behaviour. In an ideal solution, the molecules exhibit the same intermolecular interaction between all the components/constituents. Components having similar molecular structure, approximate this behaviour.

A model of paraffin wax precipitation was presented by Lira-Galeana et al. (1996). In the model, each solid phase is assumed to be a pure component (pseudo-component), determined by phase stability considerations. An EOS was used to describe properties of the vapour and liquid phases. An activity coefficient model was used for the solid phase. The wax % wt. versus temperature was estimated and found to match reasonably well with published data. The molar composition of eight crude oils was tabulated, from C_1 to C_{30+} for two of the crude oils and C_1 to C_{20} for six of the crude oils. The specific gravities were also given. Whether the paraffin wax precipitation occurs continuously, from high to low molecular weight components, or whether the precipitation occurs in a discontinuous multi-solid fashion, does not matter. The well-tried idiom, that the simplest modelling method that gives the required accuracy, should preferably be used.

A precipitation curve was measured for an API 71° heavy crude oil (specific gravity 0.953), and modelled (Lira-Galeana et al., 1996). The measured points and the model line are shown in Figure 4.8. The model predicts nicely the WAT and then overpredicts initially the amount precipitated, thereafter to shown a good match. The cloud point (WAT) was detected at 304°K (31°C). The precipitated paraffin wax was estimated to have a carbon number $C_n > 25$. The modelling and measurements suggested that the wax precipitated in a multi-solid phase precipitation process. Another way of expressing the precipitation behaviour is to state that heavier wax molecules precipitated first

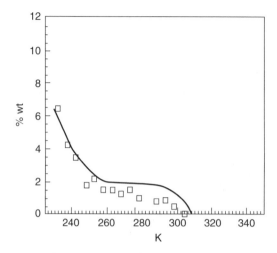

Figure 4.8 Paraffin wax precipitation curve for a heavy crude oil with a cloud point (WAT) of 304 K (=31°C), showing model (line) and data points (Lira-Galeana *et al.*, 1996).

(under cooling), before the light molecules. One way to express this behaviour is that the alkane number *n* of the precipitating molecules is proportional with temperature *T*.

4.5 SPATIAL AND TEMPORAL BUILDUP

The amount of paraffin wax that precipitates in a pipeline, depends heavily on the rate of cooling with distance. An analytical equation can be used to calculate the bulk temperature in a pipeline, from inlet to some distance downstream (see *Section 2.1 – Bulk and Wall Temperature*). The inlet temperature of a crude oil (and gas condensate) may be in the range 60–80°C, and the ambient temperature will be that existing on-land or subsea. On-land temperatures in northern climates (Canada and Russia, for example) will fall below freezing. Subsea temperatures near land in the Atlantic Ocean may be in the range 5–10°C and in South-East Asia oceans, 15–25°C. With distance, the bulk temperature decreases exponentially to approach the ambient temperature. That the WAT (cloud point) of crude oils being typically in the range 30–40°C, means that paraffin wax will precipitate both in-bulk and deposit on cooling surfaces (pipeline wall). Thermal insulation and even electrical heating may alleviate pipeline cooling.

Spatial and temporal precipitation and deposition refer to the *distance along* a pipeline and the *operational time*. The operational time may range from days to months, and even years. Precipitation does not necessarily mean that paraffin wax will form a deposit where the precipitation occurs. Detailed composition of crude oil and condensate are needed in modelling. Thermodynamic equilibrium (solubility of paraffin wax in crude oil and gas condensate) and general phase behaviour, determine the amount of paraffin wax precipitation. For a particular situation, the bulk and wall temperatures are perhaps the main variables affecting precipitation. Deposition

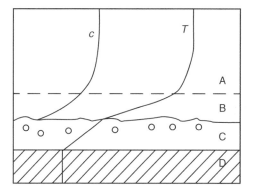

Figure 4.9 Schematic wax concentration and temperature gradients at a pipeline wall. A: turbulent convective flow, B: viscous sublayer, C: wax deposit with occluded oil and D: steel wall.

models are needed to relate the amount precipitated to the amount deposited. The general understanding of the complex chemical and physical processes involved, is that the build-up of paraffin wax deposits occurs due to a combination of molecular and particle mass transfer. The main transfer processes are presented in *Chapter 2 – Basic Phenomena* and the associated appendices.

The driving force for mass transfer by *molecular diffusion* is based on the temperature gradient and associated concentration gradient in laminar and turbulent pipe flow. In most instances, the flow is turbulent in pipelines carrying crude oils and gas condensates. The temperature and concentration gradients perpendicular to a pipeline wall, are illustrated in Figure 4.9. The pipeline wall D is the cooling surface. The wax layer C is deposited on the wall, contains a volume fraction of oil (liquid inclusion). The volume fraction of oil close to the wall is less than just below the wax-oil boundary. The viscous sublayer B extends from the surface of the wax layer to the buffer boundary where turbulent convective flow A dominates. In the figure, the wax concentration *c* and the temperature *T* are schematic; the flow is from left-to-right.

General values of thermal conductivity at 25°C are shown in Table 4.4. *Fourier's Law* describes the temperature conductive gradient through parallel layers by

$$q = -k\frac{dT}{dy}$$

where q W/m^2 is the heat flux, k W/m·K the thermal conductivity and dT/dy the temperature gradient perpendicular to the wall. For the same heat flux, the thermal conductivity determines the temperature gradient (see *Appendix C – Boundary Layer Temperature Profile*).

The thermal conductivity of paraffin wax deposits containing entrapped oil, cannot be calculated directly. Mixture density can be calculated directly; mixture viscosity cannot be calculated directly (see *Appendix Y – Two-Phase Flow Variables and Equations*). Expressions have been derived for different mixture geometries of solid-solid, liquid-liquid and solid-liquid. The expressions are reviewed by Stubsjøen (2013), who

Table 4.4 Thermal conductivity of materials in paraffin
wax deposition in pipelines; oil and wax and
two types of steel (approximate values).

Material	kW/m · K
Oil*	0.15
Wax	0.25
Stainless	16
Carbon	43

*Kerosene

suggest which would be most appropriate for the thermal conductivity of wax-oil mixtures.

The concentration gradient in Figure 4.9 is based on the solubility of paraffin wax in crude oil (gas condensate). The concentration in the turbulent core flow is practically constant. Through the viscous sublayer, the concentration decreases to zero at the surface of the wax deposit, where crystallization/solidification occur. The concentration gradient in the viscous sublayer is the driving force for mass transfer described by *Fick's Law* of diffusion

$$J = -D\frac{dc}{dy}$$

where J mol/s · m^2 is mass flux, D m^2/s the diffusivity coefficient and dc/dy the concentration gradient. Concentration will be mol/m^3, that can also be expressed by chemical activity using an activity coefficient.

An important question that arises, is at what distance from the wax surface the WAT of the oil-wax system is reached. Inside the viscous sublayer or in the turbulent core? If inside the viscous sublayer, paraffin wax crystals may move toward the wall, or away from the wall due to a lift force. If in the turbulent core, paraffin wax crystals will be precipitated in the bulk of the flow. An additional complication is the local break-up of the wax deposit, as envisaged in the deposition-release model. Turbulent bursts may hit the wax layer and carry a solid particle away from the wall. The particles mentioned are not illustrated in Figure 4.9. The several scenarios that may arise are different and too cumbersome/speculative to present in the current text. Some of the relevant mass transfer details are given in *Chapter 2 – Basic Phenomena* and associated appendices.

Nucleation (see *Appendix V – Solubility and Nucleation*) and heat of dissolution play a role in the precipitation of paraffin wax. Nucleation is easier in dirty fluids, such as crude oil, than in pure laboratory fluids. The heat of dissolution and crystallization of paraffin wax in crude oil and condensate is reported in the range 100–300 kJ/kg according to Han *et al.* (2010), based on data from Hansen *et al.* (1991). For comparison, the heat of crystallization (freezing) of water at 0.01°C is 333.5 kJ/kg (Rogers and Mayhew, 1980). The heat of solution of methane in water is reported 15 kJ/mol (Duan and Mao, 2006), equivalent to 940 kJ/kg. All of the three processes are exothermic,

Table 4.5 Molecular weight, melting temperature and heat of melting (fusion) for three normal paraffins.

n	M g/mol	T K	T °C	ΔH kJ/mol	ΔH kJ/kg
20	282	310.9	38	53.6	190
30	392	331.1	58	79.3	202
40	562	346.7	74	119.1	212

meaning that heat must be removed. In terms of modelling (see *Appendix L2 – Regular Solution Model*).

For the freezing of water (mentioned here as an example), and for the solidification of *pure paraffin wax*, measurable amount of heat must be extracted. It may be useful to separate what happens on micro-scale and macro-scale; what happens to molecules and what happens in measurable bulk. The crystallization of an individual paraffin wax molecule requires extraction of heat. The total paraffin content of crude oils is on the percentage-scale. In pipelines, the cooling of flowing crude oils will occur gradually with distance. The heat of wax solidification, will be absorbed by the surrounding crude oil. Locally, the crystallization/solidification process, will occur at near-isothermal conditions. In regular solution theory, the heat consumed/liberated in mixing and phase-change process, is included in terms of an *Interaction Parameter*. The term *athermal* is used for processes that take place without change in temperature.

The heat of melting of normal paraffins can be estimated from the empirical equation (Coto *et al.*, 2010)

$$\Delta H_m = C_1 M_i T_{m,i}$$

where C_1 is a constant, M_i the molecular weight and $T_{m,i}$ the melting temperature, the subscripts indicating wax component *i*. The value of the constant $C_1 = 0.6111 \cdot 10^{-3}$ is based on the units M_i kg/kmol and $T_{m,i}$ K. The empirical equation does not have consistent units. For the melting point $T_{m,i}$, Cato *et al.* (2010) suggested the use of an equation from Won (1989)

$$T_{m,i} = C_2 - \frac{C_3}{M_i}$$

where the constants have the values $C_2 = 382.72$ and $C_3 = 20242.59$.

Using the general formula for normal paraffins (given in *Appendix X1 – Crude Oil Composition*), for carbon numbers $n = 20$, 30 and 40, the molecular weight, melting point and heat of melting were estimated, shown in Table 4.5. Paraffin wax in crude oil and condensate has carbon number in the range 20 to 40.

Consider a long (10–100 km) subsea pipeline carrying crude oil or gas condensate in single-phase or multiphase flow. Furthermore, that the pipeline inlet temperature is well above the wax appearance temperature (WAT). The fluids will cool down with distance and paraffin wax will precipitate and deposit. For illustration purposes, the distinction between precipitation and deposition will be ignored. The spatial and temporal thickness of paraffin wax in a typical subsea pipeline is sketched in Figure 4.10.

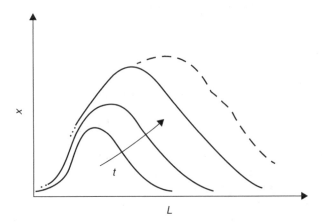

Figure 4.10 Deposition profiles showing paraffin wax deposit thickness *x* mm with pipeline length *L* km at increasing times *t* in hours, days, weeks or years.

Table 4.6 Suggested rule-of-thumb for maximum paraffin wax deposit thickness in pipelines.

Time	x mm
Day	<1
Week	1–10
Month	>10

The spatial and temporal paraffin wax deposit thickness depends on a whole host of factors, as exemplified by Zhang *et al.* (2014). Nevertheless, the following rule-of-thumb is offered here, based on days, weeks and months, shown in Table 4.6. The suggested wax thickness is the *maximum* thickness, localized at some distance from the pipeline inlet. The maximum thickness is most important because of pigging operations. Localized flow restriction gives a more profound and rapid change in total pressure drop than an evenly distributed wax deposit.

The bell-shaped curves in Figure 4.10 increase in height and width with time. The limited deposit thickness on the left-hand-side of the curves can be related to precipitation-deposition when the bulk temperature is above the wax appearance temperature. The central main bell-shapes can be related to precipitation at temperature below the wax appearance temperature. The decreasing deposit thickness on the right-hand-side of the curves can be related to the bulk temperature approaching the surrounding temperature of the ambient sea (no further cooling).

The broken line bell-shaped curve in Figure 4.10 is meant to indicate the deposit thickness when the up-stream deposit has become relatively thick. A thick deposit means that the up-stream cooling (heat transfer) has been reduced significantly due to the insulating effect of the paraffin wax deposited. The maximum deposit thickness will not increase significantly in the absence of external cooling. The deposition profile moves down-stream with time. The overall deposition process, can be described as a self-restraining process. This because as wax thickness increases, the cooling decreases and deposition reduces. The cooling decreases because a wax deposit reduces the overall heat transfer coefficient (wax adds a thermal resistance).

An important factor to consider in the cooling of crude oil pipelines, is the presence or absence of produced water. The heat capacity of crude oil is about one-half that of the heat capacity of water; respectively 2.1 kJ/kg·K and 4.2 kJ/kg·K at 25°C. The heat capacity depends on temperature. The basic consideration is that if the watercut of the crude oil is high, the cooling rate will be lower than in water-free oil flow.

High molecular weight (large carbon number) paraffin wax precipitates at higher temperature than low molecular weight paraffin wax. It means that the composition of paraffin wax depositing along a subsea pipeline will vary with distance. With time, the entrapped oil in wax deposits will diffuse away from the wall, making the deposit harder near the wall, a process called aging. The diffusion is forced by the ripening of the paraffin wax. Ripening is the process where small crystals join to make larger crystals, to reduce their *Gibbs Free Energy*. Gelling of paraffin wax in crude oil or gas condensate (solidification of pipeline content) is of major concern when pipelines are shut-down and on start-up. Paso *et al.* (2005, 2009) have published in this area.

4.6 NATURE OF DEPOSITS

Wax crystals examined under a microscope are reported to be shaped primarily as rods and needles and to a lesser extend as plates. The size of individual paraffin wax crystals is reported in the range 1–10 μm with small crystals forming in gas condensate and large crystals in crude oil (Bacon *et al.*, 2009). Such a particle size range suggests particle deposition in the diffusion and inertia regimes (see *Section 2.5 – Particle Mass Transfer*). Also, that wax crystal particles will be subject of an outward lift force in the viscous sublayer. Wax crystals can be classified as paraffinic (macroscopic) and microcrystalline (microscopic). Microscopic wax crystals in early studies were reported to be amorphous, which later studies have contradicted. Microcrystalline/microscopic wax crystals are smaller in size that the range given above.

The formation of paraffin wax crystals is likely to conform to crystallization theory in general. Factors of interest in oil and gas operations, include the presence of impurities (nucleation sites), the rate of cooling (rapid cooling giving small crystals) and the flowing conditions (laminar or turbulent flow). Understanding of paraffin wax crystallization (precipitation) is hampered by the fact that microscopic examinations are undertaken on quiescent (non-stirred) fluids, while the opposite (stirring, flowing) exists in pipelines and equipment.

The stiffness (hardness) of paraffin wax deposits in laboratory flow loops and operational pipelines ranges from mushy to hard. Mushy soft deposits (Gudmundsson, 1977) are usual in flow loops while hard deposits are found in pipelines. The inclusion of fluid (for example, kerosene and similar hydrocarbons in flow loops; crude oil and condensate in pipelines) in paraffin wax deposits is responsible for their mushy nature. The fluid content of paraffin wax deposits may range from 50–60%, expressed as volume or mass (density of oil and wax is similar, for practical purposes). A different way to view the entrapped oil volume, is to consider the minimum volume % of wax needed to support a solid structure. Some authors refer to the fluid volume fraction as porosity (Rygg *et al.*, 1998).

The hardness of paraffin wax deposits may be caused by several factors, discussed below in terms of *four* effects: fractionation, cooling rate, flow velocity and aging:

First, consider that fractionation occurs under cooling in pipelines. Heavy paraffin wax molecules are the least soluble and will precipitate (and deposit) before the more soluble light molecules. Information is lacking about whether heavy wax molecules will have the same or different degree of fluid inclusion than light wax molecules. It is not known (measured) whether the hardness of pipeline deposits varies with pipeline distance. Keeping in mind that the wax deposition profile moves downstream with time (see Figure 4.10) the heavy-light fractionation will simultaneously move downstream.

Second, consider how quickly the precipitation occurs. Factors affecting how rapidly paraffin wax precipitates (and deposits) are the presence of impurities, the rate of cooling and the wax supersaturation (depends on concentration and temperature). Crude oil and gas condensate contain impurities that facilitate nucleation, suggesting that wax deposits in flow loops without impurities would form more slowly than in pipelines. Alternatively, that the presence of impurities favours the formation of more compact deposits. It can be argued that deposits that form under rapid precipitation will be softer than deposits formed slowly. Rapid cooling and high wax supersaturation, are similarly assumed to form soft deposits. The presence of impurities may favour the formation of hard deposits. More importantly, the slow cooling in pipelines, compared to rapid cooling in laboratory flow loops, may favour the formation of hard deposits.

Third, consider the effect of flow velocity and similar phenomena (for example, Reynolds number). Flow loop experiments have shown that the deposition-release model describes reasonably well the build-up of paraffin wax deposits (see *Section 2.6 – Deposition Models*). One of the observations made (Bott and Gudmundsson 1976; Gudmundsson 1977) is that deposits formed at high velocity are harder than deposits formed at low velocity. However, laboratory flow loops experiments are usually carried out at flow velocities similar to flow velocities in pipelines. Therefore, flow velocity cannot explain why paraffin wax deposits in flow loops are softer than deposits in pipelines.

Fourth, consider the effect of aging and diffusion. The surface energy of small crystals is higher than the surface energy of large crystals. Because physicochemical systems tend to move to a lower energy state, then small crystals will have a tendency to fuse into larger crystals. The effect is called *Ostwald Ripening* (Fredrich Wilhelm Ostwald, 1853–1932). While small crystals form more rapidly than large crystals, large crystals are more stable (small crystals are kinetically favoured while large crystals are thermodynamically favoured). Simultaneous with large crystals growing at the expense of small crystals, the occluded fluid (included in the fluid volume fraction) will diffuse to outside the deposit material. *Ostwald Ripening* and fluid diffusion are processes that have been described as aging, resulting in a denser (harder) deposits with time. The ripening is probably the controlling process; the diffusion being a result of the ripening. In terms of pipeline and pigging operations, it would be useful to know the slowness of paraffin wax aging processes. Are the processes slow or fast, relative to the need for pigging operations?

It has been argued above, that *hard deposits* are formed in pipelines primarily due to slow cooling and aging; that is, compared to soft deposits formed in laboratory flow loops. Laboratory experiments take from hours to days while pipeline operations are measured in weeks and months. A practical implication may that pipeline scraping should perhaps be carried out more often than strictly required (due to reduced flow capacity). However, operators want to keep PIG-operations to a minimum, because of shut-in of wells cause loss of production. Soft deposits are more easily scraped off pipe walls than hard deposits.

The surface roughness (see *Section 2.3 – Surface Roughness* and *Appendix F – Friction Factor of Structured Deposits*) of paraffin wax deposits in pipelines is not known (measured). Assuming that paraffin wax deposits are not necessarily smooth, calculated pressure drop based on sand-grain-roughness will be lower than measured pressure drop. Rippled deposits will exhibit a much higher pressure drop than evenly rough deposits (Gudmundsson, 1977; Bott and Gudmundsson, 1977). Such deposits may form in oil and condensate pipelines when the deposition driving force is minimal; for example, in pipelines subjected to relatively slow cooling. The *minimal-condition* is likely in pipelines carrying waxy crude oils.

4.7 FLOW LOOP OBSERVATIONS

Laboratory flow loops can be used to study the build-up of paraffin wax in cooled pipes. The pipes used are envisioned to simulate pipelines. Much new knowledge has been gained in many laboratories and various aspects identified and observed. The scale-up of flow loop data to field conditions has been difficult, because of the pipe dimensions and test times involved. While laboratory pipes are of the order of meters in length, pipelines are of the order of kilometres in length. While laboratory experiments are carried out in days, pipelines are operated for months and more. The concentration and temperature driving forces in flow loops are also much larger than in operated pipelines. Paraffin wax aging/hardening is an important unknown in pipeline operations.

Observations and results from different flow loop studies, tend to be similar. Drawing on the literature and a wax-in-kerosene study reported by Gudmundsson (1975), the following *Observations* of paraffin wax deposition can be presented:

1   The deposition of paraffin wax occurs due to reduced solubility upon cooling, starting at the cloud point temperature.
2   The mass transfer mechanisms from fluid-to-wall is convection in the turbulent core and diffusivity in the viscous sublayer.
3   The nature and size of wax crystals formed in bulk flow is affected by the rate of cooling and the presence of impurities and natural surfactants.
4   Wax crystal particles formed in the turbulent core and viscous sublayer may contribute to the deposit build-up.
5   Paraffin wax deposition thickness increases more-and-more slowly with time, subject to deposition-release mechanisms.
6   The initial rate of deposition, the long-term deposits thickness and the amount of entrapped oil, all decrease with increased flow velocity.

7   The initial rate of deposition and the long-term deposit thickness, both increase with increased temperature driving force and wax concentration.

8   Paraffin wax deposits are held to surfaces by physical adsorption forces. Adsorption processes are exothermic, thus favouring cold surfaces.

9   Cohesive failure can occur in paraffin deposits, resulting in random processes of build-up and break-down (deposition-release).

The effect of aging (Zhang *et al.*, 2014), and deposit roughness (see *Appendix F – Friction Factor of Structured Deposits*), are not included in the above observations, neither the effect of flow patterns in multiphase flow (Sarica and Volk, 2004). The following three *Observations* can be added to the list, for completeness:

10   Aging of paraffin wax deposits reduces their liquid oil content (occluded/entrapped fluid) and makes the deposits harder with time.

11   The roughness of paraffin wax deposits may be structured, resulting in greater frictional pressure drop than predicted using sand-grain roughness.

12   The thickness of paraffin wax deposits is affected by multiphase flow pattern; each pattern has different contact area with the surrounding solid surface.

The work of Zhang *et al.* (2014) on wax deposition modelling in pipelines, was largely in accordance with the above observations. The *shearing effect* and the *aging effect* were considered carefully. In both cases, however, the results were unclear. The shearing effect was not exhibited in a field pipeline, where the deposit thickness increased linearly with time. The aging effect was not observable in flow-loop test and neither in field tests. Surprisingly, the wax porosity assumed was in the high range 60–90% wt.

The work of Rittirong *et al.* (2017) on paraffin wax deposition in gas-liquid flow in horizontal pipes, was also largely in accordance with the above observations. Flow-loop testing was carried out using a gas condensate and natural gas. In total, 52 deposition tests were carried out at 11 operating conditions with independent/variable test durations. Deposit thickness decreased with increasing flow velocity. The wax content of deposits increased (occluded liquid decreased) with velocity and time. In terms of C_{30+} the wax percentage weight (% wt.) was in the range 5–25%, meaning that the volume fraction of occluded liquid was high. The weight distributions of the gas condensate used and the paraffin wax deposited, are shown in Figure 4.4, based on measurements after 24 hours. Information about the testing is summarized in Table 4.7. The experimental results were presented in terms of the superficial velocities of the liquid phase and the gaseous phase (for definition of superficial velocities, see *Appendix Y – Two-Phase Flow Variable and Equations*). The testing showed that in gas-liquid slug flow, the deposit thickness at the top of the horizontal pipe was greater than at the bottom of the pipe; the thickness changed circumferentially. The result shows that flow pattern affect the distribution of deposit in a horizontal pipe, perhaps with an implication for scraping (pigging) of multiphase pipelines.

Under what conditions can paraffin wax deposits be described as a *solid* and when can deposits be described as a *gel*? Can the transition from a gel to a solid be described by a particular porosity (entrapped oil volume), analogous to the transition from an oil-continuous to a water-continuous emulsion? Gels are mostly liquid and

Table 4.7 Flow-loop dimensions, test conditions and gas condensate properties. The temperatures are the bulk and wall temperatures (Rittirong et al., 2017).

Description	Value
Length	16.2 m
Diameter	52.5 mm
Pressure	2.4 MPa
Bulk	25°C
Wall	17°C
WAT	36°C
API	(42°)
Wax	0.5% wt.
Superficial L	0.3–0.9 m/s
Superficial G	0.8–3.0 m/s
Duration	4, 12, 24 h

can behave like solids due to a cross-linked network within the liquid. Gels can be described as a dispersion of liquid molecules within a solid structure. The solid is the continuous phase and the liquid is the discontinuous phase. A related matter may be the coefficient of thermal expansion (βK^{-1}) of paraffin wax and crude oil. With time, the deposit thickness increases and its temperature near the wall becomes colder. The wax and the oil will shrink proportionally to their individual thermal expansion coefficients, expressed by the equation

$$\Delta V = \beta V_o \Delta T$$

Assuming that the coefficient of thermal expansion of paraffin wax and crude oil are different, the volume changes are likely to be different enough to reduce (squeeze out) the volume fraction of oil in the near-wall deposit. The alternative and commonly give explanation for aging, is that the liquid crude oil trapped in a pipeline deposit, will diffuse away from the wall through the wax structure.

4.8 DEPOSITION PROFILES

Schematic paraffin wax deposition profiles are shown in Figure 4.10. Oil and gas field operators and pipeline operators, want to know how paraffin wax deposits build-up along a wellbore and a pipeline, and how the build-up increases with time, spatial and temporal deposition, respectively. Awareness and knowledge of wax deposition is required in field development planning (front-end), engineering design and in early/start-up and long-term operations.

Flow-loop tests are carried out in laboratories to gain information and knowledge that can be transferred (scale-up) to actual wellbore/pipeline operations. A whole gamut of modelling need to be carried out in conjunction with flow-loop testing, to make the results applicable to field situations. The work of Zhang et al. (2014) is a

Table 4.8 North Sea gas-condensate used in flow-loop testing (Hoffmann and Amundsen, 2010).

Property	Value
Density	809 kg/m^3
API	43°
Wax	4.5% wt.
WAT	30°C
PPT	1°C
Viscosity	2–3 mPa · s

Table 4.9 Description of paraffin wax deposition testing flow-loop apparatus (Hoffmann and Amundsen, 2010).

Description	Value
Length	5.5 m
ID (=5″)	52.6 mm
Pump	3–30 m^3/h
Pressure	≅1 atm.
Oil	15–70°C
Water	10–25°C
Tank	4000 L

good example of such an approach. Detailed and accurate laboratory work is the backbone of useful modelling. Without useful laboratory data, there is no useful modelling. The *chemical* thermodynamics modelling of wax deposition, rests upon the already established approaches for hydrocarbon mixtures, in the form of Equation of State (EOS). The other requirement for modelling, is the *physical* testing using flow-loops in laboratories.

An illustrative and good example of flow-loop testing, is that of Hoffmann and Amundsen (2010), who carried out single-phase wax deposition experiments. A real gas-condensate from a North Sea field was used, with the properties shown in Table 4.8. The gas-condensate viscosity was measured in a rotating cylinder viscometer. The shear rates ranged from 100–2000 1/s. Above about 20°C, the shear rate had no influence on the measured viscosity. In the work reported by Rosvold (2008), the data analysed and modelled, was based on practically the same flow-loop, using the same gas-condensate.

A description of the flow-loop apparatus is shown in Table 4.9 (Hoffmann and Amundsen, 2010). The table shows the capability of the apparatus, not necessarily the values used in the flow testing. The pump rate can give 0.4–4 m/s average flow velocity ($A = 2.17 \times 10^{-3}$ m^2), corresponding to a Reynolds number in the range 6800–68,000 assuming a dynamic viscosity of 2.5 mPa · s. The test pipe was smooth and the cooling was counter-current. A small part of the down-stream test section could be removed occasionally to remove the deposited was for analysis.

A typical deposition profile obtained by Hoffmann and Amundsen (2010) is schematically illustrated in Figure 4.11. The deposit thickness increases at a decreasing rate with time. The shape of the deposition profile is logarithmic, based on extensive matching of flow-loop data and deposition-release models by Botne (2012). The flow-loop data of Rosvold (2008) was used, being a forerunner to the data of Hoffmann and Amundsen (2010). Derivations of an exponential and a logarithmic model are given in *Appendix K – Deposition-Release Models*. The logarithmic derivation is new, while the exponential derivation is of an older variety. *Section 2.8 – Deposition Models* can also be consulted.

In the logarithmic deposition-release model

$$\frac{dx}{dt} = k_1 k_2^{-x}$$

where k_1 and k_2 are parameters obtained from matching with flow-loop or well-bore/pipeline data. The first parameter k_1 may be called the deposition parameter and the second parameter k_2 the release parameter. The logarithmic model results in the following expression for deposit thickness x with time

$$x = \frac{1}{\ln k_2} \ln[1 + (k_1 \ln k_2)t]$$

The k-values are not that difficult to find (from the dx/dt expression above). At $t = 0$ the deposit thickness $x = 0$, such that

$$k_1 = \left(\frac{dx}{dt}\right)_{x=0}$$

Knowing the first k-value, the second k-value can be determined from any point on the deposition profile curve. Traditional curve fitting can be used for this purpose. An alternative would be to use multiple regression for both of the k-values (Botne, 2012).

The flowrate and temperature conditions of the deposition profile shown in Figure 4.11, where not explicitly given by Hoffmann and Amundsen (2010). Comparing the profile with other published profiles in the same paper, it was estimated in the present text that the flowrate was 5 m³/h, the gas condensate temperature 20°C and the cooled wall temperature 10°C. Volumetric flowrates used in the flow-loop experiments were in the range 5–25 m³/h. The initial rate of wax deposition dx/dt is quite steep in Figure 4.11, and somewhat unclear, which is not unusual. One reason being that at start-up the paraffin wax is depositing on a metal wall, not on an existing wax deposit.

The several deposition profiles presented by Hoffmann and Amundsen (2010), were used to read-off the deposit thickness after 65 hours testing. The thickness values are plotted in Figure 4.12. Clearly, the deposit thickness decreases with flowrate, as stated in *Observation 6* above. The approximate initial rate of deposition in Figure 4.11 can be estimated from 0.5 mm after 5 hours, giving a deposition parameter of $k_1 = 0.1$ mm/h. Based on a deposit thickness of 1.55 mm after 65 hours, the logarithmic deposition-release model gives a release parameter of $k_2 = 11$, with an undefined unit. These values can be used to calculate the deposition profile. Or more importantly,

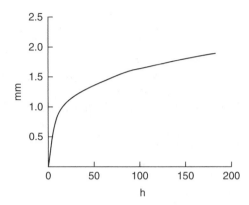

Figure 4.11 Paraffin wax deposit thickness (mm) with time in hours (redrawn after Hoffmann and Amundsen, 2010).

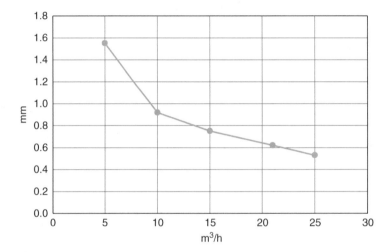

Figure 4.12 Deposit thickness mm versus flowrate m³/h after 65 hours. Gas condensate inlet temperature 20°C and cooling water temperature 10°C. Data from Hoffmann and Amundsen (2010).

they can be correlated with other variables, such as wall shear stress and temperature driving force, assuming that many deposition profiles have been measured.

The WAT in Table 4.8 is 30°C. It is an average value judged most appropriate, based on four measurement methods. DSC (Differential Scanning Calorimetry) gave 29°C, NIR (Near Infra-Red spectroscopy) gave 27°C, microscopy gave 30°C and rheometer gave 31°C. In the flow-loop experiments, it was observed that paraffin wax deposition started when the gas condensate temperature was in the range 27.5–32.5°C. Hoffmann and Amundsen (2010) measured the wax content of deposits at two flowrates, after 100 hours of operation of the flow-loop. At the low flowrate of 5 m³/h, the *wax content* was 18% wt. and at an unspecified higher flowrate, the

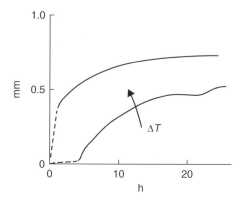

Figure 4.13 Paraffin wax deposition thickness with time for two different driving forces, WAT minus wall temperature (redrawn after Venkatesan and Creek, 2010).

wax content was 33% wt. *Observation 6* was expanded to include this effect (volume of entrapped oil in a deposit decreases with flowrate). Analysis of the paraffin wax deposits showed that they were rich in normal alkanes C_{25} to $C_{45.}$ Other results were largely in line with the *Observations* offered above.

Flow-loop experiments reported by Venkatesan and Creek (2010), exhibited the influence of the pipe wall temperature on the initial rate of deposition and an induction time, shown in Figure 4.13. The *upper curve* was obtained when 18.2% wt. wax in a crude oil deposited where the pipe wall temperature was 14°C below the bulk temperature. The initial rate of deposition the first hour is clearly much higher than the subsequent rate of deposition, indicating a deposition-release mechanism.

The *lower curve* in Figure 4.13 was obtained when 22.5% wt. wax in crude oil was deposited with the pipe wall of about $\Delta T = 4°C$ below the bulk temperature. The first four hours exhibit an initiation time (see also Figure 2.10), after which the deposit thickness increases at a declining rate with time. The difference in wax concentration (18.2 and 22.5% wt.) was not considered a significant factor in the experiments. However, the temperature driving force difference ($\Delta T = 14$ and 4°C) was considered a significant factor.

The temperature driving force in flow loop experiments ΔT is in the range 5–15°C, which is much larger than in subsea pipelines having ΔT in the range 1–2°C. It means that it is difficult to apply flow loop results to actual pipelines. The practical constraint is that flow-loop experiments with small ΔT take a very long time.

The accumulation of solids on pipe walls will increase the overall heat transfer *resistance* from inside (bulk temperature) to outside (seawater temperature). It means that the inside pipe wall temperature will decrease with time as the deposit thickness increases (self-restraining process). The decrease in inside wall temperature will enhance the Oswald ripening (large crystals grow at the expense of small crystals) of paraffin wax. Simultaneously the surface temperature of the deposited wax will increase, further reducing the temperature driving force.

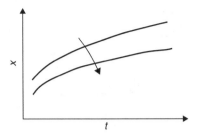

Figure 4.14 Schematic of long-term deposition-release behaviour of paraffin wax in flow-loop experiments. Flow velocity (and related properties) increases in value downwards.

The long-term effect of flow velocity on paraffin wax was deposit thickness in flow loops, as shown in Figure 4.14. The deposit thickness is sketched against time for two hypothetical experimental conditions. The pipe wall effects (initial rate of deposition) are not included. The deposit thickness x, can alternatively be expressed as an increase in the fouling resistance R_f. The average bulk flow velocity u, can alternatively be expressed in terms of volumetric flow rate q, Reynolds number Re, or friction velocity u^* (see *Appendix H – Universal Velocity Profile*). The use of universal parameters in correlating flow-loop results to modelling equations, for example Reynolds number instead of flow velocity, gives an advantage and is recommended. Meaning that flow-loop results will be more transferrable to wellbore and pipeline conditions in operational environments, when universal parameters are used.

4.9 SELECTED DEPOSITION MODELS

Selected wax deposition models used in the oil and gas industry, were reviewed by Rosvold (2008). The model reviewed were the models of Rygg *et al.* (1998), Singh *et al.* (2000, 2001) and Matzain *et al.* (2001), plus one proprietary model. Based on extensive laboratory flow loop data (gas condensate in flow-loop), it was concluded that the Matzain *et al.* (2001) model gave the best match, after tuning of parameter. It was the only model that included the effect of shear stripping (same as release in the deposition-release model). All the models included molecular diffusion. The Rygg *et al.* (1998) included the curiosity shear dispersion. The only model to include aging was the Singh *et al.* (2000, 2001) model. Particle mass transfer was not included in any of the models reviewed by Rosvold (2008).

Details of pipeline wax deposition modelling are given by Hovden *et al.* (2004). The implementation of three models (Rygg *et al.*, 1998; Matzain *et al.*, 2001 and heat transfer analogy) in a commercial computer code was described. Comparison of the models to field data gave similar results as later reported by Rosvold (2008).

Shear dispersion is a curiosity introduced to the paraffin wax literature by Burger *et al.* (1981). The idea stems from the movement of small particles subjected to a transverse velocity gradient in laminar flow. The shear dispersion is supposed to enhance the mass transfer from bulk *toward* a wall. Rosvold (2008) stated that shear dispersion was not important in the analysis of proprietary flow loop data. The concept of shear dispersion contradicts the *Saffman Lift Force* that causes small particles to move *away* from a wall due to viscous shear (Siljuberg, 2012).

The flow-loop data of Rosvold (2008), was digitized and analysed by Botne (2012). The focus was on matching the data to the deposition parameter k_1 and the release parameter k_2 in the logarithmic deposition-release model. The flow-loop and fluid data (gas condensate) were the same as reported by Hoffmann and Amundsen (2010). The data consisted of five flowrates (5, 10, 15, 21 and 25 m^3/h) at constant temperature of 20°C and five temperatures (15, 20, 30 and 40°C) at the same flowrate of 21 m^3/h. With only one exception, the experimental data followed the logarithmic increase in deposit thickness with time.

The experimental data were correlated by Botne (2012) using two parameters, the wall shear stress

$$\tau_w = \frac{r}{2}\frac{dp}{dL} = \frac{1}{8}f\rho u^2$$

and the dimensionless temperature

$$\Delta T^+ = \frac{T_{cloud} - T_{wall}}{T_{oil}}$$

taking into consideration the cloud point (=WAT), the cooling wall temperature and the inlet temperature of the oil (gas condensate). For a particular flow-loop experiment, the dimensionless temperature driving force will be constant. In a pipeline situation, the oil and wall temperature values may need to be adjusted (pipeline divided into segments).

The deposition-release parameters are shown schematically in Figure 4.15. In the work of Botne (2012), numerical values were obtained for the two parameters. The *deposition parameter* is shown on the ordinate (y-axis) versus the inverse of the wall shear stress $1/\tau_w$ and the temperature driving force ΔT^+ on the abscissa (x-axis). Low wall shear stress gives a high deposition parameter. High temperature driving force gives a high deposition parameter. The *release parameter* is shown on the ordinate versus the wall shear stress squared, and the inverse of the temperature driving force. The higher the wall shear stress, the more deposit release/removal. Low temperature driving force gives a high release parameter. The overall results, confirm *Observations 6 and 7*. The methodology confirms also the usefulness of the deposition-release approach to correlating deposit buildup with time (deposition profile).

Experimental flow-loop data of Lund (1998), Hernandez (2002) and Venkatesan (2004), and field data of Singh *et al.* (2011), were also analysed by Botne (2012). The deposition and release parameters were determined for each of the experimental sets/series. The methodology was applied with success.

Flow-loop tests were carried out by Gudmundsson (1997) and reported by Bott and Gudmundsson 1976). The data analysis resulted in an exponential deposition-release model, presented in *Appendix M2 – Wax-in-Kerosene Deposition Model*. Fully refined paraffin wax 51/54°C and commercial kerosene were used in the experimental work. The deposited wax was mushy and not hard; the entrapped liquid volume was not measured. The paraffin wax used had a relatively low melting point. Based on

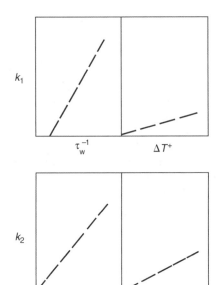

Figure 4.15 Schematic illustration of deposition parameter k_1 and release parameter k_2 as functions of wall shear stress and dimensionless temperature driving force.

Table 4.10 Experimental conditions in wax-in-kerosene flow-loop study.

Conditions	Value
Duration	24–240 h
Reynold number	6720–20,900
Bulk temperature	28–47°C
Wax concentration	4.2–26% wt.
WAT	10–32°C

the values in Table 4.5, the molecular weight was about 350 kg/kmol and the carbon number $20 < C_n < 30$. The experimental conditions are given in Table 4.10.

The analysis of the experimental data showed that the rate of deposition (expressed by thermal fouling resistance) was proportional to the wax concentration times the temperature driving force, divided by the product of fluid density and friction velocity. The asymptotic deposit thickness was proportional to the same terms, with the denominator squared. In terms of deposit thickness

$$\frac{dx}{dt} \propto \frac{c\Delta T^+}{\rho u^*}$$

$$x_\infty \propto \frac{c\Delta T^+}{(\rho u^*)^2}$$

The proportionalities agree with the trends shown in Figure 4.13, Figure 4.14 and Figure 4.15, with the addition of wax concentration. The greater the wax concentration and the greater the temperature driving force, the greater then rate of deposition and the long-term deposit thickness. The opposite hold true for the flow rate/velocity expressed as the tangential (flow direction) mass flux, readily expressed in terms of the shear stress at the wall (see Figure 4.15). An additional comment by Bott and Gudmundsson (1976), was that the transport of material to the surface is not the limited factor, but the cohesive properties of the wax deposit. The wax concentration result can be added to *Observation 7* and the cohesive comment to *Observation 9*.

4.10 MONITORING OF PIPELINES

An operator of a pipeline suffering from paraffin wax deposition, wants to know the amount and location of the deposits that form with time. Pressure drop, tracer testing and pressure pulse measurements, are among the tools that can be used to monitor paraffin wax deposition in pipelines. Monitoring of single-phase (liquid) flow pipelines is less difficult than the monitoring of two-phase (gas and liquid) pipelines. In the present context, stabilized (steady-state) flow of an oil-in-water and water-in-oil emulsions is considered single-phase flow.

Pressure drop in pipelines depends heavily on the internal diameter. At constant volumetric flowrate, the pressure drop in pipelines depends on the pipe diameter inversely to the power of five. The average flow velocity in a pipeline is related to the effective flow area through

$$u = \frac{q}{A} = q \frac{4}{\pi d^2}$$

Substituting volumetric rate q for velocity u in the common frictional pressure drop equation (see *Appendix D – Darcy Weisback Equation*), the following can be written

$$\Delta p = \frac{f}{2} \frac{L}{d} \rho u^2 = \frac{f}{2} \frac{L}{d} \rho \left(\frac{4}{\pi d^2} \right) q^2 \propto \frac{1}{d^5}$$

The above applies to a pipeline with a constant volumetric flowrate q m³/s. In real operations, the volumetric flowrate may change with time, for various reasons. If the inlet pressure is constant, the volumetric flowrate will decrease with time due to deposition. If the inlet pressure can be changed by increased pumping, for example, the total pressure drop versus flowrate will change.

The relationship between pumping and pipeline pressure drop can be illustrated in Figure 4.16. A plot of pressure drop against flowrate will not give a straight line. However, a plot of specific pressure drop $\Delta p/q$, will give a straight line. The *lower half* of the sketch shows the performance of a centrifugal pump and the pressure drop performance of a pipeline at two conditions (two deposit thicknesses). The whole curve shows the performance of a clean pipe and the dashed line shows the performance at

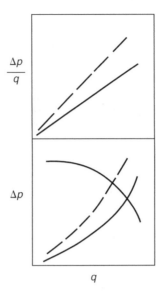

Figure 4.16 Lower diagram, performance of a centrifugal pump and a liquid carrying pipeline, clean and deposit laden. Upper diagram, specific pipeline performance, clean and deposit laden.

some later time, when a deposit has reduced the diameter (ID). The *upper half* of the sketch shows the specific performance of a pipeline

$$\frac{\Delta p}{q} = FI \times q$$

where *FI* is a flow index (gradient of straight line), that encapsulates all the variables in the frictional pressure drop equation above, the effective diameter being the most important for solid deposition. Step-rate tests (at least three) are needed to draw the FI-line. It may be difficult if not impossible to carry out step-rate tests in many field situations. For example, where oil or gas condensate from a production platform is pumped through a pipeline. Lower flowrates, means that the whole production system, from wells to processing, needs to be choked. Hydrostatic pressure head has not been considered in Figure 4.16. Of course, the friction factor and liquid density may change with time because of deposit buildup. The friction factor may change due to changes in surface roughness and liquid density and viscosity due to changes in the temperature profile of the pipeline.

An extensive paraffin wax pigging program for subsea oil export pipelines in the Caspian Sea, was reported by Stevenson *et al.* (2015). Pipeline performance plots of pressure drop versus oil volumetric flowrate were illustrated. A better illustration would have been the specific pressure drop versus flowrate. The total length of the pipelines was 190–200 km, consisting of several segments. The thickness of the wax deposits was reported in the range 2–5 mm. An interesting observation was that the volume of the wax removed by pigging was much less than expected from pressure drop modelling. It was concluded that the *roughness* of the wax deposit was a significant contributor to the pipeline pressure drops. The effect of rough/rippled deposits on pressure drop in pipes is presented in *Appendix F – Friction Factor of Structured Deposits*.

Tracer testing can give information about the volume (not the location) of paraffin wax in a pipeline. Injected at the pipeline inlet and measured at the pipeline outlet. The travel time from inlet to outlet depends on the liquid content of a pipeline. The liquid content decreases when paraffin wax deposits increase and reduce the pipeline diameter. The volume difference between a clean pipe and a deposit laden pipe, given an estimated of the volume of deposits. Perhaps the tracer profile at different times will contain information about dispersion due to local restriction and deposit roughness. The rougher the deposit, the greater the dispersion of the tracer concentration.

Pressure Pulse profiling is a new method to determine the location and thickness of deposits in pipelines (Gudmundsson *et al.*, 2001). The original intension of developing the method was multiphase flow metering in wells and pipeline (Gudmundsson and Celius, 1999). It turned out that the oil and gas industry was in a dire need to measure deposits in pipelines, for example paraffin wax depositing in gas-condensate pipelines. The method works well in pipelines carrying liquids such as crude oil and condensate. The essential features of the method are sketched in Figure 4.17. Consider a flowing pipeline where the pressure is measured up-stream of a quick-acting valve at the *outlet*. Alternative location for the pressure measurement is at the *inlet*, immediately downstream of a quick-acting valve. On quick closure of the outlet valve, the pressure will increase according to the water-hammer equation (*Joukowski Equation*)

$$\Delta p_a = \rho a \Delta u$$

where ρ kg/m^3 is fluid density, a m/s acoustic velocity in the fluid and Δu m/s the change in fluid velocity (from steady-state velocity to zero on full closure). For a 700 kg/m^3 fluid/oil, an acoustic velocity of 1200 m/s and a 2 m/s flow velocity, the water hammer pressure increase will be 16.8 bar. If the acoustic velocity is known (from model or own measurement) the flow velocity can be calculated from the water-hammer equation, assuming a clean pipe (known diameter).

Quick-acting hydraulically actuated valves in oil and gas production operations, can be closed in 0.1 s. In practice, however, the closing time of subsea valves is much slower. A rule-of-thumb, is that the closing time will be 1 s per inch in diameter. After full closure at a pipe outlet, the pressure upstream of the quick-acting valve will continue to increase after the water hammer, as illustrated in Figure 4.17. The increase is due to the frictional pressure drop in the pipeline. As the fluid comes to rest, the frictional pressure drop will be superimposed on the static pressure. This increase in pressure, propagates at acoustic velocity in the direction of the quick-acting valve.

The frictional pressure gradient in a pipeline is given by the *Darcy-Weisbach Equation*

$$\frac{\Delta p_f}{L} = \frac{f}{2d} \rho u^2$$

This is the gradient of the pressure increase after full closure of the quick-acting valve. A pressure pulse measurement on a clean pipeline will give a certain frictional pressure

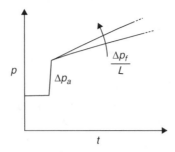

Figure 4.17 Pressure Pulse measurement. Pressure is measured immediately upstream a quick-acting valve.

gradient. A pressure pulse measurement on a pipeline with paraffin wax deposits will give a different and a higher gradient, as illustrated in Figure 4.17. The gradient need not be a straight line but could vary with time, each segment representing different deposit thickness. Thus, the thickness and location (converting measured time to distance using acoustic velocity) of paraffin wax deposits can be determined. The *Pressure Pulse* profiling method can be used to locate obstructions in pipelines, for example a stuck scraper (pig). The present description of the pressure pulse method is highly simplified. Detailed analysis of *Pressure Pulse* signals can reveal a lot of details about the inside walls of an entire pipeline.

The *Pressure Pulse* profiling method has been used on a 197 mm internal diameter (=8 inch nominal diameter), 116 km long gas condensate pipeline (Fahre-Skau *et al.*, 2013). The profiling was used to measure the location and thickness of wax deposits along the pipeline. The wax problem started when gas condensate from another field was mixed with condensate that reportedly had not exhibited any signs of deposition for years. As a part of the planning of pipeline pigging, the pipeline was profiled using the Pressure Pulse method. One of the many profiles measure is shown in Figure 4.18, illustrating the wax deposit thickness with distance. The sudden drop in deposit thickness after about 25 km, was related to various previous attempts to clean the pipeline by pigging. Otherwise, the deposition profile follows the expected trend, as illustrated schematically in Figure 4.10.

Wax deposits need to be removed from pipeline because of reduced flow capacity and increased pressure drop. One additional reason is the need to satisfy inspection requirements by the authorities. The 116 km was eventually cleaned by pipeline pigging. The approximate volume of the paraffin wax removed was 350 m^3. An interesting observation was that the pressure drop measured on-board across the pipeline scraper, corresponded roughly with the local wax thickness from the profiling measurements. A bypass scraper tool was used and its was about 0.3 m/s, which is quite low for a scraper PIG. The measured bypass flow (measured on-board) was shown to increase in proportion with the pressure drop across the scarper. This property of bypass scrapers is operationally favourable, because it provides greater liquid flow to transport the wax ahead/downstream of the scraper. The mechanical scraper was fitted with a non-return valve on the bypass, in case the scraper became jammed/stuck and had to be pumped back using high downstream pressure. Good paraffin wax management

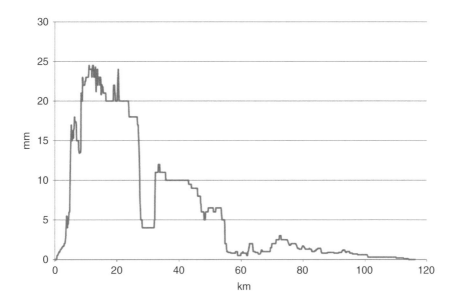

Figure 4.18 Wax deposit profile in a gas condensate pipeline (Fahre-Skaun *et al.*, 2013).

should include a profile measurement after pipeline pigging, to confirm the effectiveness of the cleaning operation. Such a practice may be challenging, because expensive inspection tools are required.

4.11 PIPELINE PIGGING

The technology of pipeline pigging is semi-empirical and rests predominantly with service companies, large and small. For an oil/gas company, production can continue (needs not necessarily stop) during pigging, which is a tremendous advantage. Pigging is the method of choice for removing flow assurance solids. The need for pipeline pigging has increased markedly in recent decades, due to subsea developments and an increase in the production of heavy and difficult to handle crude oils.

Pipeline pigging is an established part of sound field management. The following is a partial list of common pigging operations in the oil and gas industry: (1) removal of deposited solids, (2) removal of debris and liquid slugs, (3) dewatering of new subsea pipelines, (4) inspection of pipeline walls and (5) separation of different fluid products. The first point, the removal of paraffin wax deposits (scraping off), is the subject of the present section.

For a newcomer, the words pig and pigging may sound strange. The objects/projectiles/scrapers pumped through a pipeline may resemble the shape of the animal (round and long). The saying is that in olden-days, the cleaning of pipelines was carried out using straw-like materials, wound into a piston-like brush by a wire binding. During cleaning operation, the sound of the wire scratching against the pipe wall, gave a scratching/squealing noise, not that different from that of the animal. The acronym PIG (Pipeline Integrity/Inspection Gauge) has appeared in the literature, more correctly called a backronym, because the words were selected to fit the letters.

In principle, paraffin wax deposition in pipelines can be *prevented* by thermal insulation and external heating. Thermal insulation is expensive, especially in subsea pipelines. Long-distance (hundred to thousands of kilometre) pipelines on-land are commonly insulated. In subsea production, it is common to insulate pipelines feeding crude oil and condensate from satellite production sites to nearby processing facilities/platforms. External heating by either pipe-in-pipe technology (where hot fluid is circulated) or electrical heating is used for shorter distances, from production sites to processing platforms. External heating is much more expensive than thermal insulation. Electrical heating technology is limited to shorter distances. The prevention of paraffin wax deposition is akin to the prevention of natural gas hydrate deposition. Cold flow technology, as tested for natural gas hydrate, fortuitously/simultaneously includes also paraffin wax solids. The technology is presented in *Section 5.8 – Prevention by Cold Flow*.

Fine chemicals (not bulk chemicals) can be added to crude oil and gas condensate to prevent the agglomeration of paraffin wax crystals. Small crystals will not settle out and will therefore be carried with the bulk fluid. The field of chemical additives is within the realm of service companies and thus highly proprietary. Production chemicals for the oil and gas industry are presented by Kelland (2009).

Paraffin wax deposits may occur in wellbores (production tubing). The wax deposit can be removed by cutting and drilling, also chemical dissolution. However, the most common method is hot oiling. Relatively hot oil is pumped through coil tubing (the drillstring can also be used). The most difficult waxes to melt are deep in a wellbore, the waxes having highest molecular weight. The lighter waxes are at shallow depth. One challenge in hot oiling is that other flow assurance solids are likely to be mixed with the paraffin wax, likewise clay fines and sand from the reservoir formation.

Pipeline pigging is the most common method to remove paraffin wax deposits in pipelines, being much more cost effective than chemical methods. For oil/gas industry details, see O'Donoghue (2004), Hovden *et al.* (2004) and Wang *et al.* (2005). A general introduction to pipeline pigging is that of Davidson (2002). Two central technical issues that need to be addressed in flow assurance solids management, are the *deposition profile* (deposit thickness in time and space) and *pigging frequency*.

The most common method to *estimate* a paraffin wax deposition profile is modelling, as exemplified by Zhang *et al.* (2014). Commercial simulation software (Rahman and Chacko, 2013) and in-house software (Stevenson *et al.*, 2015) are other examples of deposition profile estimation. Comparisons of models in the public domain to flow-loop data were presented by Rosvold (2008). An analytical model and a numerical model were formulated and executed/programmed by Stubsjøen (2013). The analytical model was diffusion based while the numerical model was convection based. Field data from the North Sea were used as input parameters. Public domain models were used as reference points. The diffusion based model gave a much higher deposit thickness than the convection based model; in modelling, it is always a question of balance. As a part of the modelling work, Stubsjøen (2013) reviewed models for the *thermal conductivity of mixtures* of paraffin wax and entrapped oil (as in wax deposits on wall), because there is no *a priori* knowledge to derive it directly. In the present context, elements of paraffin wax modelling are presented in *Section 4.3 – Thermodynamics of Precipitation*.

Deposit thickness profile can be measured by *Pressure Pulse* profiling and by inline inspection (ILI) tools. Such tools will have a multi-arm calliper (U.S. English, caliper) and other sensors, commonly called intelligent pigs. These are direct measurements that will capture the actual deposition profile at the time of when carried out. *Pressure Pulse* measurements can be carried out at any time during on-line production (no need to stop pipeline flow) and are cost effective. Inline inspection requires a pigging operation, that is a more involved operation and costlier, although quite feasible during continuous production (no need to shut-in pipeline).

The pigging frequency is based on a synthesis of technical, operational and cost factors, which are at the heart of flow assurance solids management, including wax management. The velocity of scraping tools in pipelines is in the range 1–2 m/s, suggesting the time needed for a single pigging run. Early in the decision process, trial-and-error testing of pipeline gauges and scrapers, pigging modelling may be useful. A recommendation by O'Donoghue (2004), is that a pipeline should be cleaned no later than when the deposit thickness has become 2 mm. Another recommendation, is that the rate of bypass flow should be greater than the rate of build-up of scraped wax in front (downstream) of the scraping tool; what this means is unclear. In the case of paraffin wax pigging, one of the considerations should be the fact that liquid oil/condensate is entrapped in the deposit and that the deposit becomes harder with time (deposit aging).

A simple pipeline pigging model is presented in *Appendix M3 – Pigging Model for Wax*. Such a model can contribute to a better understanding of the main parameters involved, and to evaluate the operational pressures and flowrates required in pigging operations.

The pigging models is based on the force balance

$$F_p = F_f + F_w$$

where the subscripts signify, the **p**ressure drop across a scraper, the **f**rictional resistance between a scraper and the pipe wall and the **w**ax removal force. The following expressions were used in the model

$$F_p = \frac{\Delta p}{A}$$

$$F_f = \mu_f g M$$

$$F_w = \sigma_w A_w = \sigma_w \pi (\delta d - \delta^2)$$

$$\sigma_w = 660(1 - \phi)$$

The pressure drop force is based on the flow area of the pipe, while the wax removal force is based on the deposit annulus area (wax thickness given by the symbol δ, not to be confused with solubility parameter). The variables in the friction force expression are coefficient of friction, gravity constant and the mass of the wax scraper. In the simple pigging model, the wax removal force is assumed to depend on the compressive strength of the wax deposit, signified by sigma, which depends on the volume fraction of entrapped oil.

For a wax scraper to move through a pipeline, the pressure drop force must be greater than the sum of the frictional and removal forces. The simple pigging model was used by Galta (2014), to study wax pigging of subsea pipelines. The model was found to ignore the forces due to the bending (elastic compression) of the scraping discs. The diameter of scraping discs is typically 102–105% larger than the pipe diameter (Davidson, 2002). The scraping discs must be compressed to fit into a pipe. The force acts on the outer area of the discs can be expressed by

$$F_s = \sigma_s A_s$$

where the subscript stands for scraper. This expression is analogous to the wax removal force. The elastic compression of a scraping disk can be related to *Young's Modulus* (stiffness of a material) through

$$\sigma_s = \varepsilon E$$

$$\varepsilon = \frac{\Delta L}{L_o}$$

The disk materials used pipeline pigging are commonly polyurethane elastomers. The *Young's Modulus* for such materials is reported to be in the range 8–13 MPa (Nieckele *et al.*, 2001). It follows from the above, that the frictional force will consist of two parts, the normal gravitational force and the compressional force

$$F_f = \mu_f g M + n E_s \frac{\Delta L}{L_o} A_s$$

The number of disks n has been added to the second part of the expression. The simple model in *Appendix M3 – Pipeline Pigging of Wax*, needs to be corrected by adding the second part, the effect of bending of the scraping disks. Any up-dating of the model should take into consideration the work of Barros *et al.* (2005), on wax removal forces in pigging operations.

As pointed out by Galta (2014), the compressional force is much larger than the gravitational force in pipeline pigging. Estimating the gravitational force (first part), reasonable values would be $\mu_f = 1$ and 100 kg, resulting in about 1 kN. Estimating the compressional force (second part), reasonable values would be minimum 4 discs, $E_d = 10$ MPa, 105% diameter (ID = 200 mm, $\Delta L/L_o = 10/200$) and a disc thickness of 1 inch, resulting in about 32 kN. The wax removal force was estimated 0.6 N, assuming 2 mm deposit thickness and 30% vol. content (entrapped fluid). Therefore

$$F_{scraper} \gg F_{friction} \gg F_{wax}$$

Table 4.11 Values used in estimating pressure gradient in a typical pigging operation.

Description	Value
Length	10 km
Diameter	0.3 m
Flowrate	0.07 m^3/s
Reynolds number	110,000
Bypass area	1%
Oil velocity	1 m/s
Pig velocity	0.83 m/s
Inlet temperature	45°C
Inlet pressure	5.2 MPa
Overall coefficient	20 W/m^2 · K
Density	780 kg/m^3
Viscosity	2.14 mPa · s
Oil content wax	30% vol.

In pipeline pigging, the scraping device is moved forward by a pressure difference. An example calculation was carried out by Galta (2014), based on the values shown in Table 4.11, taking into consideration a multitude of factors. The values represent a typical liquid carrying subsea pipeline. The scraper has a bypass to wash along the removed paraffin wax to downstream.

The pressure profile in the pipeline described in Table 4.11, is shown in Figure 4.19. The inlet pressure is 52 MPa and the pipeline scraper is assumed to be in the middle, where the pressure drop across the scraper is approximately 2.5 MPa. With an outlet pressure of about 46.5 MPa, the total pressure drop, from inlet to outlet is 5.5 MPa. The pressure drop across the scraping device is about one-half of the total pressure drop. The pipeline pressure gradient downstream of the scraping tool is slightly steeper than upstream of the tool. The temperature in the pipeline decreases exponentially with distance, affecting both fluid viscosity and density. Furthermore, downstream of the scraping tool, the flowing medium is a liquid-solid slurry (see *Section 4.3 – Viscosity of Oil-Wax Slurry*). The *Einstein Equation* was used by Galta (2014), for the viscosity of a slurry

$$\mu = \mu_o(1 + 2.5\phi)$$

where the subscript o signifies pure liquid at the same temperature and pressure as the slurry (Einstein, 1906). The volume fraction is that of the paraffin wax solids. The use of the equation presumes that the removed wax particles are larger than precipitated wax crystals. To avoid confusion, note that the symbol μ is used for both the coefficient of friction above and fluid viscosity in general (also used for chemical potential). The *Einstein Equation* can also be used for dilute emulsion, in the absence of experimental data and/or more relevant correlations.

A sketch of a mandrel scraping tool is shown in Figure 4.20. The length of such tools is typically 1.5-times the pipeline diameter. The tool has four flexible scraping (sealing) disks and two support disks. The disks are held together by a central mandrel.

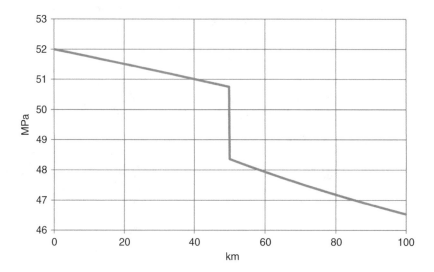

Figure 4.19 Pressure profile in a pipeline under pigging.

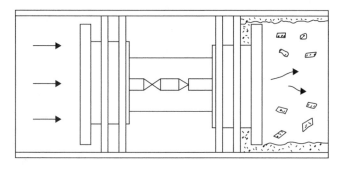

Figure 4.20 Mandrel-type scraping tool with a bypass. A valve and a non-return valve on the bypass line.

The sealing disks have a diameter slightly larger than the pipe diameter (ID) and act as the main scrapers. The support disks are rigid and have a diameter slightly smaller than the pipe diameter, their purpose being to ensure that the scraping tool remains central in the pipeline. The scraping tool is fitted with a central bypass line through the mandrel. The purpose of the line is to pass a fraction of the upstream liquid through the tool to the downstream side. The bypassed fluid acts to carry forward the removed wax, to prevent a buildup of a wax plug ahead of the scraping tool; a wax volume that the receiving facilities cannot handle. The central bypass line has a pressure drop control valve, pre-set for a particular pigging operation. The line has also a non-return valve, open during a pigging operation. In the case of the scraping tool getting jammed/stuck in the pipeline, it can be pumped backwards. A scraping tool can be fitted with special on-board instruments, to monitor variables of interest. The paraffin wax deposit is also illustrated in the figure, a bit thicker on the bottom than on the top (the case in

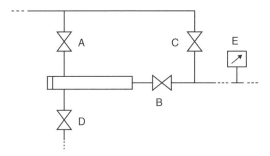

Figure 4.21 Pipeline scraper tool launching/receiving barrel. Launching, flow left-to-right. Receiving, flow from right-to-left.

multiphase slug flow). In reality, the paraffin wax will fill all the spaces between the sealing disks.

A scraping tool needs to be launched from one location and received at another location. A simplified sketch of a launching barrel (trap) is shown in Figure 4.21, the flow being from left-to-right. During normal pipeline operation, valve C is open and all other valves closed. The launch barrel has a quick opening trap closure/door, having a large enough diameter to accommodate a scraping tool. A scraping tool can be launched by closing valve C and opening valves A and B. Fitted along the pipeline are several tool indicators, marked E. The launch barrel can also be used as a receiving barrel. In such a mode, valve C is closed and valves A and B open. The scraped-off paraffin wax carried by the oil/condensate, is a resource and will enter the processing facilities. Valve D can be used as a drain valve or coupled to a high-pressure source to balance the barrel pressure to the pipeline pressure.

4.12 CONCLUDING REMARKS

The paraffin wax *deposition problem* is as old as the oil industry. Produced oil is hot and the ambient temperature is cold, leading to wax precipitation and deposition in surface and subsea pipelines. Crude oil consists of the whole spectrum of hydrocarbon molecules, from the lightest to the heaviest. Gas condensate consists of a lighter spectrum of hydrocarbon molecules, including the normal paraffins. The deposition of paraffin wax is almost inevitable in normal field operations world-wide. Thermal and chemical methods to control wax deposition, can be used in certain limited situations, but at exorbitant costs. The most cost effective solution to the paraffin wax deposition problem, still remains pipeline pigging in most situations.

Over the years, *observations* of paraffin wax deposition in flow-loops and field pipelines tend to be consistent. Twelve observations have been listed in the present chapter, ranging from the established to emerging. Among the established observations, is the dissolution and precipitation of paraffin wax in crude oil and gas condensate. Among the emerging observations, is the aging and hardening of wax deposits with time, often formulated in terms of the transfer of entrapped oil in a wax crystal structure. Large crystals are more thermodynamically stable than small crystals.

Vapour-liquid-solid *equilibrium models* have been developed and tested for paraffin wax (normal alkanes) in hydrocarbon liquids, such as crude oil and gas condensate. The models are based on chemical thermodynamics, formulated in terms of equations of state (EOS) supplemented by a range of physical property correlations. Activity coefficient models and regular solution models have been used with success. Any and all models have only been tested against the precipitation curves of different crude oils and gas condensates. The models are equilibrium based and cannot capture time-dependent behaviour.

Paraffin wax *deposition profiles* shown the buildup (thickness) of deposits with time. The apparently most common type of buildup can be described as logarithmic, increasing steadily but at a slower and slower rate. The logarithmic behaviour seems to be most common in flow-loops using natural fluids. In flow-loops using synthetic fluids (mixing of hydrocarbon liquid and refined paraffin wax), the buildup behaviour seems to fall into the exponential category, reaching relatively rapidly a constant asymptotic deposit thickness with time. The build-up of paraffin wax deposits in real pipelines occurs much more slowly than in laboratory based flow-loop testing. The time-depending buildup is not really documented, but here are indications that it may be linear.

Empirical *deposition-release* models have been derived and tested for exponential and logarithmic buildup of paraffin wax deposits. The two most important parameters were shown to be the wall shear stress (flowrate, flow velocity, friction velocity) and the temperature driving force, formulated as a dimensionless temperature difference. The effects of the mentioned parameters were in line with the relevant flow-loop and pipeline wax summary observations. The empirical deposition and release parameters can be related to theoretical properties and parameters, making scale-up from laboratory to field conditions more reliable.

An improved *knowledge and understanding* of paraffin wax precipitation and deposition, will contribute to improved solids management practices. Best practices need to penetrate front-end planning, engineering design/development and the operations of wells, pipelines and processing facilities. A key element of effective paraffin wax management will always be pipeline pigging. Oilfield operators need to avoid potential total blockages of pipelines, that could shut-down production wells and facilities and cause a substantial loss of revenue.

Chapter 5

Natural gas hydrate
The ice-like solids

Subsea production of oil and gas is totally dependent on reliable methods to prevent the formation (precipitation) of natural gas hydrate blockages and plugs. Long tie-backs (flowlines and pipelines) cannot be operated without gas hydrate prevention methods. Pipeline cooling is the main culprit. Hydrates can also form in on-land pipelines and equipment. Natural gas must be processed to give a low water dew-point to avoid water condensation in abrupt pressure drop situations (Joule-Thomson cooling).

The most common hydrate prevention method is diluting the liquid water phase 20–50% with antifreeze chemicals at high investment and operating costs. Antifreeze bulk chemicals are technically called thermodynamic inhibitors. Pipeline insulation and/or heating can be used for short distances in some field developments. Low-dosage (fine) specialty chemicals are beginning to be used in some oil provinces of the world. The term used is Low Dosage Hydrate Inhibitors (LDHI). A challenge in using LDHI's has been environmental concerns; ideally, such chemicals should be *green chemicals*. The low-dosage chemicals are carried by the liquid water/brine phase. Other technologies are being developed to prevent hydrate plugging, including *cold flow*.

The importance of gas hydrate flow assurance is reflected in the large volume of literature on the topic in recent decades. A review of natural gas hydrate inside and outside flowlines is that of Sloan *et al.* (2009). A review of the fundamentals of natural gas hydrate is that of Koh (2002). Extensive deposits of natural gas hydrate are found world-wide on-land and subsea at depths where both pressure and temperature are within the hydrate envelope. The voluminous deposits are considered potential future sources of natural gas for mankind (Birchwood *et al.*, 2010). With global warming, methane may be released for permafrost regions, aggregating the *greenhouse-effect*.

Oilfield chemistry conferences are held regularly and general oil and gas conferences have sessions on gas hydrate. The foremost hydrate conference is the *International Conference of Gas Hydrates*, held every three years since 1993. The conference has been held in the following cities/locations: New Palz (New York State), Toulouse, Salt Lake City, Yokohama, Trondheim, Vancouver, Edinburgh, Beijing and Denver (2017). The cities are known for active natural gas hydrate research groups. Numerous doctoral theses have emanated from the world-wide research community on gas hydrates.

An early translated book on gas hydrate is that of Makogon (1981). Another translated book is that of Berecz and Balla-Achs (1983). Later books on the subject are Makogon (1997) and Sloan (1998). Practical engineering books have been written by Sloan (2000) and Carroll (2003).

5.1 TRAPPED IN CAGES

Fresh water at atmospheric pressure forms ice at 0°C, practically independent of pressure, at least at pressures found in oil and gas operations. Water and light natural gas molecules form gas hydrate at moderate temperatures and pressures, typically at 20°C and 100 bara (=10 MPa). The crystal structure of gas hydrate is similar to that of frozen water. Water molecules, a few degrees above the freezing point of water, take on an increasingly ice-like structure when cooled. Something similar happens when water saturated with light natural gas molecules is cooled (the gas molecules must first be dissolved in the water). The presence of the gas molecules stabilizes the water molecules such than an ice-like structure forms, above the freezing point of pure water. The gas molecules are trapped inside an ice-like structure. While an ice crystal transmits light, a gas hydrate reflects light and is optically white.

Cooling is required for gas hydrate to form, just as cooling is required to form ice. The specific energy required to freeze water is 334 kJ/kg while that of gas hydrate is about 410 kJ/kg, which includes the heat of solution (enthalpy of mixing) of the gas molecules.

Gas hydrates are also called clathrate hydrates, meaning enclosed or trapped inside. The formation of gas hydrates is not a chemical reaction but a physical process. The gas molecules must be small enough to fit inside the solid water cages, also called cavities. The following molecules in oil and gas production can form a hydrate: methane, ethane, propane and iso-butane (not normal-butane), carbon dioxide, hydrogen sulphide and nitrogen.

Carbohydrates (not hydrocarbons) have the general molecular formula $C_m(HO)_n$ and were once thought to represent hydrated carbon. The values of m and n are commonly different, but in some carbohydrates $m = n$. The arrangement of atoms in carbohydrates has little to do with water molecules. Carbohydrates are found in sugars and starches in fruits and in vegetables and cereals.

Crystals are three-dimensional structures. Cubes have six sides (also called faces) and triangular pyramids four sides. The cube form is also called hexahedron and the pyramid form tetrahedron. Collectively, the various forms are called polyhedra (singular, polyhedron). A polyhedron with 12-faces is called dodecahedron. Each of the 12 faces is shaped like a pentagon. One of the crystal cages that can trap small gas molecules is shaped like a dodecahedron, shown in Figure 5.1.

A water molecule is located at each of the corners of a hydrate dodecahedron in a tree-dimensional structure. The oxygen atom of water is larger than each of the hydrogen atoms. The molecule is highly polar and the hydrogen atoms a separated by an angle of 105 degrees. The angle of each of the corners of an ideal pentagon is 108 degrees.

Unfolding a polyhedron with 12 pentagonal sides, the number of water molecules can be counted, see Figure 5.2. There are 20 corners and hence 20 water molecules. According to Jeffrey (1972), 30 of the 40 available hydrogen atoms in the $H_{40}O_{20}$ structure form hydrogen bonding. Research on gas hydrates dates back to Humphrey Davies (1778–1829) and Michael Faraday (1791–1867) and more recently Linus Pauling (1901–1994).

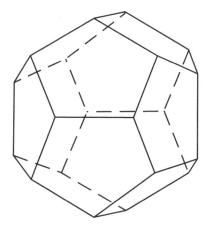

Figure 5.1 Polyhedron with 12 pentagonal surfaces.

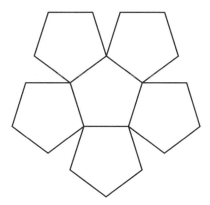

Figure 5.2 Unfolded polyhedron with 12 pentagonal sides.

The structure of hydrate cages is held in place by hydrogen bonding. Hydrogen bonding is not like chemical bonding, but arises because of the dipolar nature of the water molecule. The positive nature of the hydrogen atom is attracted to the negative nature of neighbouring atoms, namely neighbouring oxygen atom in another water molecule. Hydrogen bonding in three-dimensions keeps the hydrate cages together.

Geometrically, it is not possible to form a coherent tree-dimensional structure by 12-sided polyhedra alone, due to the angular distribution between molecules. Another polyhedron with 14-sides is required, called tetradecahedron. This polyhedron has 12 pentagons and two hexagons. A hexagon has six sides. One hexagon sits at the bottom and one at the top. Between the top and bottom are 12 pentagons. Yet another polyhedron with 16-sides, called hexadecahedron, is required to complete the structure of gas hydrate. It has four hexagons and 14 pentagons. The structures of the three types (12-sided, 14-sided and 16-sided) of polyhedra are shown in Figure 5.3.

Different combinations of 12-sided, 14-sided and 16-sided polyhedra make up the overall structure of gas hydrate. The structures are called structure I, II and H (Sloan, 1998; Carroll, 2003). Structure I consists of two small cages (12-sided polyhedra) and six large cages (14 sided polyhedra). Structure II consists of 16 small cages (12 sided polyhedra) and 8 large cages (16-sided polyhedra). Such combinations are called unit cells. Comparison of the unit cells of structure I and structure II gas hydrates are shown in Table 5.1.

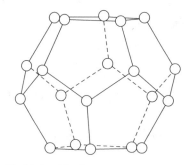

12-sided polyhedron

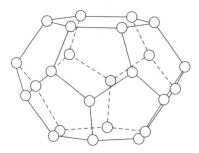

14-sided polyhedron

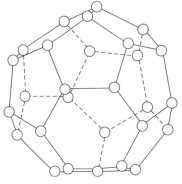

16-sided polyhedron

Figure 5.3 Structure of polyhedron forming gas hydrate (Carroll, 2003).

A widely-used illustration of the unit cell of structure II gas hydrate is shown in Figure 5.4 (Sloan, 1998; Lee and Holder, 2001). The illustration originates from Mak and McMullan (1965).

The number of cages listed in Table 5.1 is theoretical. Gas hydrate is seldom stoichiometric because not all of the cages will necessarily contain a gas molecule. Gas hydrate formed from natural gas will always have structure II. The gas molecules that form structure I gas hydrate will also fit into the cages in structure II. All of the gas molecules in Table 5.1, therefore, will take part in the formation of structure II gas hydrate in the oil and gas industry. Structure H gas hydrate is much larger than

Table 5.1 Comparison of the unit cells of structure I and structure II gas hydrates (modified from Carroll, 2003).

	Structure I	Structure II
Water molecules	46	136
Small cages	2	16
Large cages	6	8
Small cages Å	7.9	7.8
Large cages Å	8.6	9.5
Gas molecules	CH_4	C_3H_8
	C_2H_6	$i\text{-}C_4H_{10}$
	H_2S	N_2
	CO_2	

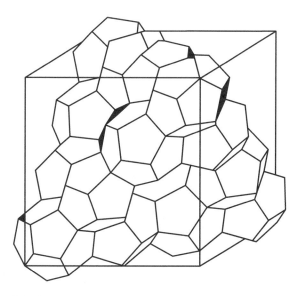

Figure 5.4 Illustration of structure II hydrate (Mak and McMullan, 1965).

Table 5.2 Composition (concentration) of a typical natural gas mixture and the ideal fraction entering structure II hydrate.

	c % vol.	c % vol.	M kg/kmol
Methane	85.828	86.179	16.0
Ethane	8.271	8.271	13.0
Propane	2.335	2.335	44.1
I-Butane	0.233	0.233	58.1
N-Butane*	0.255		58.1
I-Pentane*	0.029		72.2
N-Pentane*	0.029		72.2
N-Hexane*	0.038		86.2
Carbon dioxide	1.825	1.825	44.0
Nitrogen	1.150	1.150	28.0
TOTAL	100.000	100.000	

*Non-hydrate former

structure II, but is apparently not involved in hydrate from natural gas. Other molecules than listed in Table 5.1 can also form hydrates (see Sloan, 1998 and Carroll, 2003 for details).

The ideal formula of structure II (based on Table 5.1) gas hydrate can be written

$$24X \cdot 136H_2O$$

where X stands for the hydrate forming gas molecule. A made-up composition of a typical natural gas mixture is shown in Table 5.2. The non-hydrate formers amount to 0.351% of the total mixture, an insignificant percentage. Ignoring the non-hydrate formers and adjusting the mole percentage (same as volume percentage) of methane and the other hydrate formers proportionally, the third column in Table 5.2 represents approximately the gas molecules that will fit into the cages in structure II gas hydrate. The approximate density of this gas will be 17 kg/kmol. Using the above ideal formula, assuming all cages are filled with gas molecules, the gas molecules will be 14% of the total weight and the water molecules 86% of the total weight. Weight is important in non-pipeline technology for storage and transport of natural gas in the form of hydrate (Gudmundsson et al., 2002).

The 17 kg/kmol gas mixture will have a density of 0.73 kg/Sm³ at standard conditions. Assuming a hydrate density of 900 kg/m³ the weight of natural gas in one cubic metre will be 126 kg, corresponding to 172 Sm³, assuming all the cages are filled. Assuming 90% of the cages containing a gas molecule (non-stoichiometric hydrate), the volume value will be 155 m³. The above calculations are highly approximate. A *fractionation* of gas molecules occurs during the formation of hydrate (Levik et al., 2003). It means that the gas composition of a hydrate will not be the same as the gas composition from which it was formed.

5.2 WATER IN OIL AND GAS PRODUCTION

The oil and gas industry world-wide produces more water than oil. Water is a bigger cost item in conventional oil and gas production than is the precipitation and deposition of the big-five solids (asphaltene, hydrate, naphthenate, scale and wax). Water hampers the production rate of wells (a column of water is heavier than a column of crude oil) and must be separated in processing facilities. Separators and dehydration (drying) units are the largest pieces of equipment on offshore platforms and on-land processing facilities. The combination of water and natural gas gives rise to the problem of hydrate precipitation and deposition, causing blockages of flowlines and equipment.

Water is found in all oil and gas reservoirs. A simplified cross-section of a conceptual/representative hydrocarbon reservoir is shown in Figure 5.5 (scale exaggerated vertically). Section A represents a reservoir volume containing natural gas. The volume can be thought of as a natural gas reservoir or as a gas cap of an oil reservoir. In section B, the reservoir rock contains predominantly crude oil but also liquid water. The water saturation of the pore space may typically range from 2 to 20% vol. The liquid water saturates also the natural gas with water vapour according to the principles of mutual solubility, primarily through the saturation pressure of water. Similarly, small amounts of natural gas are dissolved in the liquid formation water. Natural gas components are also dissolved in the oil, as illustrated in Figure 3.8, showing typical solution gas ratio with pressure. The water content of natural gas is free of salt (upon cooling up a wellbore and at the surface, water vapour condenses out; the resulting liquid water phase tends to be acidic and therefore corrosive. The sour gases carbon dioxide and hydrogen sulphide play an important role in the acidity of liquid water.

It is important to make a clear distinction between *formation* brine/water (often termed free water) and *condensed* water vapour. Hydrate equilibrium curves are typically estimated based on non-saline water, meaning condensed water. This practice represents an example of conservative engineering design, because the presence of brine lowers the need for antifreeze (thermodynamic inhibitor).

Section B in Figure 5.5 represents a reservoir volume containing oil and a certain amount of liquid water. Commonly, the oil is undersaturated (reservoir pressure higher

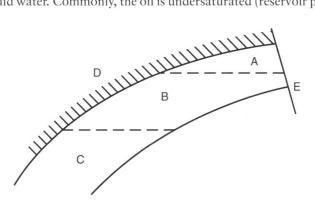

Figure 5.5 Cross-section of a conceptual reservoir. A: Non-associated gas or gas cap of oil reservoir (saturated with water vapour). B: Oil reservoir (with low to high liquid water saturation). C: Saline aquifer. D: Tight cap rock. E: Sealing fault. Interface between A and B is called gas-oil-contact (GOC); interface between B and C is called water-oil-contact (WOC).

than the bubble point pressure of the oil) such that free gas phase is not present (no gas cap). In some cases, the oil is close to the bubble point pressure such that associated gas evolves under production, for example in the near-wellbore region. A crude oil reservoir saturated with associated gas, will typically have a gas cap. With time, as the reservoir pressure decreases due to production, the reservoir volume containing oil and gas expands. The liquid formation water will saturate the associated gas with water vapour as occurs in natural gas reservoirs. Associated gas evolves from oil while natural gas reservoirs are said to contain non-associated gas. In non-associated gas reservoirs, the oil section B shown in Figure 5.5 will be very thin. With the advent of horizontal wells, thin oil zones can sometimes be produced.

The amount of liquid water produced from oil formations depends on the water saturation and on whether the formation rock is oil-wetted or water-wetted. The liquid water contains dissolved salts (see below) and is practically a brine. A common oil production strategy, especially offshore, is to inject processed (degassed) seawater and/or produced gas to maintain reservoir pressure. With time, water injection inevitably leads to *increased* watercut in the produced oil/fluids. In poorly designed water injection and/or highly fractured reservoirs, water break-through (from injection well to production well) may occur relatively quickly. As discussed in *Chapter 6 – Inorganic Scaling*, the mixing of formation water and injected water leads commonly to scale precipitation.

Section C in Figure 5.5 represents an aquifer, a subsurface rock formation containing liquid water in pressure communication (good, medium, poor) with the overlying oil. The water may contain some dissolved gas. As oil and gas are produced from above, the reservoir pressure decreases and the underlying aquifer water moves upward to fill the depleted reservoir volume. The terms water influx and water encroachment are used for water entering a depleted hydrocarbon reservoir volume. Most commonly, the aquifer water is not pure liquid water but saline water. Common seawater contains 3.5% wt. dissolved solids. The salinity of aquifer brine/water can be much higher.

Water coning describes the phenomenon of aquifer water entering production wells. Aquifer water may enter production wells at any time during the lifetime of a reservoir. Oil wells may produce from zero to more than 90% vol. water. Considerable effort is made to avoid water coning in the drilling and completion of production wells. Water coning affects not only crude oil producing wells, but also gas wells in non-associated gas fields. Aquifer water (brine) can be highly corrosive, having low pH due to dissolved carbon dioxide and hydrogen sulphide.

The interface between section A and section B in Figure 5.5 is called the gas-oil-contact (GOC). The interface between section B and section C is called the water-oil-contact (WOC). The interfaces within a reasonably homogeneous reservoir formation are practically horizontal. The symbol D in Figure 5.5 represents the cap rock of the reservoir. The cap rock is tight (low porosity and low permeability) and keeps the oil and gas in the reservoir volume. The symbol E in Figure 5.5 represents a sealing fault. Such faults can be found within reservoirs, thus dividing the reservoir into several blocks. The absence or degree of pressure communication between reservoir blocks is important knowledge for optimal reservoir production.

The thermodynamic properties of water and saline water are found in Sharqawy *et al.* (2010), both in diagrams and correlations.

5.3   WATER VAPOUR IN NATURAL GAS

Temperature has the largest (close to exponential) effect on the water vapour content of natural gas in reservoirs. Pressure has also an important effect. The presence of sour gases *increases* the water vapour content slightly; the presence of heavy gas molecules (gas gravity) and water salinity both *decrease* the water vapour content slightly.

Charts are found in handbooks and textbooks showing the water vapour content of natural gas with temperature and pressure. The best known is the chart published by the Gas Processors Suppliers Association (1998) for lean gas (gas gravity of about 0.7). A similar but a simpler chart has been presented by Rojey *et al.* (1997), shown here in Figure 5.6 (specific gravity not specified, but stated to be nitrogen free) The chart shows the water vapour content c_w mg/Nm^3 against temperature T °C and pressures from below atmospheric pressure (0.5 bara) to 300 bara.

In Figure 5.6, the *hydrate region* is at temperatures below about 20°C (demarked by dotted lines). The high pressure and high temperature region corresponds to reservoir conditions while the medium pressure and medium temperature region corresponds to dehydration/processing conditions. The low pressure and low temperature region in Figure 5.6 corresponds to conditions that may arise for example due to rapid pressure drop (Joule-Thomson expansion cooling) in pipelines and equipment.

The Nm^3 (normal cubic meter) in Figure 5.6 is referenced to 0°C while Sm^3 is referenced to 15°C (60°F) and atmospheric pressure (1.013 bara). Sm^3 is larger than Nm^3 by a factor of $(288.15/273.15) = 1.055$. Therefore, for the same milligrams of water vapour, the water content in Figure 5.6 needs to be divided by 1.055 to give c_w mg/Sm^3. In the European scientific literature, 25°C is commonly used as standard temperature. General consideration of the different units and standard states are given in *Section 1.5 – Symbols and Units*.

Methods have been developed to calculate the amount of water vapour in natural gas as discussed in *Appendix N – Water Vapour in Natural Gas*. To illustrate the effect of temperature and pressure on the water vapour content of natural gas, the method of Mohammadi *et al.* (2005) was used to calculate the water content from 20 to 120°C at three pressures, 10, 20 and 30 MPa. The results are shown in Figure 5.7. Clearly, the water vapour content decreases with pressure. Importantly, the water vapour concentration c_w mg/Sm^3 is that at *standard conditions*, not at the measured much higher pressures.

The figure shows that the water content increases approximately exponentially with temperature. The calculations show also that the water content at 90°C decreases from 6254 to 4016 to 3407 mg/Sm^3 for 10, 20 and 30 MPa, respectively; decreasing asymptotically from low pressure to high pressure. It may be worth noting that the molar volume of water remains approximately constant with temperature and pressure (within the rages of interest in oil and gas production). However, the molar volume of natural gas increases with increase in temperature and decreases with increase in pressure. Therefore, the water vapour content of natural gas has a relation to molar volume.

It can be difficult to read accurately from graphs such as that in Figure 5.6. Reading 7000 mg/Nm^3 for 90°C and 10 MPa and converting (factor 1.055, shown above) from normal to standard conditions gives about 6600 mg/Sm^3, which is 5–6% higher than calculated (shown in Figure 5.7). Phase behaviour (PVT) software calculations

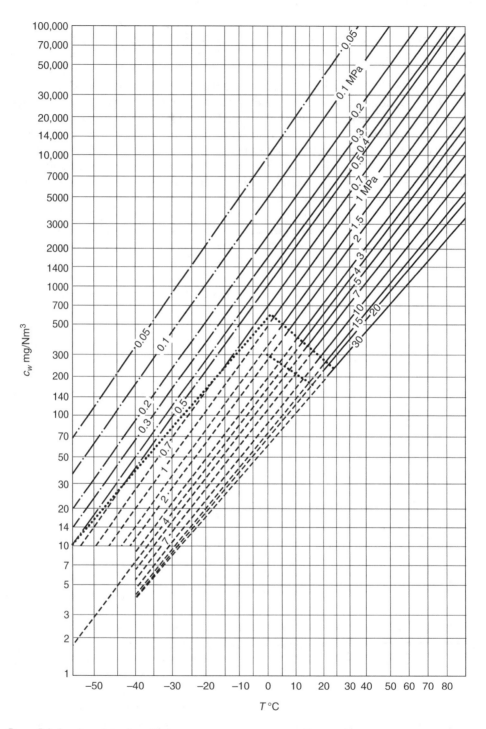

Figure 5.6 Log-log plot of equilibrium water vapour content of natural gas versus temperature at different pressures (Rojey *et al.*, 1997).

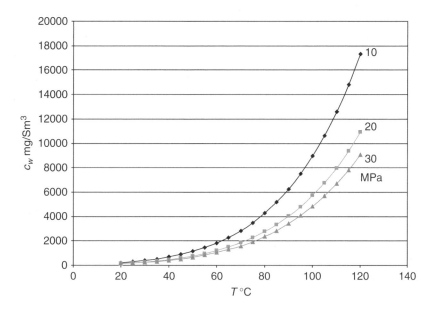

Figure 5.7 Equilibrium water vapour content of natural gas at standard conditions c_w mg/Sm^3 against temperature and pressure (top line 10 MPa, middle line 20 MPa and bottom line 30 MPa.

(reservoir and process simulations) are undoubtedly more accurate than simple chart-reading. The effect of the whole natural gas composition (sour gases and nitrogen) can be included. However, water vapour charts can be useful in off-the-cuff calculations before more involved simulations are carried out. Graphs can also be useful to control software calculations, to check garbage-in, garbage-out problems.

The effects of sour gases, gas gravity and liquid water salinity can be calculated using correlations developed by Mohammadi *et al.* (2005) given in *Appendix N – Water Vapour in Natural Gas*. The above figures and the correlations in *Appendix N* can be used in the oil and gas industry to estimate how much water will condense out from natural gas upon cooling. For example, natural gas at 10 MPa (=100 bara) and 90°C cooled to 5°C in a subsea pipeline, will condensate out about 6000 mg/Sm^3 of water. The condensed water will form natural gas hydrate unless removed or treated with chemicals (bulk or fine).

5.4   EQUILIBRIUM LINES

Equilibrium lines (dissociation curves) for natural gas hydrate show pressure against temperature. Gas hydrates form when the conditions are above the equilibrium line (to the left of the line). As discussed in *Appendix V – Solubility and Nucleation*, a certain supersaturation is required for crystallization to occur. The formation of natural gas hydrate (crystallization) in industrial systems requires a super-cooling corresponding to a temperature 4–6°C below the equilibrium line temperature (to the right of the line). The required temperature difference (super-cooling) in the formation of gas hydrate in clean laboratory systems may be even greater. Real field fluids contain a lot of fines and course particles. The fines and particles easily act as seeds for solid hydrate formation. The most accurate equilibrium line for gas hydrates can be obtained by melting solid hydrate, called the dissociation pressure (avoids super-cooling).

A number of methods are available in the literature to determine the equilibrium lines of gas hydrates (Sloan, 1998; Carroll, 2003). They include simple K-factor methods, gas gravity methods and various computer codes based on equilibrium thermodynamics. The equilibrium thermodynamics computer codes continue to be improved and represent the most accurate methods available to the oil and gas industry. In the present section, one empirical gas gravity methods applicable to natural gas mixtures will be presented. Such a method can be used in the absence of advanced computer codes; it illustrates the effects of pressure, temperature, gas gravity and the presence of carbon dioxide and nitrogen (effect of hydrogen sulphide not yet included, apparently). Again, easy-to-use empirical methods can be useful to check garbage-in, garbage-out problems.

The equilibrium lines of natural gas components that form hydrates (hydrate formers, Table 5.2) reflect the molecular weight of the components. The pure methane CH_4 equilibrium line lies above the pure ethane C_2H_6 equilibrium line. The pure propane C_3H_8 line lies further down at lower pressure. Natural gas mixtures will have different equilibrium curves depending on the gas composition. Importantly, the presence of propane causes the hydrate structure to be structure II. Structure II hydrates form at much lower pressure than structure I hydrates. In real oilfield situations, it is always structure II that forms.

The equilibrium lines for individual gas molecules are shown in Figure 5.8 (modified after Lee and Holder, 2001). Also shown are the equilibrium lines for argon, nitrogen, carbon dioxide and hydrogen sulphide. The equilibrium lines are based on the work of Holder *et al.* (1988), referenced in Katz and Lee (1990). Empirical correlations for the equilibrium line of individual hydrate formers by Holder *et al.* (1988) are also presented in Katz and Lee (1990). It is worth noting that H_2S hydrate forms at relatively low pressure. Pure carbon dioxide CO_2 hydrate forms at a lower pressure than methane hydrate. The chemical potentials of the natural gas components are different and result in a fractionation between the gas phase and the hydrate phase (Levik *et al.*, 2003).

The equilibrium lines for natural gas mixtures (structure II) exhibit a much lower dissociation pressure than pure methane (structure I). Natural gas mixtures are commonly characterized by reference to common dry air at standard conditions (atmospheric pressure and $15.6°C = 60°F$). Natural gas gravity is defined by

$$\gamma = \frac{M}{28.97}$$

where M kg/kmol is the molecular weight of the gas mixture and 28.97 kg/kmol is the molecular weight of common air. The molecular weight of methane being 16.043 kg/kmol means that its gravity is

$$\gamma = \frac{16.04}{28.97} = 0.554$$

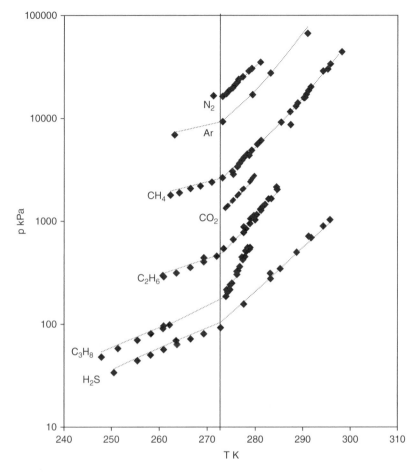

Figure 5.8 Equilibrium lines of hydrates showing dissociation pressure with temperature (after Lee and Holder, 2001).

Common natural gas mixtures (non-associated) have molecular weights 18–19 kg/kmol, thus having gas gravities of 0.62–0.66. Associated gases (dissolved in oil at reservoir conditions) have a higher molecular weight than non-associated gases.

In the oil and gas industry, gas gravity is commonly used in property correlations. The dissociation pressure (equilibrium curve) of natural gas hydrate is an example of such a correlation. An empirical correlation expressed in terms of gas gravity was developed and reported by Østergaard *et al.* (2000). The correlation was not based on experimental data but based on calculated results from a well-proven computer code based on equilibrium thermodynamics. The empirical correlation was successfully tested against experimental data with a maximum deviation of 0.1K. The gas hydrate dissociation pressure was given by the expression

$$p = \exp(aT + b)$$

Table 5.3 Constants in the Østergaard
et al. (2000) correlation.

Constant	Value
c_1	$4.5134 \cdot 10^{-3}$
c_2	0.46852
c_3	$2.18636 \cdot 10^{-2}$
c_4	$-8.417 \cdot 10^{-4}$
c_5	0.129622
c_6	$3.6625 \cdot 10^{-4}$
c_7	-0.485054
c_8	-5.44376
c_9	$3.89 \cdot 10^{-3}$
c_{10}	-29.9351

$$a = c_1(\gamma + c_2)^{-3} + c_3 F_m + c_4 F_m^2 + c_5$$

$$b = c_6(\gamma + c_7)^{-3} + c_8 F_m + c_9 F_m^2 + c_{10}$$

The pressure and temperature units were p kPa and T K. The factor

$$F_m = \frac{f_{nh}}{f_h}$$

expresses the molar ratio of non-hydrate forming hydrocarbons (pentane and heavier) to hydrate forming hydrocarbons (methane, ethane, propane and butanes). Both of the butanes are included as hydrate formers; furthermore, that the gas composition needs to be known (not enough to know the gas gravity). The constants c_1 to c_{10} are given in Table 5.3. The gas gravity used in the Østergaard *et al.* (2000) correlation is based on the hydrate forming gases only; that is, the non-hydrate formers are not taken into account when calculating the gas gravity.

An example was presented by Østergaard *et al.* (2000) where the dissociation pressure at 280.55 K was calculated based on a gas composition where the *hydrate forming* gas gravity was 0.6766. The saturation pressure, without correction for the effect of carbon dioxide and nitrogen, was calculated 1630 kPa. For details, see *Appendix O – Hydrate Dissociation Pressure.* It may be tedious, with all the constants involved, to calculate the equilibrium line (dissociation pressure) for a particular natural gas mixture. The use of spreadsheet makes such calculations more manageable.

5.5   NON-HYDROCARBON GASES AND WATER SALINITY

The dissociation pressure of gas hydrate decreases with the amount of heavier hydrate forming molecules as illustrated in Figure 5.8. The dissociation pressure of nitrogen is greater than that of methane and the dissociation pressure of carbon dioxide is lower.

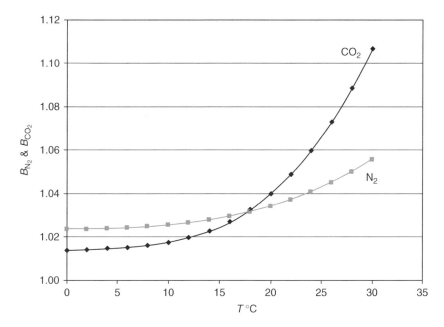

Figure 5.9 Correction factor for effect of carbon dioxide (◇) and nitrogen (□) on dissociation pressure of gas-hydrate, assuming 2% volume fraction of both gases.

However, the presence of both of these non-hydrocarbon gases increases the dissociation pressure of gas hydrate. In the Østergaard *et al.* (2000) method, the calculated dissociation pressure (based on the equations and constants given above) needs to be corrected for the presence of nitrogen and carbon dioxide

$$p_{real} = (B_{N_2} \times B_{CO_2})p_{ideal}$$

where the *B*'s are correction factors, each larger than unity (>1). The correction factors were calculated assuming 2% volume fraction (same volume fraction for comparison purposes) of both of the non-hydrate forming gases, using the same natural gas mixture as used by Østergaard *et al.* (2000). The correction factors are shown in Figure 5.9. See *Appendix O – Hydrate Dissociation Pressure*, for details. At temperatures below about 15°C the effect of nitrogen is larger than that of carbon dioxide. At temperatures above about 15°C, however, the effect of carbon dioxide is larger. At 30°C the presence of 2% vol. carbon dioxide increases the dissociation pressure by about 10% and that of nitrogen by about 5%. Clearly, the effect of the two non-hydrocarbon gases on hydrate dissociation pressure need to be included in the estimation of equilibrium curves.

The Østergaard *et al.* (2000) correlations for the correction factors are given in *Appendix O – Hydrate Dissociation Pressure*. The effect of hydrogen sulphide has not yet been included in the Østergaard *et al.* (2000) correlation, apparently. Based on an example gas composition given in *Appendix O*, the gas hydrate dissociation pressure was calculated taking the effects of CO_2 and N_2 into account, the

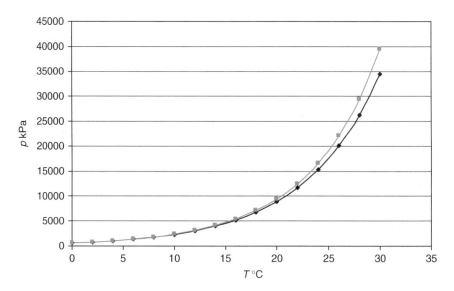

Figure 5.10 Dissociation pressure with temperature. Lower line without CO_2 and N_2, the upper line with the effect of the non-hydrocarbon gases (see mole percentages in text above). Based on correlations by Østergaard *et al.* (2000).

two non-hydrocarbon gases having concentrations of 2.38% mole and 0.58% mole, respectively. The results are shown in Figure 5.10. At about 20°C the non-hydrocarbon gases *increase* the dissociation pressure by 4% and at 30°C by 17%. These percentages may or may not be significant, depending on the accuracy of the gas concentrations and the correlations used. Nevertheless, the numbers show that the non-hydrocarbon gases should be included in dissociation pressure calculations, be it by semi-empirical correlations or full scale thermodynamic equilibrium software calculations. Based on Figure 5.8, it may be noted that pure nitrogen has a higher dissociation pressure than methane, while carbon dioxide has a lower dissociation pressure than methane.

The salinity of liquid water forming a gas hydrate affects the dissociation pressure. The presence of salt lowers the hydrate formation temperature, just as the presence of salt lowers the freezing point of water. The presence of salt increases also the boiling temperature of water. The presence of salt in water stretches the temperature scale at both ends; below the normal freezing point of 0°C and above the boiling point of 100°C, at atmospheric pressure.

An ingenious method to calculate the effect of salinity of the dissociation pressure of gas hydrate was presented by Mohammadi and Richon (2007). Recognizing the fact that salinity affects both the freezing point and the boiling point of water, and knowing that it is difficult to measure the dissociation pressure of gas hydrate in the presence of a saline solution, they turned the problem up-side-down. Mohammadi and Richon (2007) proposed to correlate the effect of salinity on the dissociation pressure of gas hydrate by the effect of salinity on the boiling point of water. The empirical correlation

$$T = T_o - 2.65 \Delta T_b$$

expresses the real dissociation temperature T K in terms of the ideal dissociation temperature T_o K and the boiling point elevation ΔT_b K, often abbreviated to BPE.

An accurate correlation for the boiling point elevation of seawater is given in *Appendix P – Boiling Point Elevation of Seawater* (Bromley *et al.*, 1974). Calculations based on 5, 15 and 25% weight of seawater salt showed that boiling point elevations (BPE) were approximately 0.5°C, 1.5–2.0°C and 3.25–4.25°C, respectively, for temperatures in the range 0–30°C. Inserting the maximum value for each of the salinities (0.5, 2.0 and 4.25) into the Mohammadi and Richon (2007) correlation, shows that the dissociation temperature needs to be lowered by 1.3, 5.3 and 11.3°C, apparently without the effect of pressure.

An interesting implication of the effect of water salinity on the dissociation pressure of gas hydrate in reservoir water, occurs in injection operations. The salinity of reservoir water is commonly much higher than that of ordinary seawater (3–5% weight). Water injection offshore is usually based on treated seawater. While high salinity formation water is produced, the dissociation pressure will be high. But, as injected seawater breaks through to production wells, the gas hydrate dissociation pressure will be lower, calling for greater use of chemicals, for example, to prevent the formation of hydrates.

5.6   PREVENTION BY ANTIFREEZE

The formation of gas hydrate can be prevented by the addition of antifreeze (thermodynamic inhibitor) to the liquid water phase. The antifreeze lowers the hydrate dissociation temperature. We know from Raoult's law (Francois-Marie Raoult 1830–1901) that the vapour pressure of solutions is the sum of the vapour pressure of the constituents of the solution. Similarly, the freezing point of solutions depends on the freezing point of the constituents.

The addition of alcohols and glycols to water lowers the freezing point of the aqueous solution. For dilute solutions (also called ideal solution) the following equation from Glasstone and Lewis (1960) describes the freezing point lowering

$$\Delta T = \frac{RT_0^2}{H_i} x_{mole}$$

where R is the gas constant, T_0 K is the freezing point of the pure solvent (here, that of pure water at 273.5 K) and H_i kJ/kmol, the molar heat of freezing ($=334$ kJ/kg $= 6008$ kJ/kmol) and x_{mole} the mole fraction of the solute (here, antifreeze). The molecular weight of the solute is not included in the equation. Using the given constant values,

$$\frac{RT_0^2}{H_i} = 103$$

Therefore, for a solute mole fraction of 0.1 the freezing point lowering will be 10.3°C, etcetera. Considering methanol and ethylene glycol as the solutes, the respective mass fractions will be 0.167 and 0.277. According to Carroll (2003) the above *ideal equation* can be used for methanol mass fractions below 0.3 and ethylene glycol mass fractions below 0.15.

Using the gas constant $R = 8.314$ J/mol $\cdot$ K in systems not involving gases may seem strange. The constant has the same unit as molar specific heat capacity kJ/kmol $\cdot$ K. The gas constant can be expressed by the product of two constants $R = N_A k_B$. The two constants are the Boltzmann constant $k_B = 1.381 \cdot 10^{-23}$ J/K and the Avogadro constant $N_A = 6.022 \cdot 10^{23}$ 1/mol. More than four significant numbers are required to get the correct value of R. The Avogadro constant is the number of atoms or molecules in one mole of substance. One mole of substance is defined equal to that of 0.012 kg of carbon 12. The international scientific community has agreed to use carbon 12 as a reference for one mole, which then determines the Avogadro constant. The Boltzmann constant, however, relates the energy of atoms and molecules to macroscopic variables such as pressure and temperature, at ideal conditions, including dilute solutions. The ideal gas law is just one of many equations where the Boltzmann constant is used. The gas constant should perhaps be called the energy of matter constant.

The legendary equation of Hammerschmidt (1939, 1969) can be used to calculate the dissociation temperature lowering of gas hydrate due to alcohols and glycols

$$\Delta T = \frac{K}{M} \frac{W}{(100 - W)}$$

where K is a constant, M kg/kmol inhibitor molecular weight and W % weight percent of inhibitor. The term $(100 - W)$ is the weight percent of water. By inhibitor means the solute, not the solvent. The equation can also be written in terms of mass fraction

$$\Delta T = \frac{K}{M} \frac{x_{mass}}{(1 - x_{mass})} = \frac{K}{M_{solute}} \frac{x_{mass,solute}}{x_{mass,solvent}}$$

Experiments have been carried out to find the constant for the bulk chemicals (solute = inhibitor) used in the oil and gas industry. Typical values are shown in Table 5.4, based in part on Pedersen *et al.* (1989) and Carroll (2003). The Hammerschmidt equation is generally considered applicable for inhibitor weight fraction below about 25%. The equation does not include the effect of pressure and gives a conservative answer, meaning higher antifreeze concentration than strictly required. Engineers like conservative answers, to be sure it will work.

As shown in Table 5.4, methanol is more effective (larger K/M ratio) than ethanol in lowering the dissociation temperature of gas hydrate. Furthermore, the table shows that the alcohols are more effective than the glycols. However, the alcohols have substantially higher vapour pressure than the glycols, meaning that they will evaporate relatively more when in contact with a natural gas mixture. Methanol, ethanol, MEG and water have the following vapour pressures at 20°C, respectively: 30 kPa, 9 kPa, 0.5 kPa and 2.4 kPa. The cost of bulk chemicals is an important factor in the overall hydrate management strategy of oil and gas companies.

Table 5.4 Hammerschmidt constants, molecular weight, density and viscosity of common gas hydrate inhibitors. Reference temperature 20°C except for glycol viscosity 25°C.

Inhibitor	K	M kg/kmol	K/M	Density kg/m³	Viscosity mPa · s
Methanol	1297	32.04	40.5	792	0.59
Ethanol	1297	46.07	28.2	789	1.2
MEG	1350	62.07	21.7	1115	16.9
DEG	2222	106.12	20.9	1118	35.7
TEG	3000	150.17	20.0	1125	49.0

MEG = Monoethylene glycol
DEG = Diethylene glycol
TEG = Triethylene glycol

An inhibitor should remain in the liquid water phase to prevent losses into the natural gas phase. MEG is probably the most used inhibitor in the oil and gas industry. MEG remains in the liquid phase and can be regenerated for repeated use. The effectiveness of MEG is similar to that of DEG and TEG but the viscosity is much lower, thus facilitating pumping in subsea applications. Furthermore, the boiling point of MEG being 198°C makes it possible to separate it from liquid water at reasonable evaporator temperature and pressure conditions. MEG reclamation and regeneration units are widely used in the oil and gas industry to reduced operating costs and environmental costs.

An equation proposed by Nielsen and Bucklin (1983) has proven more general for the estimation of the lowering of hydrate dissociation temperature than the Hammerschmidt (1939) equation. The Nielsen and Buckling equation has proved to give good results for inhibitor concentrations up to about 90 mass percent

$$\Delta T = -72 \ln(1 - x_{mole})$$

The temperature lowering/suppression is given in °C. The constant 72 was envisaged for *methanol only* but has proven reasonably accurate for other hydrate inhibitors also (Carroll, 2003). The constant must be multiplied by 1.8 when the temperature lowering is in °F.

The Nielsen and Buckling (1983) equation can also be written as

$$x_{mole} = 1 - \exp\left(\frac{-\Delta T}{72}\right)$$

to determine how much inhibitor (mole fraction) is required for a specified temperature depression.

The mole fraction of inhibitor, for example MEG, is related to the mass fraction of inhibitor by the equation

$$x_{mole} = \frac{M_{H_2O}}{M_{MEG}} \frac{x_{MEG}}{(1 - x_{MEG})}$$

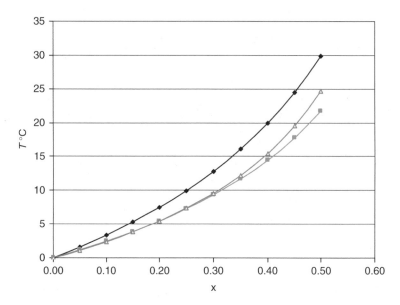

Figure 5.11 Gas hydrate dissociation temperature reduction with increasing MEG mass fraction x. Dilute/ideal solution equation (top), Bucklin and Nielsen equation (middle) and Hammerschmidt equation (bottom).

The dissociation temperature lowering of hydrate inhibited by MEG is shown in Figure 5.11. The temperature lowering is plotted with the mass fraction of MEG in water. The three lines are for a dilute/ideal solution (top line), the Hammerschmidt equation (bottom line) and the Nielsen and Bucklin equation (in the middle). At mass fractions below 0.3 the Hammerschmidt and Nielsen and Bucklin equations are similar. At greater mass fractions, the Nielsen and Bucklin equations follows the ideal equation line in shape, but moved parallel to the right in the figure. The line based on the ideal solution equation is always above the others, showing increasing difference with increasing mass fraction of MEG (as the solution becomes less and less ideal). To repeat, the dilute/ideal solution is the freezing point lowering equation of Glasstone and Lewis (1960), presented above. An inspection of Figure 5.11, shows that the Hammerschmidt equation recommends more MEG than the other equations to obtain the same temperature lowering/suppression. Hence, it is more conservative than the Nielsen and Bucklin equation. It is also reported to be more conservative than the dissociation pressures calculated using chemical thermodynamics.

The ratio between the ideal solution equation and the Nielsen and Bucklin equation is exactly 0.726 at all mass fractions such that

$$\Delta T = 0.726 \frac{RT_0^2}{H_i} x_{mole} = -72 \ln(1 - x_{mole})$$

and since

$$\ln(1 - x) \simeq -x$$

it can be concluded that the Nielsen and Bucklin equation is based on the ideal equation with the constant 72 added to fit experimental data (effect of methanol on the dissociation temperature of gas hydrate). The above approximation is strictly only applicable for $x << 1$, emphasizing the nature of dilute/ideal solutions.

Assuming a gas hydrate dissociation temperature of 20°C and a subsea pipeline that cools to 5°C, about 40% weight MEG is required to reduce the temperature by 15°C, based on the Bucklin and Nielsen equation, which is perhaps the most reliable of the three equations used in Figure 5.11. However, efforts have been made to measure the Hammerschmidt constant K for different systems. Applying the most applicable constant should give a reliable result. If the liquid water is highly saline there may be some chemical interactions with the inhibitor resulting in solids precipitation. Kelland *et al.* (1995) report that 10–60% weight of antifreeze (thermodynamic inhibitor) is typically used while Lovell and Pakulski (2002) report 20–50% weight of antifreeze is typical.

Industry experience from different projects indicates that 10–60% wt. MEG is typical and that 20–50% wt. methanol is typical. Thermodynamic inhibitors are usually delivered to field sites as 90% wt. MEG and 10% wt. water. In the case of methanol, the mixtures delivered are usually 99% wt. methanol and 1% wt. water.

Computer codes based on equilibrium thermodynamics are widely used to estimate the dissociation temperature depression due to inhibitors. The common practice in the oil and gas industry is to use such codes for calculating the dissociation pressure and temperature (equilibrium line) and the effects of inhibitors. In practical oil and gas industry settings, for a range of dissociation curves, the following empirical relationship is apparently used to extrapolate hydrate curves beyond the range of pressure and temperature for which they have been generated

$$T = a_1 + a_2 p + a_3 \ln p + a_4 p^2$$

where the a's are constants based on calculated curves (thermodynamic or empirical).

5.7 PREVENTION BY LOW-DOSAGE CHEMICALS

Natural gas wells produce water vapour that condenses to liquid and partakes in the formation of hydrates. Oil wells produce liquid water that may reach 90% or more of the total liquid/volumetric flow rate. Assuming that the liquid water phase needs to be diluted 50/50 with antifreeze to prevent the formation of gas hydrate, large volumes of glycol are required. Large volumes are expensive, require pumping and processing and pose an environmental challenge. Therefore, the oil and gas industry has looked for alternative chemicals (Kelland *et al.*, 1995; Lovell and Pakulski, 2002). Chemicals that would prevent the formation of gas hydrate 10–100 times more effectively and more safely. Effectively means lower concentrations (lower costs) and more safely means greener chemicals, posing less environmental hazard.

The term low-dosage hydrate inhibitor (LDHI) is used for chemicals that are used in concentrations below 1% weight of the liquid water phase. There are two main types of such chemicals: kinetic inhibitors (KI) and anti-agglomerants (AA). New chemical formulations are being developed in laboratories and field tested (Sun and Firoozabadi, 2014). They can be used alone or in combinations. To illustrate how the two types of low-dosage chemicals affect the formation of gas hydrate, the pressure in a hydrate forming test cell (containing liquid water and natural gas mixture) is shown schematically with time in Figure 5.12. The pressure and temperature in the test cell are within the zone of hydrate formation (below the dissociation temperature and above the dissociation pressure).

The initiation of gas hydrate crystal formation takes some time, shown as line A in Figure 5.12. Line B shows the pressure with time for a kinetic inhibitor (KI) system. The inhibitor prevents the formation of gas hydrate. However, after some time the crystal formation begins and the pressure in the test cell decreases. Line C shows the pressure with time for an anti-agglomerant (AA) system. The formation of gas hydrate crystals is not prevented. However, the anti-agglomerant prevents the gas hydrate crystals from agglomerating. The crystals remain small (less than 1 μm) and separated. Small crystals flow easily with flowing liquids (oil and water mixture).

The quality of a kinetic inhibitor (KI) depends on the subcooling (lowering of gas hydrate dissociation temperature) it can maintain and for how long. The ranges in question are 5–15°C and 1–2 hours. For an oil-dominated flowline/pipeline flowing 2 m/s the corresponding distances will be 7.2 km and 14.4 km. For a gas-dominated pipeline with some liquid phase (LDHI carried in liquid phase) flowing 10 m/s the corresponding distances will be 36 km and 72 km. However, the greater the subcooling required, the shorter the time a kinetic inhibitor can prevent gas hydrate crystallization. Clearly, in the case of pipeline shut-down for several hours, kinetic inhibitors cannot prevent the formation of gas hydrate. Antifreeze chemicals need to be injected before scheduled shut-downs of long duration. Another method is to inject (fill the flowline/pipeline) with a light hydrocarbon, for example gas condensate (dead-oil) or diesel fuel. Alternatively, pipelines can be heated electrically during schedules shut-downs. Operators must have a hydrate management strategy for both planned and un-planned shut-downs (shut-ins).

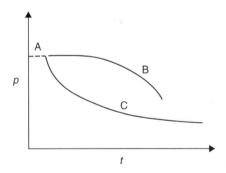

Figure 5.12 Schematic pressure in a gas hydrate test cell with time. A = initiation time, B = kinetic inhibitor and C = anti-agglomerant.

The effect/quality of an anti-agglomerant depends on the chemical composition of the flowing system; the composition of the crude oil or condensate and the salinity of the liquid water phase. Opposite to kinetic inhibitors, the quality of anti-agglomerants is not much dependent on the subcooling required and exposure time (for how long). Anti-agglomerants also offer protection in the case of pipeline shut-down. It appears that anti-agglomerants are less expensive than kinetic inhibitors and are thus favoured in oil and gas production operations. In any case, fine chemicals (AA's and KI's) are highly proprietary.

Kinetic inhibitors are synthetic polymers and anti-agglomerants are synthetic polymers and surfactants. The environmental impact of low-dosage inhibitors is therefore of concern, also their effect on downstream separation of oil, gas and water. Natural surfactants in oil and gas condensate, for example asphaltenes, may contribute to keeping gas hydrate crystals in suspension (anti-agglomerant effect). Therefore, it is necessary to test the use of low-dosage inhibitors on the oil and gas condensate in question. Hydrate inhibition and gas-oil-water separation must be viewed as a whole. Low-dosage inhibitors are offered to the oil and gas industry by service companies that are reluctant to divulge their composition and properties. Operators are sometimes complaisant about chemical selection and compatibility, relying unduly on the supplier. The use of low-dosage inhibitors is not allowed in all oil provinces of the world, even though they are reported more cost effective than antifreeze chemicals.

5.8 PREVENTION BY COLD FLOW

Research and development at NTNU (Norwegian University of Science and Technology) in the early 1990's, showed that natural gas hydrate slurry (the liquid phase was water or diesel fuel) in a circulation loop did not deposit on pipe walls in a constant temperature laboratory (adiabatic pipes). The gas hydrate particles produced in a CSTR (continuous stirred tank reactor) were small (1–10 μm) and stayed suspended in the liquid phase, even allowing shut-in for at least a day. Evidence that others studied cold flow as early as NTNU, has not be found in the literature.

In cold flow, hydrate particles suspended in the liquid phase (mixture of crude oil and brine/water) at constant temperature subsea pipelines, will not deposit on the pipe wall (the precipitated particles will not deposit). Before entering the pipeline the natural gas hydrate particles must be produced and cooled down to the surrounding (seawater) temperature. Ideally, all the gas should be converted to hydrate. A presentation of the NTNU cold flow concept is that of Gudmundsson (2002), illustrated in Figure 5.13. The gas-liquid mixture from a wellhead unit (WHU) is separated (SU = separator unit) and the liquid phase (L = oil and water) fed to a heat exchanger unit (HXU) for cooling down to close to seawater temperature. Importantly, the heat capacity of the liquid phase is much larger than that of the gas phase (G). The cooled liquid phase and the gas phase are mixed in a reactor unit (RU) where natural gas hydrate particles are formed. The total mixture then enters a cold flow pipeline (CFP), to a platform or a land-based receiving terminal.

Doctoral work by Andersson (1999) included the flow of natural gas hydrate slurries in 4 m long pipes in a constant temperature laboratory (see also Andersson and

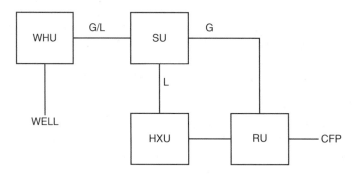

Figure 5.13 NTNU's cold flow concept.

Gudmundsson, 1999). An important result was that the frictional pressure gradient of a hydrate slurry in turbulent flow was the same as the frictional pressure gradient of the carrying liquid, as shown in Figure 5.14 (hydraulic gradient/head has unit m/m). In laminar flow, however, the frictional pressure gradient was higher, as widely reported in the slurry literature. The reason for the difference is the viscosity of the flowing mixture. In laminar flow the natural gas hydrate particles increase the bulk viscosity, while in turbulent flow the boundary layer (viscous sublayer) contains no or very few particles. The particles are small and are lifted away from the wall due to shear forces caused by the velocity gradient (see *Section 2.6 – Boundary Layer Theory*).

The Saffman lift force (Phillip Geoffrey Saffman 1931–2008) will make a small particle move away from a solid wall provided the particle is within the viscous sub-layer. The Saffman lift force is responsible for the fact that only the fluids viscosity dominates the pressure drop in turbulent pipe flow carrying small natural gas hydrate particles, in liquid water and in diesel fuel. The presence of the small hydrate particles has no effect on the pressure drop. It means that viscosity measurement made in the laboratory at viscous flow conditions (laminar flow) that show that the presence of particles increases the viscosity, cannot be applied in turbulent pipe flow (the pressure drop will be over-predicted). Therefore, it may be erroneous to apply viscosity values measured at laminar flow conditions in the laboratory in other flow assurance systems, for example paraffin wax in crude oil and condensate (see for example the model of Rygg *et al.* (1998). The same may apply to the viscosity of paraffin wax in light fuel oil reported by Abdel-Waly (1997).

The NTNU cold flow concept has been studied by Ilahi (2005, 2006) and Tvedt (2005, 2006), the former the technology and the latter its development and commercialization. The concept was compared to the SINTEF cold flow technology (see below) and the use of insulated pipelines and heated pipelines. In cold flow technology, the pipeline is bare; has no insulation. Tvedt (2005) estimated the investment costs (CAPEX) and concluded that insulated and heated pipelines are suitable for distances

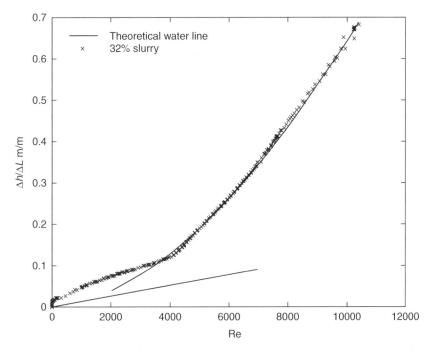

Figure 5.14 Hydraulic gradient (frictional pressure gradient, m/m) for natural gas hydrate slurry at laminar and turbulent flow conditions (Andersson, 1999). Lines show liquid water only, crosses show 32% weight hydrate in water (slurry).

less than about 100 km and cold flow technology for larger distances, as illustrated in Figure 5.15.

Two sets of lines are shown in Figure 5.15. The two lines that approach the origin are insulated pipeline (lower line) and insulated and heated pipeline (upper line). The two lines that approach CAPEX slightly below 100 million US$ at zero distance are the NTNU (upper line) and SINTEF (lower line) cold flow technologies. The insulated and heated pipeline costs are based on direct electrical heating. The general view is that the direct electrical heating technology can be used for maximum 50 km. For longer pipelines, several individual lengths need to be used (for example, four lengths of 50 km to reach 200 km). The OPEX (operating expenditures) was not included in the mentioned studies. The availability and cost of electrical power, clearly, would be important in using electrically heated pipelines. A clear advantage in using heated pipelines is the almost guaranteed trouble-free operation, on-line and during shut-in.

Traditional use of glycol to prevent gas hydrate precipitation is two orders of magnitude more expensive (CAPEX) than cold flow and insulated (and heated) pipelines (Tvedt, 2005). The reason being that the longer the pipeline the larger the liquid water inventory and hence larger volumes of chemicals are required. A chemical pipeline is required (from platform or receiving terminal to wellhead). Also, the cost of using low-dosage chemicals was estimated to cost about the same as glycols. The reason

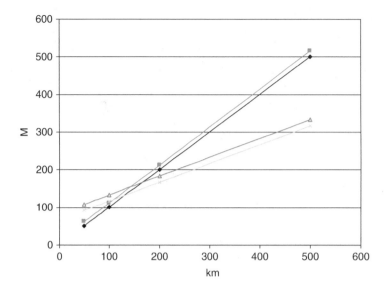

Figure 5.15 CAPEX in million $ (=M$) with distance.

being that although low-dosage chemicals are more effective than glycols, they are much more expensive.

Work carried out at SINTEF (Norwegian acronym for Foundation for Industrial and Technological Research) in Trondheim resulted in the development of a cold flow concept presented by Lund *et al.* (1998, 2010). The concept is schematically illustrated in Figure 5.16. The SINTEF cold flow concept differs from the NTNU concept in that two bare pipelines are used for heat exchange (cooling). Furthermore, precipitated natural gas hydrate particles are recycled in one of the pipelines to provide seeding. Gas-liquid flow directly from a wellhead unit (WHU) is mixed (MU = mixing unit) with a cold recycle liquid stream containing hydrate particles. The mixed stream flow in a bare steel pipeline (PL = pipe line) for cooling. At the end of the cooling pipeline the multiphase mixture is separated (SU = separator unit). A fraction of the liquid stream is recycled by the use of a pump (PU = pumping unit). The remaining liquid fraction and gas fraction enter a cold flow pipeline (CFP).

One of the advantages of the SINTEF cold flow concept is that paraffin wax will precipitate out in the bulk (not on wall) of the flow along with the natural gas hydrate. A potential disadvantage is deposition of hydrocarbon solids (paraffin wax and gas hydrate) on the cold wall of the pipelines. However, extensive laboratory testing indicates that the SINTEF cold flow concept tackles both natural gas hydrate and paraffin wax simultaneously. The test results have been reported in several technical presentations by Wolden *et al.* (2005), Larsen *et al.* (2009) and Lund *et al.* (2010).

One disadvantage of the SINTEF cold flow concept is the heat balance. The recycled liquid provides direct cooling by mixing but the pipeline cooling will be a slow process, calling for long pipelines. The temperature driving force in subsea pipelines is limited (see *Appendix B – Pipeline Wall Heat Transfer* and *Appendix C – Boundary Layer Temperature Profile*). It means that the cooling pipelines need perhaps to

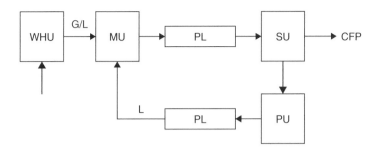

Figure 5.16 SINTEF's cold flow concept.

be mounted above the sea bottom and fitted with enhanced heat transfer fins (probably unpractical). Traditional subsea pipelines have a steel-reinforced concrete layer (anchor points can also be used) to make them heavy (prevent floating) to stay fixed at the bottom, resulting in a low overall heat transfer coefficient.

5.9   FURTHER CONSIDERATIONS

Natural gas dissociation curves are based on equilibrium thermodynamics. The passage of time is not involved in thermodynamics. An operator of production wells and flowlines is interested in time. Considerations of whether hydrate blockages and plugs will form during start-up, continuous operation and shut-down situations, are crucial to all companies/entities involved, from conceptual design to full-scale operations. The materials in *Chapter 2 – Flow Phenomena*, were compiled in an attempt to address and answer important challenges and problems. Managing gas hydrate problems requires a multidisciplinary approach, consisting of theories and experiences. Theories include chemistry, fluid mechanics and thermodynamics. Experiences include both laboratory studies and field observations. In addition to the materials presented in the above sections, mass transfer and kinetics need to be addressed. Both are difficult issues and sparsely reported in the literature.

Natural gas hydrate crystals form and grow at the gas-liquid water interface. Gas molecules that form hydrate are continuously replaced by new gas molecules that dissolve in the liquid water. The dissolved gas molecules are transported (mass transfer) to the crystal face/interface. The crystallization is exothermic with a heat of formation of about 410 kJ/kg, which is considerably higher than the 334 kJ/kg required to freeze water. The hydrate heat of formation quoted consists of both the heat of gas dissolution and the heat of crystallization.

Fortunately, the gas-liquid water interface in oil-gas-water carrying pipelines will not be at the pipe wall but in the bulk fluid. It means that gas hydrate particles are formed in the bulk of the flowing fluid, not at the pipe wall. Hydrates will form when the *in-situ* (local) pressure is above the dissociation pressure (in the hydrate forming region). Furthermore, gas hydrate solids have a tendency to form slushy clumps that compact (extrude excess water) with distance in a pipeline. With time the balls/clumps/plugs can compact further and reduce the effective gas flow area. Local flow restrictions lead to extra pressure drop that favour Joule-Thomson expansion cooling, and hence gas hydrate formation. Blockage of pipelines by gas hydrate may occur suddenly and without warning.

The oil and gas industry have developed methods to deal with gas hydrate blockages. The methods have been reported as hydrate engineering by Sloan (2000), for example. Of course, considerable efforts are made to avoid hydrate formation in pipelines. However, when a pipeline becomes blocked with gas hydrate solids, mechanical scraping and injection of chemicals are practically impossible. The most viable solution is often to depressurize the pipeline (below the equilibrium line) and allow the hydrate to dissociate. It is important to depressurize in several cycles/steps and check whether pressure communication has been established between up-steam and down-stream of a hydrate blockage/plug. Rapid and large depressurization can result in a hydrate plug becoming a high-velocity projectile that could damage equipment and pose a serious safety risk. Heat is required for hydrates to dissociate. In a subsea pipeline, the heat source will be the surround seawater. Industrial experience shows that the dissociation of hydrate in subsea pipelines is extremely slow.

The rate of hydrate formation in subsea pipelines has been reported by Mork and Gudmundsson (2001). The rate of gas hydrate formation data from a continuous stirred tank reactor (CSTR) was related to the energy dissipation W/kg in a pipeline due to frictional pressure drop. See *Appendix X3 – Energy Dissipation and Bubble Diameter* for some details. An inherent assumption is that the gas hydrate system is within the hydrate forming region and that the heat of formation is absorbed by sensible heat from the liquid phase. The following semi-empirical correlation was proposed

$$q_v = kp\alpha u_M \left(\frac{f}{2} \frac{u_M^3}{d} \right)^a \Delta T^b$$

The symbols stand for q_v m^3/s volumetric gas consumption rate at in-situ pressure, $k = 8.710 \cdot 10^{-5}$ correlation constant, pressure p bara, gas void fraction α, homogeneous mixture velocity u_M m/s, friction factor f, pipe diameter d m and temperature driving force ΔT °C (cooling below dissociation temperature). The two other correlation constants have the values $a = 0.1299$ and $b = 0.1702$. The correlation may give the maximum rate of gas hydrate formation in a subsea pipeline. The variable change along a pipeline. Normally, therefore, a pipeline can be divided into appropriate length segments. The correlation shows that the rate of gas hydrate formation increases with pressure and the temperature driving force. It depends also greatly on the mixture flow velocity. Note that the variables inside the bracket, represent the energy dissipation. The above correlation assumes a continuous formation rate of gas hydrate; not the discontinuous clumping of hydrate solids.

Finally, it may be interesting to know that gas hydrate can also have beneficial properties. Gas hydrate can be used for the storage and transport of natural gas. Early texts on this topic are Gudmundsson (1990), Gudmundsson and Parlaktuna (1994), Gudmundsson *et al.* (1994, 1995, 2001).

5.10   CONCLUDING REMARKS

Natural gas hydrates have become a *major field of study* in the oil and gas industry, in future energy resources scenarios and potential large-scale greenhouse gas emissions.

In the oil and gas industry, the importance of hydrates has been brought to the forefront in subsea production operations. In the field of energy resources, large volumes of gas hydrates have been mapped on-land and subsea, theoretically existing at temperatures and pressures within the required temperature and pressure envelope. The technology of producing gas from subsurface hydrates has *not yet* reached infancy and the economics are highly speculative. Continued global warming poses a major reason for potential melting of hydrates in permafrost regions of the world. Methane releases from permafrost in the northern hemisphere would self-propagate global warming.

Natural gas hydrates and clathrates in general, are fascinating materials. The most basic atoms/molecules of life, carbon, hydrogen and oxygen, are joint together in an ice-like structure, *above* the freezing point of water. Liquid water has unique properties; when cooled-down, the molecules prepare for becoming solid ice. However, in the presence of small gas molecules (dissolved in water), the burgeoning ice structures are stabilized enough to solidify to hydrates. That it happens at above ambient pressures is nature's way of finding the minimum of available energy, called equilibrium. Several appendices deal with free energy and equilibrium issues.

There is often a discrepancy between science and engineering. Science determines the dissociation pressure (equilibrium line) versus temperature, for pure gases and natural gas mixtures. Engineering determines what constitutes a reasonable approach to operability, costs and safety. In the oil and gas industry, the tradition has been to inject more chemicals than strictly required, just in case. The conservative engineering approach takes *unknowns and uncertainty* into consideration. As more and more field experience becomes available, the difference between the science and engineering narrows. For example, many operating companies injected enough thermodynamic inhibitor (antifreeze) to lower the dissociation pressure curve 3°C below expected lowest operating temperature. This practice is now apparently waning, with the advent of improved knowledge and experience.

Theory and practice has shown that the non-hydrocarbon gases carbon dioxide and nitrogen affect the dissociation curve of natural gas hydrates. Nitrogen tends to push the pressure upwards while carbon dioxide tends to push the pressure curve downwards. Hydrogen sulphide is present in may associated and non-associated natural gases. The dissociation pressure of hydrogen sulphide is relatively low, meaning that it may lower the pressure curve of natural gas hydrates (this hypothesis has not been tested in the present text). The use of computer models based on chemical thermodynamics may elucidate the effect of hydrogen sulphide. The presence of dissolved solids (salts) in formation water, lowers the dissociation pressure of hydrates. An important distinction needs to be made between saline (brine) formation water and condensed water vapour. Condensed water has no salinity, but may be corrosive because of dissolved sour gases. The type of reservoir and the reservoir drive mechanism, result in a range of compositions and mixtures. It means that the flow assurance issues are field dependent local issues.

The *new strand* in hydrate technology is the storage and transport of synthetic natural gas hydrates. The technology has been tested on laboratory and pilot-scale, but not yet implemented commercially. The price of natural gas will dominate in decisions to utilized hydrates for non-pipeline transport of natural gas.

Chapter 6

Inorganic scale

The salt-like solids

Chemical and thermodynamic concepts and methods are used in the analysis of inorganic scaling. The concepts and methods are used to understand and predict the *precipitation* of solids along the streamline from the subsurface to the production separator. Concepts such as mole fraction, chemical potential, ionic activity, partial pressure, fugacity and equilibrium constant (solubility product), are all important to understand solids precipitation and deposition. Whether a precipitate will form a *deposit* is currently more empirical than theoretical. Results from laboratory and field trials are valuable in understanding inorganic scaling, making inhibition and removal more manageable.

The petroleum literature on scaling overlaps with the water literature (boilers, heat exchangers, treatment, pipelines). Scaling in geothermal operations is also akin to what happens in oil and gas wells and processing facilities. A relevant book in the present context is that of Frenier and Ziuddin (2008). The book gives an overview of the science and technology of scale formation, removal and inhibition, with emphasis on the basic chemical and mechanical principles of scale control. Kelland (2009) elucidates the use of specialty chemicals in the oil and gas industry, including chemicals to abate scaling. Crabtree *et al.* (1999) present in a lucid manner field-related scale removal and prevention practices. Kan and Tomson (2010) proved both theoretical and practical aspects of scaling in the petroleum industry. A number of relevant references on inorganic scaling in the oilfield are listed in Ishkov *et al.* (2015).

Detailed theses on scaling are those of Kaasa (1998) and Sandengen (2006). The theses cover thoroughly the field of study and are extensively used in the present chapter. Both contain a considerable amount of experimental data and calculations. The fouling of heat exchangers is a problem of long standing in many industries, including inorganic scaling. Bott (1995) has been a leading proponent of fouling studies in industry. Fouling increases the heat transfer resistance of exchange surfaces (decreases heat transfer coefficient), meaning that larger heat exchangers are required at higher costs.

6.1 MAIN FACTORS AND ISSUES

A multitude of chemical and thermodynamic factors are involved in inorganic scaling in oil and gas production. The *first* and foremost, is the concentration of dissolved species in the produced water/brine. Precipitation occurs when the concentration of the species is above the solubility limit (supersaturated solution). The *second* important factor, is chemical reaction kinetics. If the reaction kinetics are fast, a precipitate will form locally, where supersaturation arises, and may form a hard deposit. If the reaction kinetics are slow, a precipitate will form gradually, after supersaturation is reached, and may form a scale from that location and onward in the remainder of the flow stream. The *third* important factors, are temperature and pressure. They affect the solubility product of precipitating species. The *fourth* factor, is the mixing of incompatible fluids; for example, one fluid high in sulphate and another fluid high in barium. The *fifth* important factor affecting scaling is alkalinity; the ability of a solution to neutralize acids. The opposite of alkalinity is acidity, the pH of a solution. The partial pressure of carbon dioxide plays an important role in controlling the acidity of aqueous solutions. The *sixth* factor, is the influence of other dissolved species; the ionic composition of a subsurface brine solution. In addition to the chemical and thermodynamic factors, the role of fluid mechanics (effect of turbulence) in inorganic scaling *cannot* be ignored, as illustrated throughout *Chapter 2 – Flow Phenomena*, and associated appendices.

Oilfield scales typically consist of different inorganic deposits (one or more types), containing debris of fines, sand and corrosion products. Organic deposits may also be involved. In oil and gas field operations, there are a number of issues involving inorganic scaling. Scaling may occur in drilling and completion operations. Water based drilling fluid (mud) may typically contain 65% wt. brine/water and 30% wt. barite salts. The remaining 5% wt. are stabilizing additives. The solid barite increases the density of the mud above that of brine/water alone. A scaling situation may arise where the mud and formation brine are mixed. The chemical equilibria may shift such that solid scale is formed downhole.

Scaling occurs in production operations. Mixing of fluids from different brine carrying layers and from injection seawater breakthrough, may lead to formation plugging (damage) in the near-wellbore region. Gravel packs, perforations and screens may become clogged by barium sulphide and other inorganic scale. Formation fines and sand may also be produced. Sand particles have a tendency to settle/deposit in downhole equipment, leading to increased pressure drop from the near-wellbore formation to the production tubing. Downhole pumps (ESP = Electrical Submersible Pump) are susceptible to inorganic scaling, curtailing their lifetime significantly. Production tubing, subsea safety valves (SSSV) and wellhead valves (gate valve and choke valve) are all susceptible to inorganic scaling. Solid particles are known to stabilize oil-water emulsions. The solids can be formation fines and precipitated inorganic solids. The stability of oil-water emulsions is a major challenge in process separation of oil, gas and water. Organic deposits, in particular asphaltenes, stabilize oil-water emulsions.

The drilling, completion and production issues have important safety implications. The main implication is perhaps clogging of a SSSV. The clogging of other valves is also relevant. An important safety issue in oil and gas production is the removal and disposal of radioactive scale, so called NORM (Naturally Occurring Radioactive Material). Radioactive scale can be harmful to personnel that handle such materials. It is an important disposal issue. The main NORM constituent is radium sulphate ($RaSO_4$). It co-precipitates with the much larger molecules of barium, calcium and strontium sulphates.

Numerous studies have been carried out on scaling in the oilfield. The thermodynamics of scale prediction was presented by Atkinson and Raju (1991), with emphasis on barium sulphate. The effect of temperature and pressure on oilfield scale formation was presented by Dyer and Graham (2002). They studied the solubility of insoluble sulphate scales of barium, strontium and calcium, and the solubility of carbonate scales of calcium and magnesium. Testing was carried out in a dynamic tube-blocking rig, where differential pressure drop is measured, including the effect of inhibitors. Temperatures and pressures relevant for petroleum reservoirs were employed in the testing, and several brine compositions and inhibitors were used. The overall conclusion reached by Dyer and Graham (2002) was that the effect of temperature was generally much greater than that of pressure. The use of near-infrared spectroscopy (NIR) to detect scale formation was presented by Ohen et al. (2004). It was concluded that NIR-technology is useful in planning application of inhibitors in oil and gas production. The kinetics of barite crystal nucleation, precipitation and the effect of several inhibitors was studied by Fan et al. (2012). Using a dynamic tube-blocking method they carried out test to shed light on the selection of inhibitors in HPHT oilfields. HPHT (acronym HTHP also used) stands for high pressure and high temperature. Deeper resources have TDS commonly above 300,000 ppm (Kan and Tomson, 2010), see also Fan et al. (2012). Similar numbers are given by Dyer and Graham (2002), but with 1000 bara as the upper pressure limit.

When scaling has occurred, there are diagnostic tools (PLT = Production Logging Tool) available to measure their location and thickness. The most common tool is perhaps multi-arm caliper, that measures by difference the scale thickness downhole (ID = Inside Diameter). A simultaneous fluid velocity measurement using a spinner tool enhances the usefulness of the caliper log. Caliper logging has several other more common applications. Recent technology for monitoring downhole, flowline and pipeline deposits, is pressure pulse technology (see *Section 4.10 – Monitoring of Pipelines*). Whatever technique is used, it is important to carry out measurements at the start of production, to know the upward flow velocity and wellbore (tubing) internal diameter before scaling occurs.

In all new oil and gas field developments, there are many uncertainties. An important uncertainty is whether scaling will occur, where and to what extent. Furthermore, how can/will scaling affect investment and operational costs/expenditures (Capex and Opex). Computer tools (scale prediction softwares) have been developed to analyse scaling potential in the whole flow stream, from reservoir to processing facilities, and beyond. The softwares available are based on established chemistry, thermodynamics and phase behaviour. The available softwares are used by oil companies, engineering companies and consultants alike. An example study is that of McCartney et al. (2014). The general statement can be made that scale prediction softwares are useful in identifying where scaling is likely to occur downhole, in flowlines, processing equipment and pipelines. However, the softwares are based on thermodynamic equilibrium and predict the maximum amount of scale that theoretically may deposit, which is useful to identify a potential problem. They predict equilibrium precipitation, not deposition. The commercial software packages lack the kinetics of scaling and the effects of multiphase oil/gas/water flow patterns.

6.2 PRODUCED WATER

The formation of inorganic scale occurs directly from the aqueous phase commingled with produced oil and gas. The aqueous phase is liquid water containing dissolved

Table 6.1 Composition of seawater and example formation water (Østvold *et al.*, 2010).

Constituent	Seawater mg/L	Formation mg/L
Na	11150	17465
K	420	325
Mg	1410	124
Ca	435	719
Sr	6	198
Ba	0	428
HCO₃	150	834
SO₄	2800	0.1
Cl	20310	28756

solids. For comparison, fresh water contains less than 1000 ppm of solids, seawater contains 30,000 to 40,000 ppm of solids and brine 40,000 to 300,000 ppm and higher. The abbreviation TDS (Total Dissolved Solids) is commonly used to express the ppm (mg/kg) concentration of inorganics salts/solids in produced waters. While scaling has been a problem in the oil and gas industry from the beginning (early times), it has increased in importance as deeper resources are exploited at higher temperatures and pressures.

There are four kinds of water that are coproduced with oil and gas, collectively called produced water: *formation* water, *condensed* water, *injection* water and *aquifer* water (see *Section 5.2 – Water in Oil and Gas Production*). Formation water is the water found in the pore spaces of oil and gas formations. Condensed water is water condensed from water vapour in the gas phase (see *Section 5.3 – Water Vapour in Natural Gas*). Injected water originates from injection of water/seawater, for reservoir pressure support, which is widely practices world-wide. This water flows from an injection well in the direction of a production well. Eventually is breaks though and is produced with the oil and gas. Aquifer water stems from water-bearing formations that encroach from the outside of hydrocarbon-bearing formations. Reservoirs with strong water-drive (strong pressure support), will eventually produce aquifer water; this happens mostly in non-associated natural gas reservoirs.

An example molar composition of a formation water is shown in Table 6.1. The water/brine is from one of the subsurface formations in a small North Sea oilfield. Also shown is the composition of normal seawater. Note the high concentration of *sulphate* in seawater and the high concentration of *barium* in the formation water/brine. Adding the concentration columns in Table 6.1, the TDS of seawater is $36.7 \cdot 10^3$ mg/L, equivalent to 3.67% wt. The TDS of the formation brine is $48.8 \cdot 10^3$ mg/L, equivalent to 4.88% wt.

Inorganic scaling can occur throughout the total flowstream from a reservoir to processing facilities, including separation. An example of the total flowstream is illustrated in Figure 6.1. Oil is produced from the main formation. A horizontal well is the production well and a vertical well is the injection well (for pressure support). The small cross-lines in both wells indicate perforations. From the reservoir to the process separator, there are several sets of valves. The first one is the downhole safety

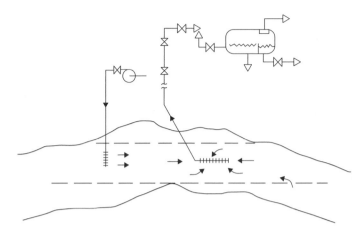

Figure 6.1 Flowstream (oil, gas and water) from reservoir to process separator. Also shown is water injection. Production well is horizontal, injection well is vertical.

valve (SSSV = Subsurface Safety Valve). The valve represents a constriction, resulting is some pressure drop. At the valve, some carbon dioxide may transfer (flash) from the aqueous phase to the gas phase. The result of this transfer may be calcium carbonate scaling in the safety valve.

The second set of valves is at the wellhead. Along the flowstream are the main valve(s), the gate valve and the choke valve. Most of the pressure drop takes place over the choke valve; the valve used to control the well flowrate. For the same reason, as in the safety valve, calcium carbonate may form, along with other solid precipitates. The flowline from the wellhead and the inlet valve to the process separator are also subject to precipitates, not only calcium carbonate but potentially a host of other materials, organic and inorganic (any of the big-five deposits). The process separator separates gas at the top and oil and water at the bottom. The stabilized oil and gas are *products* with specifications that make them suitable for transport, by shuttle-tanker or pipeline. A high-pressure pump feeds the injection fluid to the injection well.

Consider the reservoir in Figure 6.1. A gas cap zone at the top, the main oil bearing formation in the middle, and an aquifer underlying the oil zone (in the figure, separated by horizontal dashed lines). Each zone may have layers with different permeability, thickness and porosity. The horizontal well has been drilled and perforated to produce from the main oil bearing formation. Typically, the flow in the wellbore will be oil, solution gas and water. The water will be formation water initially, and then a mixture of formation water and injected water with time, and eventually a mixture of formation water, injected water and aquifer water. In production from the gas-cap zone only, the produced water will primarily be condensed water. Reservoir gas is saturated with water vapour, that condenses to liquid due to cooling in the production tubing. Although no faults are shown in Figure 6.1, they are prevalent in (all) oil and gas reservoirs. The near-horizontal boundaries between subsurface formations are seldom tight; they may leak when pressure drawdown occurs due to production by the horizontal well. The amount of produced water changes with time, as is the case in most production wells. In Figure 6.1, aquifer water is shown to flow from below toward the production well. The aquifer water may equally likely flow (encroach) from the outer boundary of the reservoir.

The fluid injected into a formation for pressure support is processed water. The water is degassed to remove oxygen and filtered to remove solid contaminants. In offshore operations, the injection water will be seawater, in most cases. In on-land operations, it may be seawater and/or groundwater, but also surface waters from lakes and rivers. Treatment chemicals are commonly added to injection waters, for biological, scaling and corrosion reasons. With time, the injected seawater will break through to the production wells. Seawater is rich in sulphate and most formation waters are rich in barium. When these water mix, the chemical equilibrium favours the formation of barium sulphate. The precipitation of barium sulphate is one of the main scaling problems in oil and gas production. The precipitation may occur in the reservoir formation, the near wellbore region and in the wellbore. The precipitation of barium sulphate in the formation and the near wellbore region may be remedied by repeated squeeze treatments; that is, the injection of chemicals that hinder the formation of the scale. Scale in the wellbore (production tubing) can be removed by chemical treatment/washing, high-pressure nozzle jetting or mechanical reaming by drilling.

6.3　SCALING MINERALS

Formation water/brine contains ionic species from the dissolution of minerals in the sedimentary formations, at the reservoir temperature and pressure. The main types of rocks in oil- and gas-bearing sedimentary formations are sandstone and carbonates. The concentrations of such ionic species as natrium, potassium, magnesium and chloride are not necessarily related to the equilibrium solubility of minerals. Formation waters are not saturated with respect to sodium chloride and potassium chloride, for example. However, formation water (and seawater) are predominantly saturated with calcium carbonate. During the flow of formation water through different formations toward production wells, changes in the concentration of dissolved species may occur, including the mixing of similar formation waters. The dominant change in water composition at reservoir and near-wellbore conditions, occurs when formation water it is mixed with injected water. The mixing leads commonly to barium sulphate deposition, a major problem in water injection for pressure support and in water flooding (pushing oil toward production wells). In oil reservoirs, composition of produced water is the result of mixing of formation water, injected water and aquifer water. In non-associated gas reservoirs, the liquid water produced, stems primarily from the condensation of water vapour.

The main minerals in oil and gas production scaling are shown in Table 6.2. The table shows the mineral name and the chemical formula. The most common scale-types are barite (barium sulphate) and calcite (calcium carbonate). An unfamiliar scale-type is strontium sulphate, precipitating for similar reasons as barium sulphate. Surprisingly, common salt NaCl can form crystalline halite in oil and gas production. Halite becomes solid when the aqueous phase evaporates in the production tubing, especially in HPHT situations.

Scales found in geothermal operations are commonly calcite, amorphous silica and pyrite. Calcite in fields with reservoir temperatures around 200°C, silica in all

Table 6.2 Main scaling minerals in oil and gas production.

Mineral	Formula
Anhydrite	$CaSO_4$
Barite	$BaSO_4$
Calcite	$CaCO_3$
Celestite	$SrSO_4$
Gypsum	$CaSO_4 \cdot 2H_2O$
Halite	$NaCl$
Pyrite	FeS_2
Siderite	$FeCO_3$

reservoirs above 200°C and pyrite in higher temperature reservoirs. Calcite and pyrite form downhole and in surface flowlines. Silica forms in surface flowlines. Silicates are also known to form in low-temperature geothermal water; for example, when different fluids are mixed.

The fundaments of the chemical thermodynamics of aqueous solutions in oil and gas production, have been presented and explained in detail by both Kaasa (1998) and Sandengen (2006). The focus of their work was calculations of inorganic equilibria in the presence of hydrocarbons. The work was an integral part of the development of a computer code for use in the petroleum industry. There are several other commercial computer codes available for such calculations.

6.4 SOLUBILITY GRAPHS

The solubility of many inorganic salts (scaling minerals) increases with temperature. For some salts, the solubility increases with temperature to reach a maximum, thereafter to decrease with increasing temperature, as shown for barium sulphate in Figure 6.2. The solubility of the carbonates and sulphates (spelled sulfate in American English), decrease with temperatures found in oil and gas production. One of the solubility rules states that all sulphates are soluble, except those of calcium, barium and strontium. These are precisely the sulphates found in oil and gas operations. The solubility of scaling minerals in water/brine solutions, depend also on the ionic strength of the medium. The presence of salt increases the solubility of inorganic minerals, in *dilute* solutions (see Figure 6.5). Acidity affects the solubility of ionic compounds, where the anions are conjugated bases of weak acids.

The solubility of barium sulphate in water with temperature at saturation pressure is shown in Figure 6.2 (Kaasa, 1998). The saturation pressure is that of liquid water. The concentration is expressed in mmol/kg H_2O, which is molality m. Appreciating that reservoir temperature increases with depth, the temperature at 3 km depth will be approximate 100°C and 130°C at 4 km depth. These reservoir temperatures are based on a 30°C/km geothermal gradient and a surface temperature of 10°C. In an offshore situation, the reference surface is the sea bottom. The solubility of barium sulphate

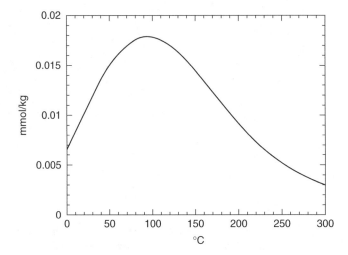

Figure 6.2 Solubility of barium sulphate in water with temperature at saturation pressure of water (Kaasa, 1998; Sandengen, 2012).

increase from ambient temperature and atmospheric pressure to a maximum of about 0.017 mmol/kg at 100°C and decreases subsequently with increasing temperature. The solubility of barium sulphate in water is very low.

The solubility of calcium sulphate with temperature at saturation pressure is shown in Figure 6.3. Two curves are shown, one for anhydrite and one for gypsum. The solubility of anhydrite decreases significantly with increasing temperature. The solubility of gypsum exhibits a concave down curve with a maximum at around 40°C. The solubility of anhydrite at 100°C is about 5 mmol/kg while the solubility of barium sulphate in Figure 6.2 is about 0.017 mmol/kg. The two orders of magnitude difference, shows that calcium sulphate (anhydrite) is much more soluble than barium sulphate (barite).

The solubility of calcium carbonate (calcite) with temperature (>100°C) is shown in Figure 6.4, at CO_2 pressures ranging from 1 atm. to 62 atm. The solubility decreases with temperature at all pressures (see *Appendix Q1 – Solubility of Gases in Water*). The solubility of calcium carbonate in water versus salinity (in terms of NaCl), at three temperatures and CO_2 atmospheric pressure, is shown in Figure 6.5. The solubility of NaCl in water at laboratory temperatures is about 6 mol/L (the figure extends to this value). The solubility of calcite increases as NaCl is added, until about 1 mol/L ($\cong$mol/kg), thereafter to decrease. The solubility decreases with increasing temperature. Reading off the diagram for 25°C and zero NaCl salinity, the solubility is approximately 1.5 mmol/kg.

Carbon dioxide is present in all petroleum fluids. Typical mole fractions in hydrocarbon reservoir fluids were presented by Pedersen *et al.* (1989), shown in *Appendix X1 – Crude Oil Composition*. The particular carbon dioxide values shown are 1.1% vol. in natural gas, 8.65% vol. in gas condensate, 2.18% vol. in volatile oil and 2.11% vol. in black oil. The unit % vol. is equivalent to % mole. The phase diagrams of these four typical hydrocarbon fluids, are illustrated in the named appendix.

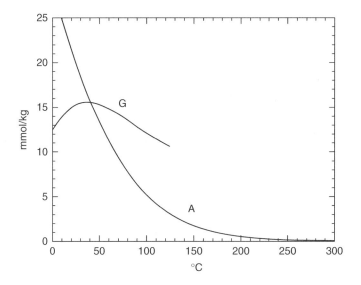

Figure 6.3 Solubility of calcium sulphate (A = anhydrite and G = gypsum) with temperature at the saturation pressure of water (Kaasa, 1998; Sandengen, 2012).

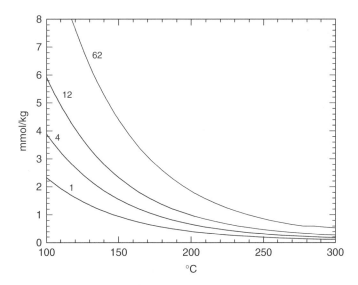

Figure 6.4 Solubility of calcium carbonate with temperature and pressure (Kaasa, 1998; Sandengen, 2012). Lines from low to high: 1 atm., 4 atm., 12 atm. and 62 atm.

The carbon dioxide content of conventional petroleum fluids, can range from 1–10% vol., perhaps more specifically within 0.1–5% vol. In exceptional gas reservoirs, the carbon dioxide concentration can be as high as 80–90% vol. The carbon dioxide exerts a partial pressure proportional to its concentration. In oil and gas production, the partial pressure is relevant to the aqueous phase (water/brine) of the production

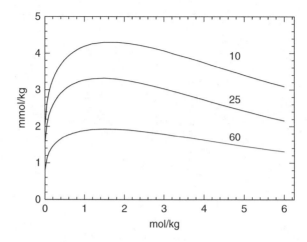

Figure 6.5 Solubility of calcium carbonate in mmol/kg versus NaCl molality mol/kg, at atmospheric CO_2 pressure and three different temperatures: top 10°C, middle 25°C and bottom 60°C (Kaasa, 1998; Sandengen, 2012).

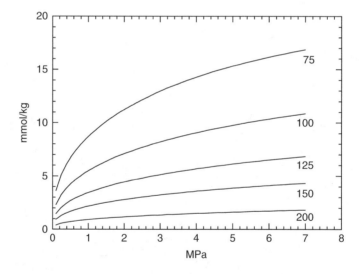

Figure 6.6 Solubility of calcium carbonate with carbon dioxide partial pressure at temperatures from top to bottom in the range 75–200°C (Kaasa, 1998; Sandengen, 2012).

fluids. The partial pressure of carbon dioxide affects directly the solubility of calcium carbonate in produced water. The relationship between the solubility of calcium carbonate ($CaCO_3$) with carbon dioxide (CO_2) partial pressure is shown in Figure 6.6 (Kaasa, 1998; Sandengen, 2012). The effect of temperature is also shown. In real situations, the partial pressure will unlikely exceed 10 bara (=1 MPa), such that the solubility will be no higher than 5 mmol/kg, taking reservoir and production temperatures

Table 6.3 Scale dependency on acidity (pH).

pH-dependent	pH-independent
Carbonates	Sulphates
$CaCO_3$	$BaSO_4$
$FeCO_3$	$CaSO_4$
Sulphides	$SrSO_4$
FeS, FeS_2	Halites
Silica	$NaCl$
SiO_2	

into consideration. Compared to Figures 6.5 and 6.6, this value seems to be within the right bounds.

Inorganic scales in oil and gas production, have been classified according to their acidity (pH) dependence. Carbonates are acid soluble while sulphates are not acid soluble. The pH-dependent and pH-independent scales are shown in Table 6.3. The table was made based on materials found in the thesis of Kaasa (1998). Scaling by sulphates and common salt are pH-independent. However, the carbonates are highly pH-dependent. The sulphides and silica are also pH-dependent. A range of acids, inorganic an organic, need to be included in pH-dependent equilibrium. That sulphates are not acid soluble, means that they cannot be removed downhole and elsewhere using an acid wash. They can only be removed by mechanical methods or by using an alkali complexing agent.

The acidity (pH) of aqueous solutions (produced water, condensed water) are highly dependent on the partial pressure of carbon dioxide (CO_2), and the whole carbonate equilibrium chemistry. The main chemical relationships are presented in *Section 6.9 Carbonate Scale* and in *Appendix W – Calcite and Silica Chemistry*. The relationship between the partial pressure of carbon dioxide in the gas-phase and its mole fraction in the aqueous phase, are presented in *Appendix Q1 – Solubility of Gases in Water*, dealing primarily with *Henry's Law*.

6.5   EQUILIBRIUM AND ACTIVITY

Several chemical and thermodynamic concepts are used in the analysis and modelling of inorganic scaling in oil and gas production. Scale is the solid material/mineral, while scaling is the process of precipitation and deposition. A most basic concept is the concentration of scale forming species. Molality m mol/kg is most commonly used because it is independent of temperature and pressure. Molarity M mol/L is also used, but has the disadvantage that the solute/water volume depends on temperature and pressure. Molality is a thermodynamically consistent unit and is preferentially used in the remainder of the present chapter. In practical applications, the unit ppm is commonly used, usually expressing a mass concentration mg/kg. Akin to concentration, is the concept of mole fraction, here shown for chemical component A

$$x_A = \frac{n_A}{\sum n_i}$$

The number of moles of component A is n_A and the summation sign encapsulates all the component in solution, A, B, C etc. By convention, x stands for mole fraction in the liquid phase, while y stands for mole fraction in the gaseous phase.

Concentration expresses the amount of substance in a solution. By substance is meant any form of atom, molecule and species. By species is meant any form of anion and cation in solution. Anions are negatively charged, for example the chloride Cl^- ion. Cations are positively charged, for example the sodium ion Na^+. Together they form common salt $NaCl$ without electrical charge. Common salt can be dissolved in a solution (brine in oil and gas production) and can exist as the solid crystal halite. In real solutions, the presence of dissolved atoms, molecules and species will influence whatever chemical interactions taking place. The concept of *activity* is used to express the effective concentration in liquid solutions. The concept of *fugacity* is used to express the effective partial pressures of gases in mixtures.

A typical chemical balance equation for reactants A and B and products C and D can be written as

$$aA + bB \leftrightarrow cC + dD$$

Further details are given in *Appendix Q2 – Chemical Equilibrium*. The lower-case letters stand for the number of molecules/species and the upper-case letters stand for the active mass of the molecules/species A, B, C and D. The active mass can be concentration in molarity M mol/L or more commonly molality m mol/kg. The equilibrium constant for the above reaction is defined from the *Law of Mass Action* as

$$K = \frac{c_C^c c_D^d}{c_A^a c_B^b}$$

The concentrations have subscripts that stand for the molecule/specie name, the superscripts stand for the stoichiometric coefficient (number of molecules). The numerical value of the equilibrium constant depends on the units used. The constant can have different subscripts, depending on the processes involved: chemical reaction, solubility relations and acid-base reaction, for example. The solubility product subscript sp is commonly used in the dissolution and precipitation of inorganic scaling minerals.

The effective concentration, to be used in the equation is the chemical activity, is most commonly expressed for component (molecule/specie) A as follows.

$$a_A = \gamma_A m_A$$

Same kind of equation can be written for the other components B, C and D in the chemical equilibrium equation above. The gamma γ_A is the activity coefficient of component A. Activity is the effective concentration of a chemical species in a reaction. The concept takes into account the deviation from an ideal solution. The effective concentration depends on the presence of other species in solution. Ionic species are especially affected by other species in solution. Divalent species ($2\pm$) are more affected than monovalent species ($\pm$). In dilute/ideal solutions the activity coefficient is approximately unity ($=1$), such that the effective molal concentration is approximately equal to the molal concentration.

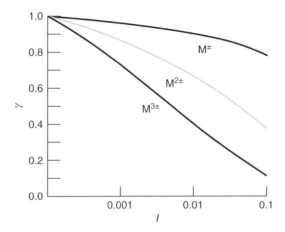

Figure 6.7 Activity coefficient in aqueous solutions, versus ionic strength for ions having different valence (Meyerhoff, 2003).

The activity (effective concentration) of a chemical species in solution is given by the general expressions

$$a_i = \gamma_i m_i$$

based on molality. The expression is for chemical component i which are the A, B, C and D above. The gamma in the activity expression is the activity coefficients. An activity coefficient must to be assigned to all the ions in solution in chemical equilibrium calculations.

The activity coefficient of ions in dilute aqueous solutions, is shown in Figure 6.7 (Meyerhoff, 2003). The capital M stands for ionic specie with valence from one to three. The coefficient decreases with ionic strength for all ions; the rate of decrease increasing with the valence of the ions. The figure shows that the activity coefficient of ionic species is highly dependent on the ionic strength of the solution. The value of the activity coefficient decreases dramatically with ionic strength. Values in the range 0.4–0.6 can easily be expected. Therefore, the activity coefficient of species results in major corrections in chemical equilibrium calculations.

Potable groundwater (drinking water) is reported to have an ionic strength in the range 0.001–0.02 (Reeves 2009). *Appendix R – Ionic Strength and Activity Coefficient*, gives details about the calculation of ionic strength and models for estimating an activity coefficient. The ionic strength of seawater and formation water, based on the compositions in Table 6.1, were calculated as $I_{SW} = 0.73$ and $I_{FW} = 0.85$, respectively. The unit of ionic strength is the same unit as the concentration used in the calculation. These values are off the scale in Figure 6.7, because it is based on the Debye-Hückel (1923) *Limiting Law*, an approximation for dilute solutions. Therefore, seawater and formation water cannot be classified as dilute solutions.

6.6 CHEMICAL POTENTIAL

Chemical equilibrium, the tendency of chemical species to change from one state to another, is encapsulated in the concept of *chemical potential*. The concept is equivalent to the *Gibbs Free Energy* per mole. In terms of symbols and units, *Gibbs Free Energy* is given by G and the unit of joule (=J), and chemical potential is given by μ and the unit of J/mol. Derivations and details are presented in *Appendix S – Gibbs Free Energy* and *Appendix T – Chemical Potential*. Any change in *Gibbs Free Energy*, represents the redistribution of energy. An example being the balance between reactants and products in inorganic scaling.

An increase in *entropy* is the primary criterion for any kind of change, including a chemical change (dissolution/precipitation) and reaction. The following relationship, derived in *Appendix S – Gibbs Free Energy*, for a constant pressure process, is relevant in the present context

$$\left(\frac{\partial G}{\partial T}\right)_p = -S$$

Chemical potential relationships are presented in *Appendix T – Chemical Potential*. The following relationships were derived/presented:
Chemical potential of pure substances

$$\mu = \left(\frac{\partial G}{\partial n}\right)_{p,T}$$

Chemical potential of ideal gases

$$\mu = \mu^o + RT \ln\left(\frac{p}{p^o}\right)$$

Chemical potential of real gases

$$\mu = \mu^o + RT \ln\left(\frac{f}{p^o}\right)$$

Chemical potential of an ideal solution

$$\mu = \mu^o + RT \ln\left(\frac{m}{m^o}\right)$$

Chemical potential of a real solution

$$\mu = \mu^o + RT \ln\left(\frac{a}{a^o}\right)$$

The symbol p stands for partial pressure, the symbol f stands for fugacity and the symbols m and a stand for molal concentration. The superscript ^{o} stands for intensive properties at standard conditions, $T = 25°C$ and $p = 100\,kPa$. The symbol R stands for the gas constant, equal to 8.314 kJ/kmol·K.

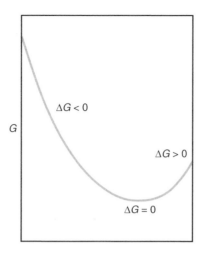

Figure 6.8 Illustration of change in *Gibbs Free Energy* from pure reactants (left-hand-side) to pure products (right-hand-side).

Consider the reactants A and B and the products C and D. The *Gibbs Free Energy* can be written as

$$\Delta G = G_C + G_D - G_A - G_B$$

expressing the energies of the products, minus the energies of the reactants. The behaviour of the *Gibbs Free Energy* in chemical reactions is illustrated in Figure 6.8. Consider the reactants, $\Delta G < 0$ and the reaction quotient Q is less that the equilibrium constant K (see *Appendix Q2 – Chemical Equilibrium*). The reaction proceeds to equilibrium as reactants are converted to products. At equilibrium $\Delta G = 0$, indicating that no further change occurs. For illustration purposes, on the pure products side of the figure, $\Delta G > 0$ and $Q < K$. Another way to explain the changes in *Gibbs Free Energy*, is to state that when $\Delta G < 0$ the process/reaction is spontaneous, when $\Delta G = 0$ the system is at equilibrium and when $\Delta G > 0$, the reverse process/reaction is spontaneous.

The spontaneity ($\Delta G < 0$) and non-spontaneity of chemical reactions and other processes, where redistribution of energy occurs, can be viewed through the change in *Gibbs Free Energy* at constant temperature

$$\Delta G = \Delta H - T\Delta S$$

If $\Delta H < 0$ and $\Delta S > 0$, the process is spontaneous at any (constant) temperature. If $\Delta H > 0$ and $\Delta S < 0$, the process is non-spontaneous at any temperature. If $\Delta H < 0$ and $\Delta S < 0$, the process is spontaneous at *low* temperature. If $\Delta H > 0$ and $\Delta S > 0$, the process is spontaneous at *high* temperature.

The heat of solution of halide (NaCl) in water at 25°C is widely reported to be about 4 kJ/mol. The heat of solution of methane (CH_4) in water at 25°C and 100 kPa pressure, is reported to be $-15\,kJ/mol$ (Duan and Mao, 2006). The first process is endothermic, meaning that heat energy is absorbed (temperature of aqueous solution decreases). The second process is exothermic, meaning that heat energy is released (solution temperature increases).

Calcium salt is used to melt ice on roads in winter, and sodium salt is added to make snow harder for cross-country skiing. One salt melts ice, while the other salt freezes water. The difference lies in their heat of solution, one being positive, the other being negative. The heat of solution of common salt (NaCl) is $+4\,kJ/mol$, while the heat of solution of calcium chloride ($CaCl_2$) is about $-80\,kJ/mol$. The sodium salt lowers the temperature, while the calcium salt increases the temperature.

6.7 SOLUBILITY OF SCALING MINERALS

The brines produced in oil and gas operations, contain dissolved minerals that form inorganic scales. The scaling occurs when the solubility of the minerals decreases due to a multitude of factors. One of the critical factors, is the solubility of the minerals. In the subsurface, the minerals dissolve to form ions in solution. Continuous dissolution and precipitation take place over thousands to hundred thousand of years, resulting in chemical equilibrium. This equilibrium is disturbed in oil and gas production operations, leading to the precipitation and deposition of inorganic scaling minerals.

Dissolved minerals in water exist as ions in solution. When a solution becomes supersaturated, the excess material precipitates and a new equilibrium is established between the solid material and the dissociated ions. In a solution system at equilibrium, there is a continuous and balanced dissolution and precipitation. In *Appendix Q2 – Chemical Equilibrium*, it was shown that the equilibrium constant, also called the solubility product constant, using effective concentrations, can be expressed as

$$K_{sp} = \frac{\{\gamma_C c_C\}^c \{\gamma_D c_D\}^d}{\{\gamma_A c_A\}^a \{\gamma_B c_B\}^b}$$

for a typical chemical reaction of the form

$$aA + bB \leftrightarrow cC + dD$$

Solids and liquid water are not included in the solubility product constant K_{sp}. This can be illustrated using the dissolution and precipitation of barium sulphate

$$BaSO_4\ (s) \leftrightarrow Ba^{2+}\ (aq) + SO_4^{2-}\ (aq)$$

where s stands for solid and aq for aqueous. The solubility product for the reaction, is given by the ionic species only as

$$K_{sp} = \{a_{Ba^{2+}}\}\{a_{SO_4^{2-}}\}$$

The curly brackets are used to clarify that the concentrations are expressed in terms of chemical species activity. Another example is where solid silica reacts with water to form silicic acid

$$SiO_2 \, (s) + 2H_2O \leftrightarrow H_4SiO_4 \, (aq)$$

the solubility product is simply

$$K_{sp} = \{a_{H_4SiO_4}\}$$

In other words, solids and liquid water are not used in solubility product constant expressions. However, this statement needs to be qualified by adding that it applies to pure salts. When activity coefficients are introduced into equilibrium relationships, the activity of water needs to be included, at least. An example would be the solubility of gypsum $CaSO_4 \cdot 2H_2O$, which includes crystal water. The solubility of gypsum depends on both the activity of water and the activity of halide (NaCl). Another example would be the water vapour content of natural gas, that depends on both salinity and the presence of sour gases (see *Appendix N – Water Vapour in Natural Gas*).

The solubility product K is constant at equilibrium. At non-equilibrium, a variable called the reaction quotient, symbol Q, is given by the same expression as the solubility product. The reaction quotient is also called the ion activity product *IAP*, based on the actual/instantaneous activity. Meaning the activity of an ionic specie(s) in transition to equilibrium and eventually at equilibrium. For the barium sulphate reaction above, the ion activity product becomes

$$IAP = \{a_{Ba^{2+}}\}\{a_{SO_4^{2-}}\}$$

The *Saturation Index* is defined by

$$SI = \log\left(\frac{Q}{K_{sp}}\right)$$

The Q/K_{sp} ratio (also called the *Saturation Ratio*) and the saturation index *SI* for equilibrium, supersaturation and unsaturation are shown in Table 6.4. For the present purposes, crystallization is synonymous with precipitation. It should be noted that crystallization is a kinetic process as is precipitation. A kinetic process can progress quickly, moderately or slowly, meaning that precipitation and deposition can occur locally (*in situ*) and/or distributed throughout a reservoir, production tubing, flowline and process equipment.

The solution level of solids in liquid water can be described as Soluble, Slightly Soluble and Insoluble, as shown in Table 6.5. Other levels and terms are also used in the literature. For the present purpose, the term soluble is for concentrations $>1 \, g/100 \, g$ H_2O, the term slightly soluble for concentrations $0.01–1 \, g/100 \, g \, H_2O$ and insoluble $<0.01 \, g/100 \, g \, H_2O$. This concentration unit is commonly used in the literature. There is no agreed standard on the solution levels in thermochemistry.

To find the appropriate solubility level/term for the common inorganic mineral scales in oil and gas production, approximate concentration values were read from the

Table 6.4 The ratio of the reaction quotient Q (=IAP) over the solubility product constant K_{sp} and the saturation index SI for three solution conditions.

Condition	Q/K_{sp}	SI
Equilibrated	$=1$	$=0$
Supersaturated	>1	>0
Unsaturated	<1	<0

Table 6.5 Solubility terms and levels.

Term	$g/100\,g\,H_2O$	$g/kg = g/L$
Soluble	>1	>10
Slightly soluble	0.01–1	0.1–10
Insoluble	<0.01	<0.1

Table 6.6 Solubility/concentration of commons scales at 25°C and atmospheric pressure, read-off graphs provided by Kaasa (1998) and Sandengen (2012).

Solid	c mmol/L	M g/mol	c g/100 g H_2O
$BaSO_4$	0.01	233	0.02
$CaCO_3$	3	100	0.03
$CaSO_4$	15	136	0.20
$CaSO_4 \cdot 2H_2O$	21	172	0.36

graphs in Figure 6.2, Figure 6.3 and Figure 6.5 at a temperature of 25°C. The approximate concentration values are shown in Table 6.6. The density of water was assumed 1000 kg/m³ (=1 kg/L). Table 6.6 shows that barium sulphate is the least soluble. Using the solubility levels/terms from above, all the four solids can be considered slightly soluble.

The solubility of oilfield scales in cold water has been given by Mackay (2008), in the following increasing order: barium sulphate 2.2 mg/L, calcium carbonate 14 mg/L, strontium sulphate 113 mg/L, calcium sulphate 2090 mg/L and calcium sulphate as gypsum 2410 mg/L. These values agree reasonable well with the approximate values in Table 6.6.

Numerous tables are available in the literature giving the solubility product constant K_{sp} for various solids dissolved in aqueous solutions. Example values for inorganic scales are shown in Table 6.7. The general statement can be made, that the larger the solubility product constant, the more soluble the solid material is. The smaller the solubility product constant, the less soluble is the solid. The solubility of barite < calcite < celestite < anhydrite < gypsum. The solubility of common salt (halite) is relatively high as shown in Table 6.7, making it highly soluble.

Table 6.7 Solubility product constant K_{sp}, heat of formation/reaction ΔH^o and *Gibbs Free Energy* ΔG^o for the main minerals forming scale in oil and gas production, all at standard conditions. The N/A in the table means not available.

Mineral	Formula	K_{sp}	ΔH^o kJ/mol	ΔG^o kJ/mol
Anhydrite	$CaSO_4$	9.1×10^{-6}	-1433	-1309
Barite	$BaSO_4$	1.1×10^{-10}	-1473	-1362
Calcite	$CaCO_3$	2.8×10^{-9}	-1207	-1129
Celestite	$SrSO_4$	3.2×10^{-7}	N/A	N/A
Gypsum	$CaSO_4 \cdot 2H_2O$	2.4×10^{-5}	-2024	-1707
Halite	$NaCl$	3.6	-411	-384
Pyrite	FeS_2 (FeS)	6×10^{-19}	-178 (-102)	-167 (-104)
Siderite	$FeCO_3$	3.2×10^{-11}	N/A	N/A

The solubility product constant K_{sp} in the literature, is commonly given at 20–25°C in temperature and 1 atmosphere (=101.325 kPa) in pressure (see *Appendix U – Solubility Product Constant*). In oil and gas operations, both the temperature and pressure are much higher. The general statement can be made, however, that *pressure* in the range 5–10 MPa does not significantly affect the solubility product constant in liquids (=aqua). For reservoir pressures in the range 50–100 MPa, the effect of pressure on the solubility constant may need to be considered. The reason being that the liquids produced will decrease in pressure throughout the flowstream, down to near-atmospheric pressure. The effect of pressure on the solubility constant in hydrocarbon gases needs to be accounted for indirectly, through the standard change in *Gibbs Free Energy*. The solubility product constant is greatly affected by *temperature*.

The *van't Hoff Equation* gives the relationship between the solubility product constant and absolute temperature, at *moderate* temperature changes where the change in heat of reaction is small. The equation is derived in *Appendix S – Gibbs Free Energy*, assuming constant pressure

$$\ln K_2 - \ln K_1 = -\frac{\Delta H^o}{R}\left(\frac{1}{T_2} - \frac{1}{T_1}\right)$$

The values $K_2 = K(T_2)$ and $K_1 = K(T_1)$ refer to the *Solubility Product Constant* at the specified temperatures. A plot of ln K on the y-axis and $1/T$ on the x-axis, gives the slope $-\Delta H^o/R$. For an endothermic reaction, the slope is negative, less than zero (<0) and for an exothermic reaction the slope is positive (>0). When the slope is less than zero, the solubility product constant increases with temperature, and when the slope is larger than zero, the constant decreases with temperature. The *van't Hoff Equation* should not be used to estimate the heat of reaction at temperatures more than 20°C different from a known value, in most cases the value at standard conditions.

In the *van't Hoff Equation*, ΔH^o is the change in enthalpy of reaction (heat of reaction at constant pressure with reactants and products at standard states). All the chemical components taking part in a reaction must be included. In some texts the heat of formation/reaction has the subscript f for formation and r for reaction. The heat of formation/reaction should *not* be confused with the heat of solution.

In oil and gas production, the pressure and temperature changes are more than *moderate*, such that the standard heat of reaction cannot be used. Empirical correlations need to be developed based on experimental data. The *van't Hoff Equation* can be simplified to

$$\ln K_{sp} = A + \frac{B}{T}$$

where A and B are constants, based on experimental data. Such an equation can be used for the data range in question. A similar but more detailed empirical equation, is commonly used to match solubility product data

$$\ln K_{sp} = A + BT + \frac{C}{T} + D \ln T$$

where A, B, C and D are empirical constants, but different from the simpler equation above. This empirical relationship is equivalent to that used by Kaasa (1998), presented below. The same empirical relationship has been recommended by Carroll (1999) to correlate *Henry's Law* constant for particular systems (see *Appendix Q1 – Solubility of Gases in Water*).

The heats of formation/reaction for the common inorganic scale types is shown in Table 6.7. Such values are found in tables in the literature. All the inorganic scales in oil and gas production have a negative heat of reaction, meaning that the reaction is exothermic. A test plot (not shown here) was made to confirm the overall behaviour of the solubility product constant with increasing temperature. The plot did indeed have a positive gradient and the value of the constant decreased substantially with temperature.

It is important to calculate the solubility product constant at temperatures of relevance in oil and gas operations. Literature data have been compiled by Kaasa (1998) in the appropriate temperature ranges at nominally constant pressure. The data for the minerals of interest were fitted to an empirical polynomial function of the form

$$\ln K_{sp} = \frac{\alpha_1}{T} + \alpha_2 + \alpha_3 \ln T + \alpha_4 T + \frac{\alpha_5}{T^2}$$

The coefficients α_1 to α_5 were determined from the available data and are shown in Table U.1, along with the valid temperature range in Celsius $T\,°C$. The solubility product constant K_{sp} is dimensionless. The solubility product constants in Kaasa (1998) and Sandengen (2006) are called the *thermodynamic* solubility product constants because the semi-empirical equations are based on chemical activity, not simple concentration. The above polynomial functional form was in-part based on theoretical considerations.

At this juncture, *Hess's Law* (Hess, 1840) should be mentioned/presented. The law allows for the indirect calculation/estimation of the heat of formation/reaction of any reaction at standard conditions

$$\Delta H^o = \sum n \Delta H^o \,(products) - \sum m \Delta H^o \,(reactants)$$

In other words, the standard heat of formation/reaction is equal to the sum of all the standard heats of formation of the products, minus the sum of all the standard heats of formation of the reactants. The symbols n and m are integers (here, stoichiometric coefficients) because each heat of formation must be multiplied by its stoichiometric coefficient in the chemical balance equation. It is important that all the components taking part in a reaction are taken into consideration. *Hess's Law* states also that the enthalpy of a given chemical reaction is constant, regardless of the reaction taking place in one step or many steps. It should be noted that the heat of formation/reaction of a pure element, as in the *Periodic Table*, is always equal to unity ($=1$). The same holds true for liquid water in aqueous solutions, in most instances.

6.8 SULPHATE SCALE

The sulphates are one of the two important group of inorganic scale in oil and gas operations, including the minerals anhydrite, barite, celestite and gypsum. They are sulphate salts of alkali earth metals. The other important group are the carbonates (see the next section). Barium sulphate $BaSO_4$ will be used in the present section to *illustrate* the calculation of solubility in water/seawater/brine.

The solubility of barite is shown in Figure 6.2. It increases to a maximum at around 100°C and then decreases with temperature. The *thermodynamic* solubility product constant of barite is given by the empirical equation

$$\ln K_{sp}(BaSO_4) = \frac{37588}{T} - 747.61 + 119.28 \ln T - 0.16283T - \frac{2.88 \times 10^6}{T^2}$$

based on constants in Table U.1 in *Appendix U – Solubility Product Constants*. The equation gives the natural logarithm of the solubility product constant ($\ln K$ has negative values) with temperature, shown in Figure 6.9. At temperatures below 100°C (boiling point of water) the pressure is atmospheric. At temperatures above 100°C, the pressure is the saturation pressure of water. The temperature used in the above equation is absolute temperature T K. Similar solubility product graphs can be made for most scaling minerals, including other sulphates, sulphides, carbonates and oxides.

The chemical balance for barite is given by

$$BaSO_4 \ (s) \leftrightarrow Ba^{2+} \ (aq) + SO_4^{2-} \ (aq)$$

and the corresponding equilibrium constant (solubility product constant) is given by

$$K_{sp} = \{a_{Ba^{2+}}\}\{a_{SO_4^{2-}}\}$$

where the activity $a_n = \gamma_n m_n$. The subscript n indicates any chemical component. The use of concentration unit, here molality, must be consistent throughout any calculations. Solid barite does not enter into the equilibrium constant equation, only the aqueous species are involved. The solubility of barite at about 60°C, can be calculated using the value $\ln K_{sp} = -22.5$ (dimensionless quantity), read from Figure 6.9. Taking the exponential gives $K_{sp} = 1.7 \times 10^{-10}$. Rewriting the equation immediately above, using x as an arbitrary variable (not to be confused with mole fraction in liquid phase)

$$x^2 = 1.7 \times 10^{-10}$$

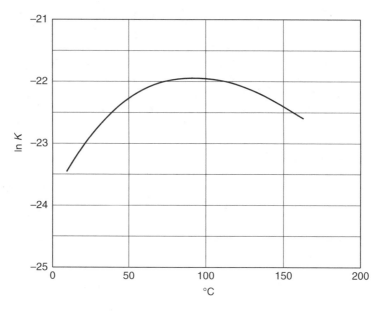

Figure 6.9 Natural logarithm of thermodynamic solubility product constant K_{sp} of barite, using values from Table U.1.

Taking the square root, means that the concentration (activity) of both the ions is 1.3×10^{-5} mol/kg (=0.013 mmol/kg H_2O) Molality m should be used because the Kaasa (1998) correlations are based on molality, not molarity M. This solubility value compares reasonably well with the graph in Figure 6.2, showing the solubility of barite with temperature. At higher temperatures than 25°C, the molal concentrations will be different (the density of water decreases with temperature). To use the example above, water density at 25°C is 997 kg/m³ and at 60°C 982 kg/m³. The solubility of barium sulphate, and all other inorganic mineral scales, will also be affected by the salinity of the aqueous phase (produced water).

The molecular weight of $BaSO_2$ (barite) is 233.30 g/mol such that the mass solubility of barite at 60°C was calculated 3 mg/L equal to 0.3 g/100 g H_2O. The solubility value of barite in Table 6.5 is 0.23 g/100 g H_2O at 25°C. The value quoted by Mackay (2008), was 2.2 mg/L in cold water. Immediately above, the calculated solubility of barite was based on molarity, not molality. For the sake of simplicity, the density of water was assumed 1000 kg/m³.

The solubility of barite calculated above, was based on the solubility product constant at 60°C. No correction was needed/made for the activity coefficient $a_n = \gamma_n c_n$. The solubility calculated was reasonably close to the value in Table 6.6, and the value reported by Mackay (2008). The barium and sulphate ions are divalent. The ionic strength of seawater I_{SW} was calculated as 0.73 mol/L in *Appendix R – Ionic Strength and Activity Coefficient*. The square root of this value is 0.85 such that the activity coefficient from Figure R.2 is approximately 0.25, based on the *Debye-Hückel Extended Law* and/or *Pitzer's Model* (they show the same value, approximately).

Other sulphates, include anhydrite ($CaSO_4$), gypsum ($CaSO_4 \cdot 2H_2O$) and celestite ($SrSO_4$). Sulphates are slightly acid soluble (Mackay, 2008). The chemical balance equations of these scaling compounds are analogous to that of barium sulphite

$$CaSO_4\ (s) \leftrightarrow Ca^{2+}\ (aq) + SO^{-4}\ (aq)$$

$$CaSO_4 \cdot 2H_2O \leftrightarrow Ca^{2+}\ (aq) + SO^{-4}\ (aq) + 2H_2O$$

$$SrSO_4\ (s) \leftrightarrow Sr^{2+}\ (aq) + SO^{-4}\ (aq)$$

The methodology illustrated above for barium sulphate, can be used for both calcium sulphates and strontium sulphate. The mineral celestite is also called celestine. The gypsum chemical balance includes water, called crystal water. Another calcium sulphate that contains crystal water is hemihydrate, having one-half water molecule per $CaSO_4$. Calcium sulphates are not common inorganic scales in oil and gas production.

The solubility of mineral is affected by the presence of salts and alcohols. In oil and gas production, halite salt NaCl and mono ethylene glycol (MEG) are commonly present. Salt is present because the water phase produced is brine; glycol is present because of hydrate inhibition in flowlines and pipelines. Both lower the solubility of inorganic minerals (Sandengen, 2006). Glycols are low molecular weight alcohols that are totally miscible with liquid water. Typical concentrations of MEG in liquid water for hydrate control are 30 to 40% wt. Methanol can also be used, but because it is more volatile than MEG, it will readily vaporize and lost into the natural gas phase. For comparison, the boiling point of methanol is 65°C and that of ethylene glycol is 197°C, both at atmospheric pressure.

6.9 CARBONATE SCALE

The carbonates are one of the two important groups of inorganic scales. The other important group are the sulphates, discussed in the previous section. Calcium carbonate is the most important and most studied inorganic scale in industry, including the oil and gas industry. The myriad of publications has progressed from the established thermodynamics, to laboratory studies of the ambiguous kinetics (induction time and rate of formation). The chemistry of calcium carbonate in water is furthermore important in a wide range of studies, including the oceans, surface water and groundwater. Carbon dioxide in the atmosphere is one part of the carbon cycle, therein the dissolution and precipitation in the hydrosphere. The chemistry fundamentals are the same in oil and gas production, as in the mentioned fields of study.

The focus, in many of the mentioned fields of study, is the role of calcium carbonate in the carbon cycle. In oil and gas production, the focus is on the precipitation and deposition in wells, flowlines and processing equipment/facilities. Mineral scaling in the oil and gas industry is due to the consequences of pressure and temperature changes in the flowstream, from a subsurface reservoir to surface facilities. In processing facilities (topside), various heat exchangers may be subject to inorganic scaling. The chemical equilibrium equations of calcium carbonate are presented in

Appendix W – Calcite and Silica Chemistry. The chemical thermodynamics of inorganic scaling are well expounded by Kaasa *et al.* (2005).

Calcium carbonate and other carbonates are pH-dependent scales (see Table 6.3), including iron carbonate, which is also vital in the corrosion of iron and steel. Carbon dioxide gas plays a most important role in carbonate chemistry and scaling (see *Appendix Q1 – Solubility of Gases in Water*). The following chemical balances are considered relevant in calcium carbonate scaling

$$CO_2\ (g) \leftrightarrow CO_2\ (aq)$$

$$CO_2\ (aq) + H_2O \leftrightarrow H_2CO_3\ (aq)$$

$$H_2CO_3\ (aq) \leftrightarrow HCO_3^- + H^+$$

$$HCO_3^- \leftrightarrow CO_3^{2-} + H^+$$

$$Ca^{2+} + CO_3^{2-} \leftrightarrow CaCO_3\ (s)$$

$$Ca^{2+} + CO_3^{2-} \leftrightarrow CaCO_3\ (aq)$$

First, carbon dioxide gas dissolves in water. *Second*, the dissolved gas reacts with water to form carbonic acid. *Third*, the carbonic acid dissociates into a bicarbonate ion and a hydrogen ion. *Fourth*, the bicarbonate ion dissociates into a carbonate ion and a hydrogen ion. *Fifth*, the carbonate ion combines with a calcium ion to form calcium carbonate solid/scale. The last balance represents the solubility of calcium carbonate in water, which is added here for the sake of completeness, notwithstanding its solubility being very low. An alternative chemical balance for the formation of calcium carbonate, used in other fields of study, is the following

$$Ca^{2+} + 2HCO_3^- \leftrightarrow CaCO_3\ (s) + CO_2\ (aq) + H_2O$$

where the carbon dioxide on the right-hand-side can be written

$$CO_2\ (aq) \leftrightarrow HCO_3^- + H^+$$

such that when calcium carbonate deposits, the pH will decrease, limiting further precipitation. The bicarbonate to carbonate reaction above can also be written as

$$HCO_3^- + H_2O \leftrightarrow H_3O^+ + CO_3^{2-}$$

which is an acid-base reaction with water as a base. The acid-base chemistry of aqueous systems has been clearly presented by Hunter (1998).

The formation of a hydrogen ion in the third and fourth balances, lowers the pH of the aqueous solution because

$$pH = -\log[a_{H^+}]$$

Note that the hydrogen concentration is expressed in terms of activity. The minus sign means that a low pH-value corresponds to a high concentration of the hydrogen ion. In pure (double distilled) water, the molal activities of the hydrogen ions H^+ and hydroxyl ions OH^- are equal at 1×10^{-7} mol/g, giving a pH-value of 7. The logarithmic expression is used because the hydrogen ion concentration can vary over several orders of magnitude. A fundamental detail of no practical consequence, is that the hydrogen ion is couplet to a water molecule, forming the hydronium ion H_3O^+. By convention, the hydronium ion is not included in acid-base chemistry.

The carbonate balances above can be combined/reorganized to give the overall balance

$$CO_2\,(aq) + H_2O \leftrightarrow CO_3^{2-} + 2H^+$$

The concentration of $CO_2\,(aq)$ depends on *Henry's Law* (see *Appendix Q2 – Solubility of Gases in Water*), formulated in terms of mole fraction as

$$x = \frac{p}{k_H}$$

such that the unit of the constant will be pressure (see Figure 6.2). Instead of mole fraction x, the solubility law can be expressed using molality m, which is more common. The value of the constant will be functionally the same, but numerically different.

Dalton's Law states that the total pressure of a gas mixture is the sum of the partial pressure of the gas components

$$p = p_A + p_B + p_C + \cdots$$

where A, B and C represent different gases. Therefore, if the total pressure decreases, the individual partial pressures will also decrease. Considering a flowstream from a reservoir through a wellbore (production tubing) to the surface, the pressure will decrease. The concentration of carbon dioxide in the produced water/brine will also decrease. *Le Châtelier's Principle* states that a system in chemical equilibrium, subjected to a disturbance, will change to opposes the disturbance. Formation water/brine contains alkaline species such that produced water has $5 <$ pH < 7 typically. In this acidity range, the bicarbonate molality is fairly constant, as shown in Figure 6.10. When carbon dioxide bubbles out of produced water, the molality of the hydrogen-ion H^+ will decrease, such that the pH-value increases. An increasing pH-value means that the molality of the carbonate ion CO_3^{2-} will increase (steep-part of its curve), whilst the molality of the bicarbonate ion remains fairly constant (flat-part of its curve). Provided Ca-ions are present, calcium carbonate will form/precipitate and potentially deposit on the wall of the conduit. The precipitation/deposition processes depend on temperature and the supersaturation in the water/brine solution.

Alkalinity is the most important parameter controlling pH in aqueous solutions (Kaasa *et al.*, 2005). An aqueous solution will have both negatively and positively charged species. The two charges must balance each other, to make the solution charge-neutral. For an aqueous solution containing sodium chloride NaCl, sodium bicarbonate $NaHCO_3$ and acetic acid CH_3COOH (=HAc), the neutral-charge balance can be written as

$$m_{Na^+} + m_{H^+} = m_{OH^-} + m_{Cl^-} + m_{HCO_3^-} + 2m_{CO_3^{2-}} + m_{Ac^-}$$

Table 6.8 Typical mole percent carbon dioxide in four hydrocarbon reservoir fluids (Pedersen *et al.*, 1989).

Reservoir fluid	CO_2 %
Gas	1.10
Gas Condensate	8.65
Volatile Oil	2.18
Black Oil	2.11

The electrical charge of the molal cations on the left, equals the electrical charge of the molal anions on the right. The *m* expresses molal concentration in mol/kg. The definition and measurement of alkalinity, have been presented by Kaasa and Østvold (1997). Acetic acid is the simplest of the carboxylic acids in crude oil, as discussed in *Chapter 7 – Naphthenate*. Consider what species in the above balance are pH-dependent, and what species are pH-independent. The halide ions Na and Cl are not pH-dependent. The remaining and pH-dependent species, define what is called *total alkalinity*

$$A_T = m_{HCO_3^-} + 2m_{CO_3^{2-}} + m_{Ac^-} + m_{OH^-} + m_{H^+}$$

The addition of NaCl to an aqueous solution, will not change the pH and will therefore not change the alkalinity. Adding or removing carbon dioxide CO_2 from the solution, will change the pH-value and hence all the molal concentrations. The alkalinity of aqueous solutions such as oilfield waters, can be measured by titration using hydrochloric acid HCl and/or caustic soda NaOH.

Carbon dioxide is found all hydrocarbon fluids produced (see *Appendix X1 – Crude Oil Composition*). Typical mole percentages of carbon dioxide in four typical hydrocarbon fluids, are shown in Table 6.8. The fluids are the same as illustrated in the phase diagrams in Figure X1.3. In addition of the carbon dioxide in hydrocarbon fluids produced, carbon dioxide will also be dissolved in produced water. The origins of reservoir carbon dioxide are many and complex. The concentrations of carbon dioxide in hydrocarbon fluids and produced water vary greatly from asset to asset (field to field), and with time, as the watercut increases. Within an asset, the carbon dioxide concentration in produced water varies from well to well. The chemical make-up of produced fluids (hydrocarbons and water), must be sampled and analysed to make possible evaluations and predictions of inorganic scaling.

The effect of pH on the distribution of carbonic acid, bicarbonate and carbonate can be calculated from their mole fractions derived/presented in *Appendix W – Calcite and Silica Chemistry*. The plotted mole fractions are shown in Figure 6.10. The figure is illustrative for the solubility product constants used in the calculations of the carbonate species. The constants may or may not be correct in practical situations. The curves will shift left-right, depending on the constant values used. A log-linear plot would have shown the mole fraction more clearly. The common pH-range for natural waters is considered 5–8, perhaps more correctly 5–7. Where the curves cross each other, the acid-base conjugates are equal such that the fractional values are 0.5. The equilibrium

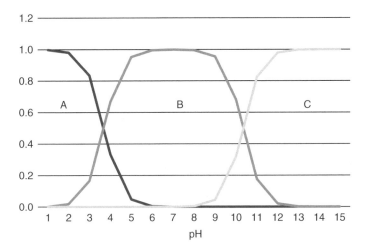

Figure 6.10 Dissociation diagram, showing mole ratios of carbonic acid (A), bicarbonate (B) and carbonate (C) in pure water, at pH ranging from 1 to 15.

constants K_1 and K_2 are given in *Appendix W*. At the conjugate points $pK_1 = 6.35$ and $pK_2 = 10.33$.

In the production of oil and gas, several important pressure drops occur from downhole pressure to surface pressure. The pressure drops depend on the rate of production and the degree of depletion (reservoir pressure decreases with time, unless pressure support is provided for by injection). A quantification of the pressure drops belongs to the realm of classical petroleum engineering and will not be expounded upon here. Details of wellbore deliverability (reservoir performance, inflow performance, tubing performance) would need more space than is available in the present text.

Abrupt pressure drops are relevant in inorganic scaling, because the accompanying flashing tends to degas the water/brine, meaning that carbon dioxide goes from being aqueous to being gaseous. When carbon dioxide degasses, the chemical balance determines whether calcium carbonate scale precipitates and/or deposits. There must be enough calcium ions and enough carbonate ions for the saturation ratio to be greater than one (SR > 1 and SI > 0, shown in Table 6.5). Calcium carbonate scale is considered to form relatively quickly, compared to other flow assurance solids. Analogous to industrial crystallization, calcium carbonate scale (on walls) will form more easily and quickly if small particles/fines are present. The same applies to calcium carbonate crystals and crystal fragments in bulk flow. The kinetics of calcium scale formation depends on temperature, degree of supersaturation, turbulence and other factors know to affect crystallization.

Four example situations are sketched in Figure 6.11. In example A, the scale forms in the production tubing (wellbore) immediately above the bubble point depth. It can be the bubble point of the crude oil or the carbon dioxide bubble point of the flowing water/brine. The process has self-propagating properties; a deposit increases the local pressure drop further, leading to more scale formation. In example B, the scale forms in the crossover from a smaller tubing diameter to a larger tubing diameter.

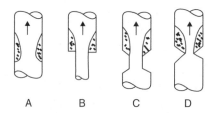

Figure 6.11 Abrupt pressure drop in a flow stream from downhole to surface. A = Bubble Point, B = Tubing Crossover, C = Safety Valve and D = Choke Valve.

The flow stream experiences a pressure drop that leads to flashing and degassing. In example C, a subsurface safety valve (SSSV) represents a restriction to flow and hence an extra pressure drop, with the consequence of flashing and precipitation/deposition. In example D, considerable pressure drop occurs over a wellhead choke valve (less severely also in other valves), with the above described consequences. Under normal oil and gas production, other wellhead valves are fully open.

In terms of chemistry, as pressure is reduced, carbon dioxide will evaporate/flash from the water phase. The pH increases and more carbonate forms, leading to calcite precipitation and deposition. Precipitated calcite crystals/particles will be carried along with the bulk fluid, and some may be transported to the confining wall. The wall will initially be a metal wall, changing to a deposit wall with time. Calcite molecules will be transported to the confining wall by turbulent convection, and will form a scale deposit by local crystallization. In general, solid particles (fines, sand, big-five solids), provide nucleation sites for crystal growth. Embryonic bubbles will also provide nucleation sites.

It is well known that the water cut in oil wells increases with time. The watercut is expressed in terms of the volumetric flowrates by the relationship

$$WC(t) = \frac{q_w}{q_o + q_w}$$

where oil and water are identified by the subscripts o and w. The water cut at the start-up of an oil field may be small and with time it may reach as much as 0.9, indicating that 90% of the produced fluids is water/brine. The water cut is important in inorganic scaling. One of the reasons being that oil-water flows can be oil-continuous or water-continuous emulsions, presented in *Appendix X2 – Emulsions of Crude Oil and Brine*. In an oil-continuous emulsion, the water/brine exists as droplet surrounded by oil. The opposite is true in water-continuous emulsions. Studies of crude oils and reservoir waters/brines show that there is an inversion point, typically at water cut of about 60% (water volume fraction of 0.6), as shown in Figure X2.2. The figure shows the ratio of mixture viscosity divided by the viscosity of the continuous phase. Laboratory measurements have shown that the inversion point increases (volume fraction of water) with crude oil density.

It was stated by Kohler *et al.* (2001), that calcium carbonate scale formation is usually generated through up-hole degassing and loss of CO_2 in producers exhibiting more than 70% water production. By up-hole they mean upper sections of the production

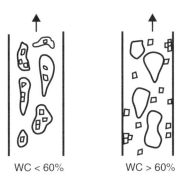

WC < 60%   WC > 60%

Figure 6.12 Upward flow of an oil-continuous mixture (WC < 60%) and a water-continuous mixture (WC > 60%) with calcium carbonate crystals shown as small squares (cube-shaped).

tubing (perhaps at depths immediately above the bubble point), not down-hole close to the perforations. Should this observation apply to other oil producing fields, then the inorganic scaling problem should be at minimum in the early-life of the field. Similarly, the inorganic scaling problem should become worse in the late-life of a field.

The picture that emerges is that calcium carbonate crystals form in two different emulsion phases. In oil-continuous flow, the crystals form in water droplets in the bulk of the flow. In water-continuous flow, the crystals form in the water phase in contact with the flow conduit (in wells, tubing wall). The two situations are illustrated in Figure 6.12, for water cut below and above 60% (= water fraction 0.6). When the continuous phase is crude oil, the cube-sized (rhombohedral structure) calcium carbonate crystals form in the water droplet and are primarily carried with the bulk flow. When the continuous phase is water/brine, the calcium carbonate crystals are formed/precipitated and carried with the bulk flow and potentially deposited at the wall. An important implication of such an emulsion-controlled precipitation and deposition process, is that not all the precipitated crystals form a deposit. Laboratory studies of the inorganic scaling in aqueous solutions (not oil/water mixtures), may only give a limited view of the overall problem. As with all hypotheses, the emulsion-controlled scaling process needs to be investigated, to find out in what situations it may be applicable. Logically, the hypothesis may also apply to other inorganic scaling minerals.

Calcium carbonate scales found in wells, flowlines and oil and gas facilities, are a mixture of different scale materials. The literature is not overflowing with analysis of inorganic scales in oil and gas operations. An exception is the elemental analysis presented by Sumestry and Tedjawidjaja (2013), shown in Table 6.9. Elements measured above one percent are included in the table, summing to about 93%. An X-Ray Fluoresces (XRF) method was used. Calcium dominates, indicating that calcium carbonate is the main constituent that deposited. The second element is magnesium. The 55.6% mass calcium and 16.6% magnesium has a ratio of 3.3 while on a mole basis the ratio is 2.0. The concentration ratio in formation water (Table 6.1) of calcium/magnesium is 5.8. Magnesium is a divalent ion (same as calcium) and exists in aqueous solutions as magnesium carbonate, and forms either magnesium carbonate or magnesium hydroxide scale. The standard solubility products are 3.5×10^{-8} for the carbonate

Table 6.9 Elemental analyses of calcium carbonate scale, based on XRF (Sumestry and Tedjawidjaja, 2013).

Element	Weight %
Ba	5.170
Sr	7.115
Ca	55.641
Na	6.590
Mg	16.631
S	1.604
Σ	92.751

and 1.8×10^{-11} for the hydroxide. Magnesium carbonate is more soluble than calcium carbonate (see Table 6.7) and the hydroxide is less soluble. There is evidence that the presence of the magnesium ion Mg^{2+} retards calcium carbonate crystallization (Østvold and Randhol, 2002). The same applies to the sulphate ion SO_4^{-2}. Also, the presence of the hydrate inhibitor MEG (Mono Ethylene Glycol), according to Flaten *et al.* (2008) and Sandengen (2012), can retard the formation of calcium carbonate scaling.

6.10 SULPHIDE SCALE

Iron sulphide scale is found in several combinations of molecules and structures, both crystalline and amorphous. The iron ion Fe^{2+} and the sulphur ion S^{2-}, make up a range of non-stoichiometric iron sulphides. Pyrite FeS_2 is one of the main scaling minerals, shown in Table 6.2. The sulphides are pH-dependent; the general formula for sulphide FeS has been added in Table 6.3. The solubility product constant of pyrite is the lowest in Table 6.7, making it less soluble than slightly soluble. Pyrite is not listed in *Appendix U – Solubility Product Constants*, but FeS is included. Unfortunately, literature data relevant for sulphide scaling in oil and gas operations are scarce.

For comparison, iron oxides are relevant in oil and gas operation. The most common oxidation states of iron are +2 and +3, the respective oxides being FeO and Fe_2O_3 (hematite = rust). The combination of the two oxidation states is found in Fe_3O_4 (magnetite), consisting of $FeO \cdot Fe_2O_3$. Iron has several other oxidations states, which is also reflected in the many forms of iron sulphides found. Iron oxide rust is found co-precipitated in many flow assurance solids. Magnetite is uncommon in industry, but is found in power cycles based on high-specification boiler water (Gudmundsson, 1997).

The formation, removal and prevention of iron sulphide scale has been reported by Nasr-El-Din and Al-Humaidan (2001). The reported scales are shown in Table 6.10, in terms of mineral name, chemical formula and density. Also shown is a *qualitative* evaluation of the effect of mineral acids on their dissolution rate, for the purpose of cleaning/removal. The acid effect is relevant when iron sulphide scale needs to be removed downhole, and in flowlines and processing equipment. It was found, that the chemical and physical characteristics of iron sulphite scale were a function of temperature, pressure, acidity and scale age. The solubility of iron sulphide was studied

Table 6.10 Iron sulphides common in oil industry, including *dissolution rate* in mineral acids (Nasr-El-Din and Al-Humaidan, 2001).

Mineral	Formula	Density	Rate
Mackinawite	Fe_9S_8	4.30	Fast
Marcasite	FeS_2	4.875	Slow
Pyrite	FeS_2	5.013	Slow
Pyrrhotite	Fe_7S_8	4.69	Moderate
Troilite	FeS	4.85	Rapid

by Pohl (1962). It was reported that at low temperature, 25°C the mineral troilite was stable; in the temperature range 40 to 150°C, the mineral pyrrhotite was stable and; in the temperature range 155 to 210°C, the mineral marcasite was most stable.

It was stated by Ning *et al.* (2015), that mackinawite was FeS and pyrite was FeS_2. The formula for pyrite agrees with Table 6.10, but not the mackinawite formula. There appears to be some discrepancy between the formulas reported by different authors, perhaps because of different disciplines and foci in the various studies. The non-stoichiometry of the pyrrhotite group can be expressed by $Fe_{1-x}S$ where x is a parameter with values in the range 0 to 0.17 (Ning *et al.*, 2015). Other authors (Sun *et al.*, 2008), state that many types of iron sulphides occur, such as amorphous iron sulphide (FeS), mackinawite ($Fe_{1+x}S$), cubic iron sulphite (FeS), troilite (FeS), pyrrhotite (FeS_2). The x in the mackinawite formula is reported in the range 0.057 to 0.064. Other ranges, for example $0.0 < x < 0.07$, are found in the literature (Lennie *et al.*, 1997).

Usually, mackinawite forms as a precursor to other types of sulphides. It suffices to state that the non-stoichiometric (polymorphous) nature of iron sulphide, makes it difficult to specify its structure. In dynamic oil and gas situations, iron sulphide scale may be a mixture of amorphous and crystalline solids. It was reported by Nasr-El-Din and Humaidan (2001), that analysis of iron sulphide scale in the oilfield shows a gradient in composition across deposits. FeS dominates at the interface at the pipe metal wall, while FeS_2 dominates at the scale surface facing the flowing fluid.

Iron sulphide is important in the corrosion of mild steel, similar to its importance in scaling. The solubility of hydrogen sulphide H_2S in an aqueous phase affects the acidity (pH), exemplified by the balances

$$H_2S \ (g) \leftrightarrow H_2S \ (aq)$$

$$H_2S \ (aq) \leftrightarrow H^+ \ (aq) + HS^- \ (aq)$$

$$HS^- \ (aq) \leftrightarrow H^+ \ (aq) + S^{2-} \ (aq)$$

$$H_2O \ (aq) \leftrightarrow H_3O^+ \ (aq) + OH^- \ (aq)$$

The above chemical balances can be illustrated in a dissociation diagram (fraction of species versus pH), as done for carbonates in Figure 6.10 and silicic acid in Figure 6.14.

The other sour gas that affect the acidity, is carbon dioxide. Data show that hydrogen sulphide is about three-times more soluble than carbon dioxide. The *Henry's Law*

Table 6.11 Thermodynamic solubility product constant K_{sp}^o for amorphous iron sulphide and crystalline iron sulphide minerals (Davison, 1991).

Mineral	$\log K_{sp}^o$	K_{sp}^o
Amorphous	-2.95 ± 0.1	1.12×10^{-3}
Mackinawite	-3.6 ± 0.2	2.51×10^{-4}
Greigite	-4.5 ± 0.1	3.16×10^{-5}
Pyrrotite	-5.1 ± 0.1	7.94×10^{-6}
Troilite	-5.25 ± 0.2	5.75×10^{-6}
Pyrite	-16.4 ± 1.2	3.98×10^{-17}

Constant at 20°C and 100 kPa pressure are quoted as 480 MPa for hydrogen sulphide and 1450 MPa for carbon dioxide, based on mole fraction. Solubility of gases is inversely proportional to Henry's constant. The constants are given in *Appendix Q1 – Solubility of Gases in Water*.

Consider the reaction

$$FeS \ (s) + H^+ \ (aq) \leftrightarrow Fe^{2+} \ (aq) + HS^- \ (aq)$$

where FeS represents the various forms of iron sulphide, both amorphous and crystalline. The solubility product constant can be written

$$K_{sp}^o = \frac{[Fe^{2+}][HS^-]\gamma_{Fe^{2+}}\gamma_{HS^-}}{[H^+]\gamma_{H^+}}$$

The square brackets indicate molal concentration and the gammas the respective activity coefficients. Because the activity coefficients are included, the solubility product constant is the thermodynamic solubility constant. Experimental work of Davison (1991), for the above expression, resulted in the solubility product constants of several of the iron sulphate scales, at standard conditions/state, shown in Table 6.11. The value for pyrite is two orders-of-magnitude greater than its value in Table 6.7. This difference underlines how difficult it can be to find solubility product constants in laboratory experiments. What is presently interesting is that pyrite is by far the least soluble of ferrous iron sulphides and amorphous iron sulphide the most soluble. It supports the general perception that pyrite is the solid mineral that precipitates and deposits most readily in oil and gas operation.

6.11 AMORPHOUS SILICA

Silica is not common in oil and gas scaling, except perhaps in small amounts in mixed inorganic scales. In geothermal operations, however, amorphous silica problems are common and represent a major constraint on efficient operations of flowlines and facilities (Gudmundsson and Bott, 1977; Gunnarsson and Arnórsson, 2003). The silica problem is as old as pipeline transport/use of geothermal energy. The word *deposition* is used to express direct scaling on surfaces, and the word *precipitation* to express the formation of particles in flowstreams and retention tanks. Particles in bulk are formed by polymerization. Silica can be described as slightly soluble, having a normal solubility (increasing with temperature). Most other common scaling materials in oil, gas and geothermal operations, exhibit inverse solubility (decreasing with temperature).

The silica concentration in subsurface water/brine, is normally found to correlate with the solubility of the mineral quartz, at least in high-temperature reservoirs. The mineral-water balances in the subsurface are complicated and are likely to involve a range of silicates, for example feldspar and mica. It is beyond the scope of the present work to discuss/present the intricacies of the geochemical balances of minerals in subsurface reservoirs.

The subsurface temperature in oil and gas reservoirs, increases with depth according to the geothermal gradient, resulting in temperatures in the range 80–120°C (non-HTHP reservoirs). The subsurface temperatures in exploitable low-temperature geothermal reservoirs are typically in the same range. An interesting phenomenon in geothermal reservoirs, is that the subsurface temperature tends to follow the boiling-point-curve of water (see for example, Wilkinson, 2001), due to natural convection. The subsurface temperatures in high-temperature geothermal reservoirs are in the range 200–300°C. In geothermal reservoirs having temperature in the range 180–240°C, calcium carbonate may be the main scaling problem. In the temperature range 240–290°C, silica may/will be the main scaling problem. Above 290°C in reservoir temperature, iron sulphide may be the main scaling problem, in addition to silica. Numerous papers on deposition of solids in geothermal systems are found in Gudmundsson and Thomas (1988).

The solubility of amorphous silica and crystalline quartz in fresh water temperatures from 0°C to 250°C, are shown in Figure 6.13. The pressure is atmospheric pressure up to 100°C and the saturation pressure of water at higher temperatures. The figure is stylized from D'Amore and Arnórsson (2000). Such/similar solubility plots are widely available in the literature, including Fournier and Rowe (1977) and Dietzel (2000).

The solubility of amorphous silica at 50°C and 100°C are 186 mg/kg and 306 mg/kg, respectively (see Table 6.12). The solubility values are in the slightly soluble range (see Table 6.5). In the relevant temperature range for commercial exploitation, the solubility of amorphous silica is around six-times greater than that of quartz. The common inorganic scales in oil and gas, are all in the slightly soluble range 0.1–10 g/L.

The dominant form of silica in water/brine is silicic acid H_4SiO_4, a weak acid. It is the hydrated form of silica. Silica is a pH-dependent scale (see Table 6.3). An acid-base reaction involves a transfer of protons from one substance to another. The

Table 6.12 Solubility of amorphous silica in water, based on Figure 6.13.

T °C	c mg/kg	ρ kg/m^3	c g/L
50	186	988	0.184
100	306	958	0.293

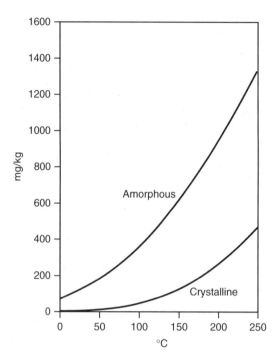

Figure 6.13 Solubility of amorphous silica (upper line) and crystalline quartz (lower line) in geothermal water.

acid is the proton donor and the base is the proton acceptor. Walther (2009) presented the following five silica balances

$$SiO_2 \ (amorphous) \leftrightarrow SiO_2 \ (aq)$$

$$SiO_2 \ (quartz) \leftrightarrow SiO_2 \ (aq)$$

$$SiO_2 \ (aq) + 2H_2O \leftrightarrow H_4SiO_4$$

$$H_4SiO_4 \leftrightarrow H_3SiO_4^- + H^+$$

$$H_3SiO_4^- \leftrightarrow H_2SiO_4^{2-} + H^+$$

The *first* balance for the solubility of amorphous silica has $pK_1^o = 2.71$ ($K_1^o = 1.95 \times 10^{-3}$). The *second* balance for the solubility of quartz has $pK_2^o = 4.00$ ($K_2^o = 1.0 \times 10^{-4}$). A comparison of the two solubility product constants show that amorphous silica is more soluble than crystalline quartz, as borne out by Figure 6.13. For clarification,

$$pK_{sp} = -\log[K_{sp}]$$

Table 6.13 First and second dissociation constants of calcite and silica in water at standard conditions.

Constant	$CaCO_3$	SiO_2
K_1^o	2.25×10^{-4}	1.5×10^{-10}
pK_1^o	3.64	9.83
K_2^o	5.15×10^{-11}	7.9×10^{-14}
pK_2^o	10.29	13.1

The neutrally charged SiO_2 (*aq*) is frequently written as H_4SiO_4. Walther (2009) argued that the solubility product constant for the first reaction, could also be used for the *third* balance, stating that the heat of formation of two water molecules will not affect the equilibrium significantly. Therefore, $K_3^o \simeq K_1^o$. The *fourth* balance has $K_4^o = 10^{-9.82}$ and the *fifth* balance $K_5^o = 10^{-13.10}$, according to Walther (2009). The first three balances express the solubility of amorphous silica and crystalline quarts. The fourth and fifth balances, are not solubility products, but *dissociation constants*. Other similar solubility product and dissociation constants (equilibrium constants) are found in the literature. Slightly different formulations of the silica-water chemical balances above, are found in the literature. The advantage of the above balances, is that they are consistent and have specific solubility product and dissociation constants.

There exists a chemical balance *analogy* between the components/species in calcite chemistry and silica chemistry (see *Appendix W – Calcite and Silica Chemistry*). Both are pH-dependent scales. The following analogies are relevant: silicic acid H_4SiO_4 (=A) and carbonic acid, once-dissociated silicic acid $H_3SiO_3^-$ (=B) and the bicarbonate ion and twice-dissociated silicic acid $H_2SiO_3^{2-}$ (=C) and the carbonate ion. One benefit of the analogy, is that the mole fraction equations presented for calcite (calcium carbonate), can also be used for silica (silicic acid). By replacing the calcium carbonate equilibrium constants with the silicic acid constants, the fractions of the silica species with pH can be calculated. The analogous dissociation constants are shown in Table 6.13. The first and second dissociation constants of silicic acid are for the fourth and fifth chemical balances shown above. The same kind of dissociation scheme can be written for iron sulphides; the dissociation constants are found in the literature.

The calculated silica *relative fractions* (not mole fractions) with pH are shown in Figure 6.14, based on relationships derived in *Appendix W – Calcite and Silica Chemistry*. A numerically similar figure (species fractions versus pH), was presented by Dietzel (2000), such that the dissociation constants used in the present work are reasonable. The A-ion dominates at pH below about 9, the B-ion dominates in the range $9 < pH < 14$ and the C-ion at higher pH-values. The pH of fresh water/groundwater is about 7, such that silicic acid is not ionized. Aqueous solutions are termed acidic when $pH < 7$ and alkaline when $pH > 7$. Commercially produced high-temperature geothermal waters, are reported to have $8 < pH < 10$ (Gunnarsson and Arnórsson 2003). The silica dissociation diagram (Figure 6.14) shows that an increase in the hydrogen ion concentration (lower pH) shifts the reaction/balance to the left and increased ratio of silicic acid. A decrease in the hydrogen ion concentration (higher pH), shifts the reaction to the right and increased silica solubility.

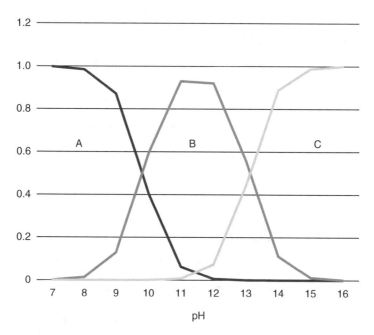

Figure 6.14 Dissociation diagram for relative fractions of silica species, $A = H_4SiO_4$ left (low pH), $B = H_3SiO_3^-$ middle and $C = H_2SiO_3^{2-}$ right (high pH).

Silicic acid can also be written as $Si(OH)_4$ to indicate a Si atom in the centre, connected to four surrounding hydroxyl groups OH. The first dissociation product can be written as $SiO(OH)_3^-$ (fourth chemical balance above) and the second dissociation product as $SiO_2(OH)_2^{2-}$ (fifth chemical balance above).

Silica as quartz does not precipitate during production (steam-water flow in wells). If that was the case, high-temperature geothermal resources could never be produced economically, because the production wells would become blocked by deposited silica. Silica deposition does not occur until the solubility of amorphous silica is reached, which has a six-times higher solubility than quarts. This fact encapsulates the conundrum of thermodynamics versus kinetics. The kinetics of quartz crystallization are much slower than the time it takes to produced aqueous fluids/solutions (water/brine), from a subsurface reservoir to surface facilities. Chemical thermodynamics are equilibrium thermodynamics, regardless of time and kinetics.

Water and brine will flash and cool down when flowing from downhole to wellhead, and further in surface flowlines and facilities. Flashing will increase the concentration of silica in the aqueous solution. The concentration will increase in direct proportion to the steam fraction generated. Considering Figure 6.13, picking any point on the quartz solubility line, the increase in silica concentration will follow an upward 45° line, approximately (Gudmundsson and Bott, 1977). Express in another way, the flashing line from quartz solubility to amorphous solubility, will follow a line perpendicular to the tangent of the quartz solubility line. If no steam flashing takes place, direct cooling will follow a horizontal line from the quartz line to the amorphous silica line.

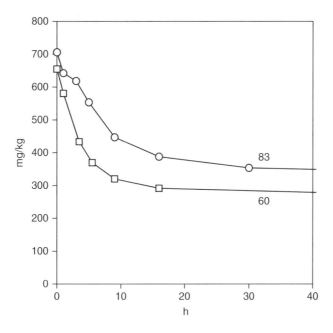

Figure 6.15 The effect of temperature on the rate of silica polymerization. The two curves (upper 83°C, lower 60°C) approach the solubility limit asymptotically (Gunnarsson and Arnórsson, 2003).

When an aqueous solution becomes supersaturated with amorphous silica, two overall processes are simultaneously effectuated. *One*, ionized silica can deposit on a solid surface forming a scale and *two*, ionized silica can polymerize, forming suspended particles. The solid silica in both processes is amorphous. Terms like monomeric silica and polymeric silica are commonly used. The anionic polymerization occurs through the formation of multiple Si-O-Si bonds. The polymeric structure can be linear or cycling and without a uniform size. Through the polymerization process, silicic acid is removed from the solution and a monomer progresses to a dimer and further to a trimer, *etc.*, each step releasing a proton and thus decreasing the pH. The following are the molecules: dimer $Si_2O_3(OH)_4^{2-}$, trimer $Si_3O_5(OH)_5^{-3}$ and tetramer $Si_4O_8(OH)_4^{-4}$ (Dietzel, 2000).

The chemical equilibria between the three main silicic acid species, the hydrated silica, the once-dissociated and the twice-dissociated anions, are presumably achieved relatively rapidly. On the other hand, the more intricate polymerization process is relatively slow. Logically, the polymerization process will depend on the degree of supersaturation, solution ionic strength, temperature and acidity-alkalinity. Measurements of silica concentration as silicic acid with time, illustrate the slow polymerization process, shown in Figure 6.15 (Gunnarsson and Arnórsson, 2003). Exact pH was not specified, but may have been in the range 9.15 to 9.55. The silica concentration decreases exponentially, to an asymptotic value of amorphous silica solubility. Two curves illustrate experiments carried out at 60°C and 83°C. Conventional knowledge would predict that the rate of polymerization increased with temperature. But that is

Table 6.14 Some inverse solubility inorganic scales (Bott, 1995).

Inorganic Scale	Chemical Formula
Calcium carbonate	$CaCO_3$
Calcium hydroxide	$Ca(OH)_2$
Calcium silicate	$CaSiO_3$
Calcium sulphate	$CaSO_4$
Magnesium hydroxide	$Mg(OH)_2$
Magnesium silicate	$MgSiO_3$

not the case in Figure 6.15. The experiment carried out at higher *supersaturation* gave a higher rate of polymerization, making it a more significant variable than *temperature*.

Barium sulphate scale occurs when sulphate-rich seawater is mixed with a magnesium-rich reservoir brine (see *Section 6.8 – Sulphate Scale*). An analogous scaling phenomenon occurs when a magnesium-poor low-temperature water is mixed with magnesium-rich fresh water. Magnesium silicate scaling occurs in district heating pipelines, when such waters are mixed (Hauksson *et al.*, 1995; Gunnlaugsson, 2012). Magnesium hydroxide and magnesium silicate are two of the inverse solubility inorganic scales listed by Bott (1995).

The chemical formula of magnesium silicate can be written in several ways. It is common to include a water of hydration, a simple version being

$$MgOSiO_2 \cdot H_2O = MgSiO_3 \cdot H_2O$$

The magnesium silicate found in low-temperature geothermal district heating pipelines has repeatedly been represented by the formula

$$MgSiO_3 \cdot H_2O \leftrightarrow Mg^{2+} + H_3SiO_4^- + OH^-$$

with a solubility product constant of

$$\log K_{sp} = -12.9 + 0.00262T - 6.212 \times 10^{-5}T^2$$

applicable in the temperature range 60–120°C. The temperature in the empirical equation is in Celsius T °C. The low-temperature geothermal water had a silica concentration of 86 mg/kg and the fresh water 21 mg/kg (Hauksson *et al.*, 1995; Gunnlaugsson, 2012). Modifications of the pH have been used to alleviate the both amorphous silica and magnesium silicate scaling problem in geothermal transport and district heating pipelines. It may be worth noting that seawater is relatively rich in magnesium, compared to formation water (see Table 6.1).

6.12 CONCLUDING REMARKS

The control of inorganic scaling in the oilfield should be included in early studies, engineering design, construction, operation and maintenance; that is, the entire field management cycle. The risk of inorganic scaling needs to be evaluated early-on in the planning and subsequent implementation of investment decisions. Flow assurance know-how should be applied throughout to reduce uncertainty (risk), capital expenditures and excessive operating expenditures. A production chemist/chemical engineer should coordinate the flow assurance solids issues, from the early-start of oil and gas projects (fluid sampling, analysis and properties, selection of chemicals/vendors, compatibility of materials and products). Understanding and analogies from the other big-five flow assurance solids, may contribute to finding cost effective and safe procedures in inorganic scaling. Integrated *flow assurance management* is required where more than one of the big-five solids pose a challenging problem.

Of the big-five flow assurance solids, inorganic scale is the only non-organic solid. Publications on the chemical thermodynamics of inorganic compounds in water (brine) abound in the literature. The problem of scaling, is found in a wide range of technical applications; wherever water is transported and used in processes. The universality of water (*aqua*) has resulted in several appendices: Solubility of Gases in Water, Chemical Equilibrium, Ionic Strength and Activity Coefficient, Gibbs Free Energy, Chemical Potential, Solubility Product Constants and Calcite and Silica Chemistry. That so many of the appendices are related to inorganic scaling, testifies to its importance in industry, here the oil and gas industry.

Carbonate chemistry takes a central place in inorganic scaling. The interaction between carbon dioxide and water in terms of solubility and pH, affects the scaling tendency of not only *calcite*, but also other scalants (see Table 6.3). The other gas that affects the pH of water is hydrogen sulphide. Dissociation diagrams can be constructed for both of the gases, showing the distribution of mole ratios (not mole fraction) of the different species from low-pH to high-pH. The dissolution of sulphates is not affected by pH. However, the mixing of formation water (rich in barium) and injection water (rich in sulphate), leads to local (*in situ*) precipitation and deposition of *barite* in near-wellbore formations and production tubing. Calcite and barite are the two *enfants terribles* of flow assurance managers.

Corrosion is a major challenge in oil and gas production operations. The topic has not been covered in the present text, because it is so enormous. The prevention of *corrosion*, however, may affect the deposition of inorganic solids. Carbon steel corrosion is more prevalent at low-pH than high-pH aqueous solutions. One method to reduced corrosion in flowlines and other oil and gas pipes and equipment, is to inject an alkaline to increase the pH-value. The basic agents NaOH (sodium hydroxide, caustic soda) and/or $NaHCO_3$ (sodium bicarbonate, baking soda) can be added to water-streams, to increase the pH-value. However, high-pH increases the ratio of the carbonate ion (see Figure 6.10), encouraging the precipitation of calcium carbonate (calcite). Therefore, there will be a trade-off between good corrosion protection (high pH) and scale control (low pH). This aspect was pointed out by Sandengen (2006), who also demonstrated that the addition of MEG (Mono Ethylene Glycol) for hydrate control, decreased the solubility of inorganic minerals in water. Another aspect worth considering is the *common ion* effect. The dissolution of sodium bicarbonate in water, irrespective of changes in pH, will increase the concentration of the bicarbonate ion. According to *Le Châtelier's Principle*, the chemical balance will adjust, and more calcium carbonate will precipitate.

Superimposed on the chemical thermodynamics of inorganic scaling, is the formation of *emulsions*. Apart from knowing that the big-five solids tend to stabilize oil-water and water-oil emulsions, the deposition of precipitated solids may be affected by the watercut. It was postulated above that inorganic scaling will be more prevalent in oil-in-water emulsions than in water-in-oil emulsions. The reason being that the inorganic chemistry processes (precipitation) occur in the water phase, not in the oil phase. The *emulsion barrier* to deposition will be less dominating in oil-in-water emulsions, compared to water-in-oil emulsions. The central point being, that solids formed in water droplets, will be carried along with the flow, without depositing directly on solid surfaces, at least not much. Because the watercut increases with time in oil production, the flowing stream will change from oil-continuous to water-continuous when the watercut increases above typically 60% vol. (see *Appendix X2 – Emulsions of Crude Oil and Brine*). The practical implication of the hypothesis is that inorganic scaling may be more prevalent after phase inversion, meaning in the mature and tail-end phases of a fields lifetime (see *Section 1.1 – Oilfield Production*).

The precipitation of inorganic minerals in water depend on temperature, acidity and salinity. Pressure affects the precipitation indirectly, through the dissolution of acidic gases such as carbon dioxide and hydrogen sulphide. Chemical additions are commonly used to prevent the precipitation of scaling molecules, and to prevent the deposition of precipitated solids. A whole industry has been developed to both prevent the formation of solids and the deposition of solids. The specialty chemical used are largely proprietary, and are employed by companies, serving the oil and gas industry. The same service companies, provide services to remove deposits throughout the flowstream, from reservoir to processing facilities.

Chapter 7

Naphthenate

The soap-like solids

Deposition of naphthenates has become a major problem in the production of acidic crude oils. The problem increases capital and operating costs/expenditures and reduces the regularity and safety of production and processing facilities. Since the 1960s it has been known that surface active naphthenates give rise to stable oil-water emulsions, making separation difficult and requiring the use of fine/specialized chemicals. Since the 1990s it has emerged that naphthenates also give rise to troublesome solid deposits in three-phase (oil, gas, water) production separators, and desalting equipment. Remedial actions include pH regulation (acid injection) and use of chemicals. Naphthenate deposits originate from dissolved naphthenic acid in crude oil.

The naphthenate problems arise when oil, gas and water/brine are produced. The cascading reduction in pressure in the flow stream, from the reservoir to surface facilities, result in CO_2 degassing of the water/brine phase and hence an increase in pH. Surface active naphthenic acid at an oil-water interface dissociates, and bonds with dissolved metal ions in the aqueous phase. The result is the formation of heavy calcium naphthenate and lighter sodium naphthenate, the metal ions being divalent and monovalent, respectively. The terms CaN and NaN are short-hands for calcium and sodium naphthenates.

Academia and industry have cooperated to understand the precipitation and deposition of naphthenates in the oil and gas industry, ranging from laboratory experiments to field observations and trials. The theoretical and laboratory studies by Havre (2002), Brandal (2005), Håvåg (2006), Shepherd (2008), Hanneseth (2009), Nordgård (2009), Ahmed (2010) and Shafiee (2014), have contributed much to the understanding of the naphthenate problem.

A review of the chemistry of tetrameric acids in petroleum, has been published by Sjöblom *et al.* (2014). The naphthenates problems are more severe in heavy and extra heavy oil than in conventional oil (light and medium). The problem of calcium naphthenate deposition has been addressed by Brocart and Hurtevent (2008), Passade-Boupat *et al.* (2012), Simon *et al.* (2012) and Juyal *et al.* (2015). The composition of crude oil related to flow assurance solids are presented in *Appendix X1 – Crude Oil Composition*.

7.1 ACIDITY OF CRUDE OIL

Crude oils contain a very large spectrum of chemical compounds. The compounds can be organized into the well-known fractions of Saturates, Aromatics, Resins and

Asphaltenes (SARA, see *Chapter 3 – Asphaltene*). Included in the fraction are various oil-soluble naphthenic acids. When the acids come into contact with degassing (depressurizing) water/brine during production, solid deposits of naphthenates are formed. The degree of solids formation and deposition correlates generally with the acidity of crude oil. Heavy crude oil is presumably more acidic than light crude oil. The water/brine pH increases upon carbon dioxide degassing, allowing for the formation of naphthenate solids. Hitherto, the effect of the water/brine chemical composition on naphthenate deposition has not been correlated. However, it seems plausible that brine salinity affects the formation of naphthenate solids. The mixing of brines from different productive formations (commingled streams) may contribute to the problem.

The severity of naphthene problems can be gauged by measuring the acidity of crude oils. The acidity results from a spectrum of molecules. There is not necessarily a correlation between the measured acidity and the concentration of the molecules responsible for naphthenate deposition (Lutnaes *et al.*, 2006). Acidity of aqueous solutions is well known in terms of pH. Crude oil is not an aqueous solution; consequently, its pH cannot be measured. The method used instead is Total Acid Number (TAN), measured using a ASTM (American Society for Testing and Materials) standardized method. The standard is based on potentiometric titration with a mixture of a hydrocarbon and KOH (strong base, potassium hydroxide, completely ionized in water). The method involves dissolving an oil sample in toluene and isopropanol with a minute amount of water. A glass electrode and a reference electrode are placed in the solution and connected to a voltmeter/potentiometer. The measurement result is reported as mg KOH/g; milligrams of potassium hydroxide required to neutralize one gram of crude oil. TAN can be defined as the amount of strong base that is required to neutralise the acidity of a crude oil sample. Examples of reported TAN-values and API gravity are Sjöblom *et al.* (2003), Barth *et al.* (2004), Simon and Sjöblom (2013), Elnour *et al.* (2014) and Shafizadeh *et al.* (2003).

A schematic potentiometric *titration curve* is shown in Figure 7.1, where the voltage between two electrodes is measured while the analytical base is added. The vertical axis shows the electromotive force in mV and the horizontal axis is the volume of KOH used. The inflexion point (ERC, Equivalence point Recognition Criteria) of the titration curve at *circa* 1.5 mL is defined as the TAN point, easily located from a gradient curve (not shown). A chemical mass balance is then used to calculate the TAN value. The TAN value is a measure of the acidity of a crude oil. Similarly, the TBN (Total Basic Number) is a measure of the basicity of a crude oil, using a standardized ASTM method. Both of the numbers in crude oils are discussed by Barth *et al.* (2004). In industry, the TAN values are important in refinery operations (due to corrosion) and the TBN number in lubricant oil degradation.

Napthenate precipitation and deposition occurs in both heavy and light crudes, and both high and low TAN value crudes. A comprehensive explanation has not yet been given. The chemistry of naphthenic acids in crude oil is elusive, and the interactions between crude oil and produced water/brine is similarly elusive. Comparable deductions have been presented by Simon and Sjöblom (2013), who presented a table with TAN and API gravity for ten stock-tank crude oils. Data also found in Aske (2003) and Sjöblom *et al.* (2003). Deposits were found in three fields and no deposits in seven fields. High molecular weight naphthenic acids (see following sections) were detected in all the stock-tank crude oils. No naphthene solids were reported for

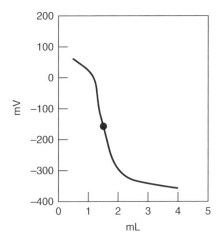

Figure 7.1 Typical potentiometric (mV) titration curve of a crude oil, showing an inflexion point at circa 1.5 mL KOH.

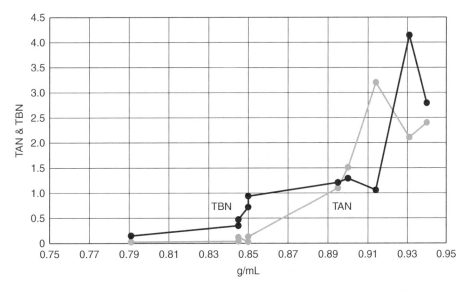

Figure 7.2 TAN and TBN for ten North Sea crude oils against density ρ_o g/mL (=g/cm^3) (Barth *et al.*, 2004).

values 0.38 < TAN < 0.66 (26.5 < °API < 34) while naphthene solids deposition were reported for values 0.11 < TAN < 2.7 (20 < °API < 34). Indeed, inconclusive results in terms of TAN-values. The no-deposition crudes are classified as light-to-medium and the deposition crudes light-to-heavy (see *Appendix X1 – Crude Oil Composition*).

TAN and TBN values were measured/reported by Barth *et al.* (2004) for ten stock-tank North Sea crude oils. The measured values are shown in Figure 7.2. The TAN and TBN values are seen to be similar and to follow the same trend, which is natural. Acids and bases must be in balance in any solution. The values are observed to increase

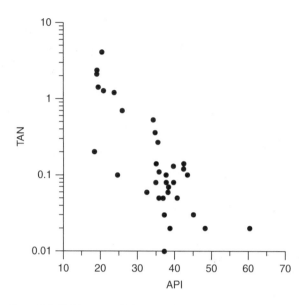

Figure 7.3 TAN against API gravity (Katz and Robinson, 2006).

exponentially with stock-tank oil density. When the crude oil density is above 0.9 g/cm³ in Figure 7.2, the API classification is heavy. At this density, TAN = 1.5 and increases to about 3.3 at greater densities.

The quality of crude oil at great reservoir depth was investigated by Katz and Robinson (2006). They reported the TAN-values for 34 crude oil samples, plotted in Figure 7.3 against API gravity: 25 or the points can be classified as light crude, three as medium crude and six as heavy crude oil. There was no mention of naphthenes deposition because the focus of the investigation was on the anaerobic biodegradation of crude oil at reservoir conditions, which increases the density (gives lower API gravity). The measurement results were plotted on a log-linear scale, confirming the observation and exponential/logarithmic relationship above.

TAN and API gravity values have been reported for 218 crude oil samples from wells in the Campos Basin in Brazil (Oliveira *et al.*, 2013). The values are plotted in Figure 7.4. Most of the points are for medium crudes, using the API classification. A few points are for light crudes and several points for heavy crudes. There are no points for extra heavy crudes.

TAN data presented by Elnour *et al.* (2014) indicate the same log-linear behaviour as shown in the figures above. The 12 data points covered the ranges 0.05 < TAN < 9.5 and 18 < °API < 37. The ranges are shown in opposite order; the lowest TAN has the highest API degree. Shafizadeh *et al.* (2003) presented a table of TAN and API gravity for five Venezuelan heavy acidic crudes. The ranges were 1.03 < TAN < 3.65 and 14.5 < °API < 23.5, also in opposite order. The following values were reported for offshore West Africa crude oils (Rousseau *et al.*, 2001): 0.87 < TAN < 4.10 and 23 < °API < 33.

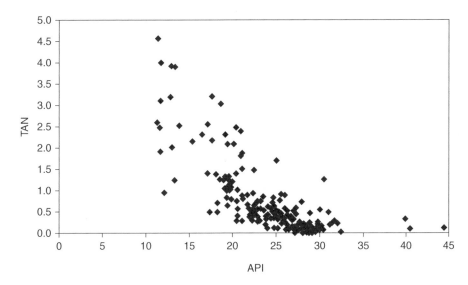

Figure 7.4 Crude oil TAN versus °API gravity for 218 samples from the Campos Basin (Oliveira *et al.*, 2013).

To summarize, the above three figures and literature values show a range of TAN-values, indicating

$$0.01 < TAN < 4$$

as the most plausible range for stock tank crude oils, although higher TAN-values have been reported. High acid crudes are stipulated to have TAN > 1.0 and acid crude TAN > 0.5 (Shafizadeh *et al.*, 2003). For comparison, crude oils with a TAN number greater than 0.5 were classified as highly acidic by Zhang *et al.* (2005).

The TAN value of crudes represents all organic acid molecules in the crude, both light and heavy acids. The most common organic acids are the carboxylic acids whose acidity depends on the carboxyl group -COOH. Light acids in crude oil include formic acid (HCOOH) and acetic acid (CH_3COOH), both of which are classified as weak acids (partially dissociated in water). The heavy acids in crude oil are the naphthenic acids. More than a 1000 species of the naphthenic acids are reportedly found in crude oils (Shafizadeh *et al.*, 2003).

There is no evidence in the literature that naphthenate deposition correlates with the TAN value of a crude oil. Deposits can occur in both low-TAN and high-TAN crudes. However, it was concluded by Shafiee (2014), that the concentration of carboxyl acids in crude oil correlated well with the TAN value. The figures and literature values presented above show a reasonable correlation between the TAN-value and API gravity.

SARA-analyse was used by Shafiee (2014), to investigate the contribution of the asphaltene and maltene fractions to the acidity of crude oils. The results, based on nine crude oil samples, are shown in Table 7.1. The TAN of the crude oils had a maximum value of 7.58 and the asphaltene fraction a maximum value of 21.24. The maltene

Table 7.1 Analysis of nine crude oils (Shafiee, 2014).

Fraction	Weight %	TAN
Crude	100	0.11–7.58
Asphaltene	2–3	3.13–21.24
Maltene	87–98	0.04–6.85

Table 7.2 Analysis of one crude oil (Zhang et al., 2005). S+A stands for Saturates plus Aromatics. The mass balance "loss" was due to the analysis methods used.

Components	Mass g	Mass %	TAN
Asphaltene	1.60	8.0	8.32
S+A	12.14	60.7	0.14
Resin	5.26	26.3	2.34
Crude oil	20.0	5.0 (loss)	4.35

fraction constituted maximum 98% wt. of an oil sample and measured a maximum TAN value of 6.85.

Work reported by Zhang et al. (2005), for one crude oil sample, indicated a similar result as that of Shafiee (2014). The asphaltene fractions had the highest TAN value, followed by the crude oil sample and then the resin fraction, shown in Table 7.2. The components' polarity, molecular weight, aromaticity and heteroatom content of the fractions increase from Saturates to Aromatics to Resins and highest for Asphaltenes.

7.2   NAPHTHENIC ACIDS

Naphthenic acids are a part of the greater carboxylic acids family found in petroleum (crude oil). Acids in crude oils are predominantly carboxylic acids having the functional group O=C–OH, shown in Figure 7.5. The general formula can be written R–COOH. The carbon atom has a *double* covalent bond to the oxygen and a *single* covalent bond to the hydroxide. The R stands for the remainder (rest) of the molecule, not participating in or affecting its main chemical reactivity. In naphthenate deposition, it is the hydroxide ion that releases (donates) the hydrogen ion when dissolved in water/brine.

The general chemical formula of carboxylic acids can be written as

$$C_nH_{(2n+z)}O_2$$

where n stands for the number of carbon atoms and z for a hydrogen deficiency factor. For $n=1$ and $z=0$, the chemical formula becomes HCOOH, which is formic acid.

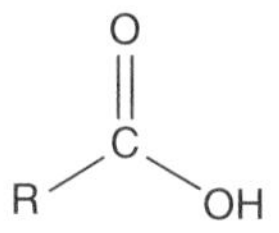

Figure 7.5 Carboxylic acid functional group (also called moiety). R stands for rest/remainder (*reliqua*) of a molecule.

For $n = 2$ and $z = 0$, the chemical formula becomes CH_3COOH, which is acetic acid. The factor z refers to hydrogen deficiency which is determined by the number of cyclic rings and double bonds present in a molecule. The factor has values from zero to negative even numbers $(-2, -4, -6, \ldots, -n)$. The above two simple acids in crude oil are not directly relevant for naphthenate deposition, but are relevant for the measured TAN-value and the injection of chemicals to mitigate naphthenate deposition. Acetic acid is an example of an organic acid injected to lower the pH of produced formation water/brine.

Naphthenic acids (NAs) are an important fraction of the carboxylic acids family in crude oils. NAs are characterized by the carboxylic acid functional group attached to a hydrocarbon molecule (see Figure 7.10). The following generalized chemical formula can be applied to NAs in crude oil

$$R(CH_2)_n COOH$$

where R is a cyclopentane ring and n typically greater than 12 (Zhang *et al.*, 2005). A cyclopentane ring has five sides (pentagon) while a cyclohexane ring has six sides (hexagon). Naphthenic acids include one or more such saturated ring structures. A multitude of other acidic compounds are also present in crude oil. The chemistry of petroleum related naphthenic acids has yet to be completely characterized. The presence of NA compounds contributes to the acidity of crude oils and is one of the major sources of corrosion in oil pipelines and refineries. Corrosion has always been a concern in the petroleum industry. The last decades, naphthenate deposition has become a major unresolved concern. An additional concern is the environmental impact of NAs dissolved in surface waters in production areas of heavy oil, extra heavy oil and bitumen.

Naphthenic acids are dissolved in crude oil, made soluble by the surrounding lighter molecules (solvation effect). They are sparsely soluble. In general, naphthenic acids are C_{10} to C_{50} compounds with 0–6 fused saturated rings. The carbon number and ring content distribution depends on the crude oil source (Kumar, 2012). Six relatively light naphthenic acids (straight chain saturated and unsaturated) were studied by Dyer *et al.* (2003), with molecular weight in the range 116–280 kg/kmol. The number of carbon atoms was 6–18. They found that the amount of solid formation increased with molecular weight and aqueous phase pH value.

An important aspect of naphthenate deposition is knowing the concentration of naphthenic acids in crude oils. 17 stock tank North Sea crudes were found to contain 0.19–9.5 mg/g with an average of 3.5 mg/g acids per gram oil, equivalent to 0.35% weight (Barth *et al.*, 2004). Separately, it was stated by Hanneseth (2009) that the

Table 7.3 Weight percentages of naphthenes components in crude oil.

Components	Weight %
Naphthenes	30–60
Naphthenic Acids	1.5–3.0
Naphthenates	0.075–0.15

naphthenic acid content in crude oils ranged from a few ppm (=mg/kg) to 3% wt. One part per million corresponds to 0.001% wt. Baugh *et al.* (2005) reported 2% wt. naphthenic acid and Håvåg (2006) reported 2 ppm naphthenate in the crude oil for Field H in Table 7.9.

Crude oil was reported to contain 2–3% wt. naphthenic acids (Moreira and Teixeira, 2009) based on Jones *et al.* (2001). Quoting another study, Dyer *et al.* (2003) stated that crude oils can contain up to 2% wt. naphthenic acids, consisting of a wide range of individual components. In an attempt to generalize, presently, the weight percentage of naphthenic acids in crude oil and the percentage involved in naphthenate deposition, a starting point was the 30–60% weight of naphthenes in crude oils (see Table X1.2 in appendix). Dividing this number arbitrarily by 20 (to get dissolved acid) and again dividing arbitrarily by 20 (to get precipitated solid), the concentrations weight percentages in Table 7.3 resulted.

The naphthenic acid concentrations are factual, while the naphthenate concentrations (the percentage of the acid that leads to solids formation) are speculative. The values in Table 7.3 can easily be described as guestimates. Knowledge of existing and future measurements are required to adjust the concentrations to more correct values. It was stated by Ahmed (2010) that the formation of calcium naphthenate in field operations was quite small, around 10 mg/L (solid matter per litre oil). This value is equivalent to 0.11% wt., which is within the naphthenates range suggested above.

Understanding what the components/compounds naphthenate, naphthenic acids and naphthenes mean/signify, may be problematic. The terms have been used in the oil and gas industry from olden times to the present. The ancient word naphtha is resilient in the petroleum industry, including oil refining. The world is used in many languages for the automotive fuel gasoline. Venturing to define the word naphthenate, one answer may be that it is a salt of naphthenic acid. Venturing to define the word naphthenic acid, one answer may be that it is a monocarboxylic acid derived from naphthene. Venturing to define the world naphthene, one answer may be that it is a cyclic aliphatic hydrocarbon found in petroleum. In organic chemistry, the naming of the naphtha-related compounds is more precise and molecule-based.

7.3 TETRACARBOXYLIC ACIDS

It was reported by Baugh *et al.* (2005), that the naphthenic acids responsible for naphthenate deposition are high molecular weight fractions in the range 1227–1235 kg/kmol. The structure of the acids was elucidated by Lutnaes *et al.* (2006)

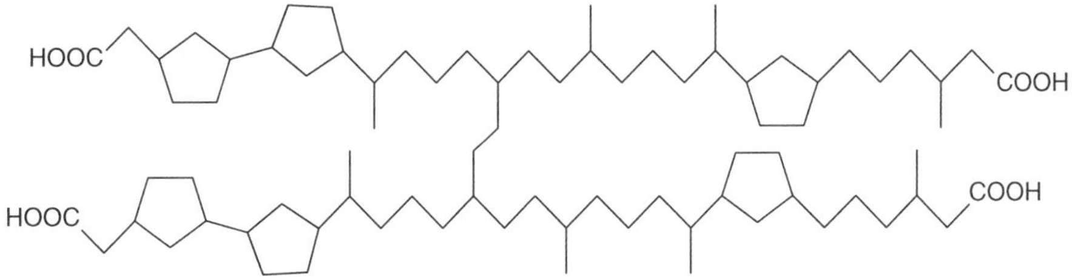

Figure 7.6 A representation of the predominant structure of the naphthenic acid responsible for naphthene deposition (Lutnaes *et al.*, 2006).

Table 7.4 Molecular weight of tetracarboxylic acids as a function of hydrogen deficiency factor.

z	M kg/kmol
−10	1238
−12	1236
−14	1234
−16	1232
−18	1230

and has been referenced thereafter in the literature. The predominant NA structure contains 80 carbon molecules (C_{80}), shown in Figure 7.6. The acid is a carboxylic acid consisting of saturated chains and six pentagonal ring structures. The acid has four carboxylic functional groups and is therefore classified as tetraprotic. The term tetrameric is also used (Sjöblom *et al.*, 2014).

Other similar structural models of the predominant naphthenic acid have been reported (Smith *et al.*, 2007). Five models containing 4–8 ring structures ranging in molecular weight from 1314–1306 kg/kmol were presented. The greater the number of ring structures, the lower the molecular weight. All the structures had four carboxyl functional groups.

Drawing upon the tradition of writing organic chemistry formulas, the following formula can be used for carboxylic acids (Baugh *et al.*, 2005)

$$C_{80}H_{160+z}O_8$$

where z stands for the hydrogen deficiency factor. Using z-factor values from −10 to −18 the molecular weight was calculated and shown in Table 7.4. The range of calculated values are about 80 kg/kmol lower than the values reported by Smith *et al.* (2007). The reason for this discrepancy is not known, but the reported values may be based on a methyl ester derivative of the carboxylic acids, perhaps related to the sample handling and chromatographic methods used.

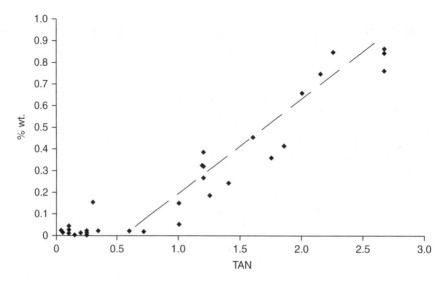

Figure 7.7 Naphthenic acid concentration % wt. against TAN (Meredith *et al.*, 2000; Shepherd, 2008).

Experimental data of Meredith *et al.* (2000) was replotted by Shepherd (2008) to focus on the relationship between TAN-values and NA concentration % wt. Data from 33 crude oil samples were plotted as shown in Figure 7.7. High acid oils have TAN > 0.5. The figure shows that below 0.5 the NA concentrations are well below 0.1% wt. and close to zero. The accuracy of the TAN measurements was reported ±0.1 units, which is of greater significance for low-TAN crude oils than above TAN 0.5 crude oils. The accuracy of the chromatographic methods used was not reported. General considerations would suggest that a complicated chromatographic measurement have no better accuracy than ±10%. The data shown that the NA concentration importantly increases linearly with TAN at values greater than about 0.6. In the present work, the straight line in Figure 7.7 was drawn by free-hand, from TAN $\cong$ 0.6, represented by the approximate equation

$$NA = 0.45(TAN - 0.60)$$

7.4 COMPOSITION OF DEPOSITS

Detailed compositions of naphthene deposits in oil and gas facilities, are seldom found in the literature. Among the few known examples, are the analyses of several organic field deposits presented by Shepherd *et al.* (2006). The elemental analyses (XRD = X-Ray Diffraction) of four deposits are shown in Table 7.5. The concentrations are in % wt. Fields X and Z are in the North Sea, field Y in Africa and field W in Asia. Fields X, Y and Z are low in Na but high in Ca. Field W is high in Na, but low in Ca.

Table 7.5 Elemental analysis (XRD) of naphthenate deposits in % wt. from two North Sea fields (X and Z), one African field (Y) and one Asian field (W), based on table in Shepherd *et al.* (2006).

Element	X	Y	Z	W
C	85.77	85.86	91.41	85.02
O	5.54	6.71	5.39	3.85
Na	0.71	1.28	0.45	3.65
Al	0.25	0.21	–	0.49
Si	0.30	0.17	–	2.41
S	0.55	0.55	–	0.28
Cl	1.06	1.70	0.38	3.50
Ca	3.72	2.42	2.37	0.23
Fe	0.76	1.10	–	0.58
Mg	–	0.21	–	–
P	–	–	–	–
Ba	1.34	–	–	–

The conclusion that can be drawn from the elemental analysis, is that the deposits from fields X, Y and X are CaN and that from field W is NaN. The several other elements shown in Table 7.5 were ascribed to sand, salt and clays. Mass spectroscopy showed that the CaN-dominated solids exhibited an intensity peak at circa 1230 kg/kmol, and a spectrum of much lower molecular weight naphthenic acids. The NaN-dominated deposit showed a spectrum of only the much lower molecular weight naphthenic acids.

Another solid composition example, is that of Vindstad *et al.* (2003), from Field H (see assay in Table 7.9). The % wt. values shown in Table 7.6 are average values from several samples. It is assumed here that the organic component includes all organic material in the samples. The samples contained 43% wt. naphthenate having an average molecular weight of 330 kg/kmol, which is within the range 300–400 kg/kmol reported by Rousseau *et al.* (2001) for interface active acids, meaning NaN-based surfactants. The average molecular weight reported by Vindstad *et al.* (2003), therefore, does not indicate CaN deposits. But calcium was reported 2.6% wt., much higher (order of magnitude) than the other inorganic compounds, except silica. One deduction can be that the deposits were CaN but that the molecular weight reported was too low, at least compared to the range 1227–1235 kg/kmol reported later for the same field (Baugh *et al.*, 2005). However, the reason for the low molecular weight may be as explained by Shepherd *et al.* (2006), that the CaN was masked by the much larger concentration of lower molecular weight naphthenic acids. Interesting that the silica value was 2.4 kg/kmol, suggesting that amorphous silica was a part of the mixed deposit; the silica may also stem from formation sand particles.

The identification and control of calcium naphthenate deposits in two offshore West Africa fields, were reported by Igwebueze *et al.* (2013). Solids were observed in the processing facilities after water breakthrough, in separators, hydrocyclones and coalesce vessels. One of the fields produced an API 34° crude oil with a TAN-value of 0.2. The produced water (mixture of formation water and condensed water) contained 50 mg/L of calcium and had pH = 8.15. The deposited solids were analysed

Table 7.6 Average composition of naphthenate* deposits from Field H (Vindstad et al., 2003).

Component	Weight %
Organic	69
Naphthenate	43
Residue	14
Ca	2.60
Fe	0.25
Ba	0.44
Na	0.40
Si	2.40
S	0.33

*Average molecular weight 330 kg/kmol.

Table 7.7 Composition (XRF) of CaN in a representative deposit in a second-stage separator in an offshore West Africa facility. Cations expressed as oxides (Igwebueze et al., 2013).

Component	% wt.
N	0.48
H	8.58
C	64.63
Na_2O	0.29
MgO	0.17
Al_2O_3	0.71
SiO_2	3.85
P_2O_5	0.46
SO_3	2.13
Cl	0.29
K_2O	0.11
CaO	7.94
TiO_2	0.15
MnO	0.10
Fe_2O_3	6.74
ZnO	1.08
Br	0.00
SrO	0.25
ZrO_2	0.02
BaO	2.03
PbO	0.02

using XRF (X-Ray Fluorescence) and MS (Mass Spectrometer). The MS showed a peak at 1224 kg/kmol and 616 kg/kmol, indicating calcium naphthenate CaN. The amount of CaN in the crude oil flowstream was a few mg/L. A representative solid deposit from a second-stage separator was analysed by XRF and is shown in Table 7.7, where the cations are expressed as oxides. Of the oxides, CaO has the highest % wt., indicating a CaN deposit (organic calcium salt). The high concentrations of hematite (Fe_2O_3) indicates corrosion products. The high silica (SiO_2) value and the various metal components, were reported to indicate the other solids in the deposit/system. It was reported that some of the wells produced formation sand, which stuck to the CaN deposits. The deposition problem was largely solved by injecting 5–10 ppm of a proprietary chemical inhibitor. The BS&W of the treated crude oil was within the specification of 0.5% wt. and the oil-in-water separated water overboard was under the required limit of 29 ppm.

7.5 DEPOSITION PARAMETERS

The front-end designers of new field developments and operators of existing production and processing facilities, need to know the parameter and factors that affect naphthenate formation and deposition. Useful case studies have been presented by Turner and Smith (2005) and earlier by Rousseau *et al.* (2001). The parameters of main interest are: the crude oil composition, the acidity of the formation brine and the overarching values/effects of temperature and pressure throughout the flow steam. Production time dependent parameters such as watercut are also important. Equipment related effects may likewise play a role in naphthenate deposition: downhole pumps, choke valves and other flow restrictions. Local pressure drops increase the shearing of flowing oil, gas and water mixtures, affecting/decreasing the size of bubbles and droplets in emulsions.

Naphthenic acid (NA) is soluble in crude oil and slightly, slightly soluble in brine. In the interface between a water droplet in oil (water-in-oil emulsion) and an oil droplet in water (oil-in-water emulsion), NA will be affected by the acidity of the brine/water. Due to the cascading reduction in pressure from a reservoir to production facilities, the brine-phase and the oil-phase, will undergo a reduction in the concentration of volatile gases (hydrocarbon and non-hydrocarbon). The gas molecules will transfer from the liquid phase to the gas phase. The non-hydrocarbon gases include carbon dioxide and hydrogen sulphide. Carbon dioxide is the gas controlling significantly the acidity of brine/water. The following chemical balances are representative of the carbon dioxide degasification

$$CO_2 \ (aq) + H_2O \leftrightarrow H_2CO_3 \ (aq)$$

$$H_2CO_3 \ (aq) \leftrightarrow HCO_3^- + H^+$$

Carbon dioxide exist in water as carbonic acid. Removal (degassing) of carbon dioxide, decreases the carbonic acid concentration. The second chemical balance shows that when this happens, the concentrations of the bicarbonate ion and the hydrogen ion also decreases. A decrease in the hydrogen ion means an increase in pH through the definition

$$pH = -\log[H^+]$$

The approximate solubility of hydrogen sulphide in water/brine is about 3500 ppm, compared to 1500 ppm for carbon dioxide and 20 ppm for methane, all at standard conditions 25°C, 100 kPa pressure and neutral pH. The solubility of gases in water depends on the *Henry's Law* constant k_H that depends on temperature, salinity and pH (see *Section 6.3 – Scaling Minerals*). Both calcium carbonate and hydrogen sulphide are sour gases. The dissolution of hydrogen sulphide will have the same qualitative effect on the pH as the dissolution of carbon dioxide, through the chemical balance

$$H_2S \, (aq) \leftrightarrow HS^- + H^+$$

The inorganic ions that partake in the formation of naphthenic acid salts are primarily divalent Ca^{2+} and monovalent Na^+. The respective naphthenate salt can be abbreviated CaN and NaN. Other ions that may perhaps be involved are divalent Mg^{2+} (Juyal *et al.*, 2015) and monovalent K^+ (Turner and Smith, 2005). Solid-like deposits CaN and emulsion-like deposits NaN are formed by the interaction of naphthenic acids and divalent (Ca^{2+}, Fe^{2+} and Mg^{2+}) or monovalent ions (Na^+ and K^+) present in produced waters (Mapolelo *et al.*, 2009).

The following chemical balances show the formation of CaN and NaN in formation water/brine

$$RCOOH \leftrightarrow RCOO^- + H^+$$

$$RCOO^- + Na^+ \leftrightarrow RCOONa \, (s)$$

$$2RCOO^- + Ca^{2+} \leftrightarrow (RCOO)_2Ca \, (s)$$

If acid is added to the water/brine, the first reaction will balance to the left. There will be less carboxylic ion available in the brine and therefore less naphthenate salts (CaN and NaN) will form.

A postulate put forward here, is that NaN is more easily formed than CaN. Both of the salts are practically insoluble in brine/water and both are amorphous (not crystalline). Consider the simplified structure of tetracarboxylic acid shown in Figure 7.5 and Figure 7.10. The figures show one carboxylic functional group connected to the remainder of the molecule, R. The three other connections represent also carboxylic functional groups (left out in figure, for simplicity). All of the four groups donate a proton ($=H^+$) to the aqueous formation brine, resulting in $RCOO^-$. The postulate states that it must be easier for Na^+ to connect with *one* negatively charged carboxylic groups, than for Ca^{2+} to connect simultaneously with *two* groups. Geometrically, it can further be argued that CaN should be a denser molecule than NaN. The properties of the two salts are reported to be different, despite a lack of hard evidence. NaN is soft and acts as a surfactant, stabilizing emulsions. CaN is a solid that deposits on metal surfaces, restricting the flow through and the operability of flowlines and surface facilities (Turner and Smith, 2005).

The concentration of Na in seawater and formation water is about 25-times that of Ca (see Table 6.1). The concentrations of Na in seawater and formation water range from 11,150 ppm to 17,465 ppm and for Ca from 435 ppm to 719 ppm. Na ions are therefore readily available to form NaN and Ca ions are available to form CaN, but to a lesser extent. The heat of formation of the simple calcium salt $CaCl_2$ is -796 kJ/mol compared to -411 kJ/mol for NaCl. The availability of ions and the heat of formation, alludes that the formation of CaN is more difficult than the formation of NaN, as already suggested in the paragraph above.

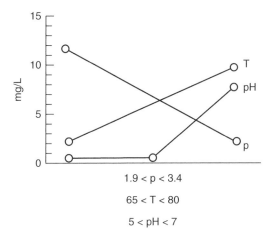

Figure 7.8 Naphthenate separated on-line from several crude oil streams, depending on temperature °C, acidity pH and pressure barg, based on data presented by Turner and Smith (2005) for a West Africa offshore field.

Five case studies were presented by Turner and Smith (2005). Two exemplified solid CaN deposits and three exemplified surfactant NaN deposits. Solid CaN was found in a North Sea field and in a West Africa field. The surfactant NaN deposits were found in a West Africa field and two fields in Asia. The influence of temperature, pressure and pH were presented for the West Africa field suffering solids deposition. In the present work, the graphical data presented were generalized/replotted in Figure 7.8. The figure shows the concentration of naphthenates mg/L in a crude oil stream in a processing facility. The concentration decreases with operating pressure, being in accord with degasification of carbon dioxide. The concentration increases with operating temperature, being in accordance with the heat addition requirement of saponification (soap making). A plausible effect may also be that interfacial tension decreases with temperature (*Appendix X3 – Energy Dissipation and Bubble Diameter*).

The total solids deposited depended strongly on the pH of the produced water/brine, shown in Figure 7.8. The line for the effect of pH has a sharp bend at pH ≃ 6. At pH < 6 the amount of naphthene solid formation is very low, practically zero. At pH > 6 the amount of solid increases linearly with increasing values. Dyer *et al.* (2003) have reported a similar behaviour. In *Section 6.9 – Calcium Scale* the mole fractions of carbonic acid, bicarbonate ions and carbonate ions are illustrated against the pH of water at standard conditions, including 25°C. At pH ≃ 6 the concentration of carbonic acid has decreased to almost zero. The concentration of the bicarbonate ion increases from near-zero at pH ≃ 2 to a maximum at pH ≃ 7 and returning to near-zero at pH ≃ 12. The pH line in Figure 7.11 depend on the equilibrium constants that in-turn depends on temperature according to the *van't Hoff Equation*. Similarities may be sought by looking through *Appendix W – Calcite and Silica Chemistry*.

The partitioning and dissociation constants of carboxylic acids from a crude oil sample and several naphthenic acid model compounds, were reported by Havre *et al.*

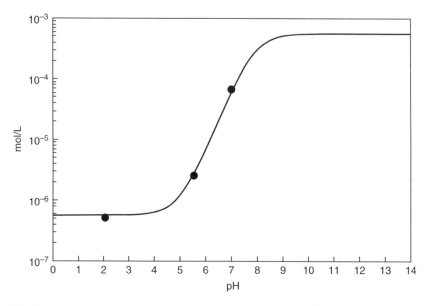

Figure 7.9 Concentration mol/L of naphthenic acids (total) in water at various pH values (Havre *et al.*, 2003 and Sjöblom *et al.*, 2003).

(2003). The naphthenic acid content of the crude oil was 2% wt. The acid extracted from the crude oil was determined to have 12 carbon atoms and two rings, based on a combined GC/MS analysis (Gas Chromatograph/Mass Spectrometer). The crude oil naphthenic acid was mixed with ordinary water having different pH. The results were fitted to an equilibrium model. The measurements and the model are shown in Figure 7.9. The concentration of the total acid content in mol/L is shown against pH. The dissociation constant was found to have the value of pK = 4.9 and the partitioning constant a value of 8.4×10^{-4} mol/L. The shape of the curve in Figure 7.9 follows the same shape as the lines in Figure 7.8. It was shown by Havre *et al.* (2003), that there was a deviation between the model and experimental data at high concentration. It was suggested that the deviation was due to the formation of normal and reverse micelles.

There may exists an analogy between the behaviour of carbonic acid and tetracarboxylic acid in water. The equilibrium constants are well known for the three form of carbonic acid. However, the equilibrium constant(s) for tetracarboxylic acid(s) are only known for one crude oil and six model compounds (Havre *et al.*, 2003). Experimental values have been obtained for model systems using carboxylic acids with molecular weight in the range 478–983 kg/kmol (Sundman *et al.*, 2010). The pH line in Figure 7.8 suggests that carboxylic acid dissolved in water starts to dissociate at pH $\simeq 6$ and that the dissociation increases with increasing pH-value. The two following chemical balances are considered to support the suggested analogy

$$H_2CO_3\ (aq) \leftrightarrow HCO_3^- + H^+$$

$$RCOOH\ (aq) \leftrightarrow RCOO^- + H^+$$

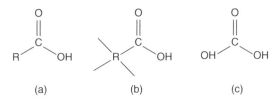

Figure 7.10 Functional groups of (a) carboxylic acid, (b) tetracarboxylic acid and (c) carbonic acid. R stands for rest/remainder of molecule.

The analogy is further strengthened by considering the molecule structures in Figure 7.10. Furthest to left (a) is the structure of any low-to-medium molecular weight carboxylic acid. In the middle (b) is a simplified structure of high molecular tetracarboxylic acid. To the right (c) is the structure of carbonic acid. All the structures have the characteristic double covalent bond between carbon and oxygen. All the structures have a hydroxyl ion bonded to the carbon atom. The structures have a mixture of nonpolar covalent bonds, polar covalent bonds and ionic bonds. At sufficiently high pH the molecules dissociate to release a proton (hydrogen H^+), leaving the rest of the molecule(s) negatively charged as shown above. The formation of calcium carbonate is presented in *Section 6.9 – Carbonate Scale* and the formation of CaN and NaN are presented above.

Apart from offering a useful analogy, carbonic acid dissociates into not only bicarbonate but also carbonate. The chemical balances leading to a calcium carbonate solid are

$$H_2CO_3\ (aq) \leftrightarrow HCO_3^- + H^+$$

$$HCO_3^- \leftrightarrow CO_3^{2-} + H^+$$

$$Ca^{2+} + CO_3^{2-} \leftrightarrow CaCO_3\ (s)$$

It was suggested by Rousseau *et al.* (2001) that calcium carbonate co-precipitated with CaN One consequence is that simulation models overpredict the amount of calcium carbonate precipitation because they are not able to account for the precipitation of CaN. An observation offered was also that naphthenate deposits were associated with high carbonated brines. The prevalent existence of mixed deposits in oil and gas production was presented by Frenier and Ziauddin (2010), including remediation methods.

7.6 INTERFACE PROCESSES

Transport phenomena take place at interfaces: heat transfer, mass transfer and momentum transfer (see *Appendix E – Transfer Equations*). Adding chemical reactions/ precipitations and surface active agents (surfactants), the simultaneous processes occurring as interfaces, makes understanding naphthenate deposition challenging.

Superimposing multiphase flow on the processes mentioned, renders a quantitative understanding even more challenging. The second-best to quantitative analysis would be semi-empirical analysis. The third-best would be qualitative analysis, to understand the main trends. What follows is a mixture of the three methods, with focus on naphthenate formation in oil and gas production situations.

Oil production involves multiphase flow of oil, gas and water with various solids. The solids can be sand particles and fines (clays and silts) from the productive formations, they can be corrosion products (aqueous oxides/sulphides and particles) from the entire flow stream and they can be any of the other big-five flow assurance solids (asphaltene, paraffin wax, gas hydrate and inorganic scale). Water-in-oil and oil-in-water emulsions are discussed in *Appendix X2 – Emulsions of Crude Oil and Brine*, including the inversion point. The ease/difficulty of the separation of crude oil and formation water/brine, depends on the temperature and the gravity of the crude oil, plus whatever surfactants and particles are present. In on-land operations, free-water knockout is conventionally the first separation stage, followed by horizontal-tank separation. In off-shore operations, the crude oil brine emulsion is most often directed to the first-stage tank separator and further too the second-stage horizontal tank separator. The general rule-of-thumb is that > 20° API oil can be separated by gravitation. Crude oil 15° to 20° API is more difficult (marginal) to separate in conventional tank separators. For crude oil gravities < 15° API, separation is difficult if not impossible, without special measures.

Naphthenate deposits form at oil-water interfaces in water-in-oil emulsions and oil-in-water emulsions. The emulsions may also be *mixed emulsions*, with oil droplets in water droplets and water droplets in oil droplets. Normal emulsions and mixed emulsions are schematically illustrated in Figure 7.11. Three main interaction *forces* between emulsion droplets have been identified: (1) van der Waal attraction (2) electrostatic repulsion and (3) steric repulsion. The steric repulsion forces are classically due to the arrangement of atoms in a molecule. In oil-water emulsions, steric repulsion can be due to non-ionic surfactants and polymers, that cover/coat the surface of a droplet. The range of factors affecting heavy crude oil emulsions, have been presented by Silset (2008).

The interfaces between oil-gas-water are illustrated schematically in Figure 7.12. The three phases are identified by (o), (g) and (w) and the interfaces by the two sloping lines. The gas-phase is in the upper left-hand corner and the oil-phase in the upper right-hand corner. The water-phase is below the other two phases in the diagram. The lines represent the boundary between the phases (water bubble in oil and oil bubble in water). The boundaries are interfaces with boundary-layers on each side, affecting the heat, mass and momentum transfer processes. Carbon dioxide and light alkane hydrocarbons (methane, ethane, propane) leave the liquid phases to enter the gas phase, as the flow steam cascades from the reservoir to top-side (surface) facilities. The TAN-square indicates the acidity of the oil phase and the pH-square indicates the acidity of the water/brine phase. Carboxylic acid RCOOH in the oil-phase is shown to cross the oil-water interface to enter the water-phase, called partition. At the interface the acid dissociates and reacts with either the Ca^{++} (aq) and/or Na^+ (aq) ions to form CaN and/or NaN solid naphthenate. The precipitation of CaN forms a *solid deposit* that can block flowstreams in processing equipment. The precipitation of NaN forms a *surfactant sludge* that can reduced the interfacial tension of droplets and bubbles.

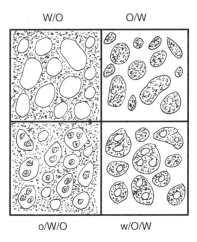

Figure 7.11 Four types of emulsions, oil-in-water, water-in-oil and mixed emulsions, water-in-oil droplets and oil-in-water droplets.

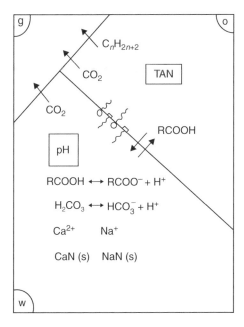

Figure 7.12 Interfacial processes in oil-gas-water (o, g, w) systems illustrating naphthenate precipitation (CaN solid and NaN sludge).

To emphasize a possible analogy between the carboxylic ion and the bicarbonate ion, the chemical balance for carbonic acid is shown in Figure 7.12.

Four surface active molecules (surfactants) are shown at the oil-water interface. The molecules with a circle-head are hydrophilic and the molecules with a square-head are lipophilic. Naphthenates are reported to be amphiphilic, meaning that each molecule is hydrophilic at one end and lipophilic at the other end. Micelles can be created in oil-water emulsions. An oil droplet in water may have the tail of a hydrophilic (water loving) molecule inside and the head of the molecule on the outside. A water droplet in oil will have the opposite, the hydrophilic head inside and the tail on the outside. Surfactants can form micelles and reverse micelles. The complex behaviours of naphthenate solids, micelles and reverse micelles stabilize emulsions, making oil-gas-water separation difficult in off-shore and on-land processing facilities.

Consider a simplified representation of oil-in-water and water-in-oil emulsions, where naphthenates form. A practical aspect is whether an oil-in-water emulsion will behave the same as water-in-oil emulsion, from the point of view of operating production wells, flowlines and processing facilities. A simplified drawing of a water droplet (on the left) in oil and an oil droplet (on the right) in water/brine, is shown in Figure 7.13. The droplets have the interfaces illustrated in Figure 7.12 (and Figure 7.14). The water droplet will contain acid-base hydrogen ions and be covered with hydrophilic surfactants. The head is positively charged and the tail neutral or lipophilic. Naphthenes are considered amphiphilic, meaning that one end likes water and the other end likes a hydrocarbon phase. For a water-continuous emulsion, an oil droplet will have surfactants with a hydrophilic head outside the droplet.

The oil-continuous phase will contain naphthenic acids (NA) that enter the electrical double layer of the water droplet. The dissociation of the acid occurs in the double layer and subsequently it reacts with calcium and/or natrium to form naphthenate

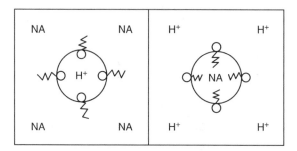

Figure 7.13 Simplified representation of a water droplet (on the left) in continuous oil and an oil droplet (on the right) in continuous water/brine. NA stands for naphthenic acid.

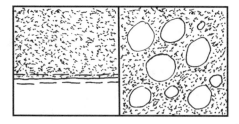

Figure 7.14 Formation (left-to-right) and breakup (right-to-left) of water-in-oil emulsion, based on two immiscible bulk liquids (oil = black, water = white).

solids. The solids, assuming they are much smaller than the droplet diameter, stabilize the droplets and hence stabilize the emulsion. The water-continuous phase will contain the whole gamut of acid-base components, including importantly hydrogen ions. However, the positively charged surfactants will be a barrier for the ions and thus reduce the potential for naphthenate formation. If this argument holds, it means that naphthenates form more easily in oil-continuous flow than in water-continuous flow. There is anecdotal evidence for naphthenate problems diminishing when the watercut increases sufficiently to make the flowstream water-continuous.

7.7 BASIC THERMODYNAMICS

The formation and breakdown of emulsions can be formalized in terms of free energy (*Gibbs Free Energy*). The system can be configured as shown in Figure 7.14, showing two bulk phases (immiscible oil and water) on the left and water droplets in oil (W/O emulsion) on the right. The bulk phases have uniform thermodynamic properties. However, at the dividing line between the phases, there is a boundary layer with properties different from the two bulk phases (Tadros, 2013). In an emulsion formation process, the free energy change can be expressed as

$$\Delta G = G_B - G_E$$

where the subscripts B and E refer to the bulk and the emulsion. The general statement can be made, that a process will be spontaneous if the change in free energy is negative. The process exchanges energy with its surroundings. In the case of emulsions, the free energy of the emulsified phase is greater than the bulk phase. To make an emulsion, energy must be provided from the surroundings (stirring and mixing), such that the change in free energy will be positive. Stated in terms of symbols, $\Delta G \leq 0$, represents a spontaneous process (spontaneous emulsion) and $\Delta G > 0$, represents a process receiving/requiring external energy.

The free energy of a constant temperature process (see *Appendix S – Gibbs Free Energy*) is given by the expression

$$\Delta G = \Delta H - T\Delta S$$

An important assumption is that the process contains a constant number of components. The enthalpy of the system can be considered as the binding energy in terms of the energy input needed to achieve a certain average droplet size. Assuming no volume change during an emulsification process, the enthalpy corresponds to the work required to expand the interfacial area. The increase in the energy of an emulsion, can be considered as a measure of the thermodynamic instability of an emulsion. The free energy of the interface can be expressed as

$$\Delta H = \sigma \Delta A$$

in terms of the interfacial tension σ mN/m^2 and change in area. The free energy of the interface corresponds to the reversible work brought permanently into the system during an emulsification process. An emulsion is prone to coalescence, leading to a decrease in interfacial area and therefore also enthalpy. As should be apparent, the enthalpy is actually the change in enthalpy, ΔH. Stability against coalescence occurs when the interfacial tension approaches zero. The entropy of an emulsion process is a measure of the extent of disorder in the system, also called configurational entropy. It measures the extent of size reduction of the dispersed phase (increase in number of droplets). The increasing disorder during the formation of an emulsion means a positive entropy. The entropy is actually the change in entropy, ΔS.

In petroleum emulsions, the actual values lead to the inequality

$$\sigma \Delta A \gg T \Delta S$$

showing that the *binding energy* is much greater than the *disorder energy*. Therefore, the key to emulsion formation and breakdown, is the interfacial tension. Surface active agents (surfactants) lower the interfacial tension ($\approx$surface tension) of droplets in emulsions. Surface and interfacial tension in petroleum emulsions are presented in *Appendix X3 – Energy Dispersion and Bubble Diameter*.

7.8   SIZE OF BUBBLES AND DROPLETS

The diameter of bubbles d_B m in liquids is presented in *Appendix X3 – Energy Dissipation and Bubble Diameter*. The two main challenging properties are the surface/interfacial tension σ mN/m and the energy dissipation P_E W/kg. The theoretical turbulence equation used was

$$d_B = \left(\frac{4\sigma}{\rho}\right)^{0.6} P_E^{-0.4}$$

The liquid density ρ kg/m^3 is that of the continuous phase. It was assumed (intellectual leap) that the same equation can be used for liquid bubbles in liquids as for liquid bubbles in gases (the theory was developed for this case). The interfacial tension values were based on a well tried semi-empirical method not involving all the bells-and-whistles of multivariable curve fits. Surface active agents (surfactants) decrease interfacial tension. Therefore, the values used in the calculations in *Appendix X3 – Energy Dissipation and Bubble Diameter*, are conservative. Knowing the liquid phase density was straight forward.

The energy dissipation term P_E can be derived logically for single phase turbulent flow in wells and pipelines, starting with the wall shear stress τ Pa/m^2

$$\tau = \frac{d}{4} \frac{\Delta p}{\Delta L}$$

where d m is flow channel (well, pipeline) diameter and $\Delta p/\Delta L$ the pressure gradient. Wall shear stress is highlight here because of its importance in naphthenate formation

(Turner and Smith, 2005). Using the *Darcy-Weisbach Equation* (see *Appendix D*), energy dissipation in pipe flow becomes

$$P_E = \frac{f}{2}\frac{1}{d}u^3$$

where f is the dimensionless friction factor and u m/s the average fluid velocity. Even though the equation has velocity-cubed, the energy dissipation for single-phase flow in wells and pipelines due to wall friction will be on the low side.

Pressure drop in multiphase flow is one of the most studied fields in oil and gas production. What is known is that multiphase pressure drop is greater than single phase pressure drop. Readily available textbooks should be consulted for details. Multiphase pressure drop can typically be ten-times greater than single phase pressure drop. Developing energy dissipation for multiphase flow was beyond the scope of the present work. Consulting *Section 2.4 – Two-Phase Flow in Pipelines* may be relevant.

The size of bubbles and droplets in multiphase flow is important in many petroleum engineering production operations. The size of bubbles is highly relevant in the flow of oil-in-water and water-in-oil emulsions. For the same watercut, the smaller the bubble, the greater the interface between oil and water. In general, the interfacial area affects the formation of inorganic scale crystals and the formation of naphthenates in both oil-in-water and water-in-oil emulsions. In gas-dominated flow, the size of liquid droplets affects the gas-liquid interfacial area and hence the formation of natural gas hydrate. The diameter of a bubble depends on the interfacial tension and energy dissipation. The details are presented in *Appendix X3 – Energy Dissipation and Bubble Diameter*.

The bubble diameter was calculated using the bubble dimeter equation, the interfacial tension equation and the energy dissipation equation for flow restrictions. The following values were used: interfacial tension $\sigma = 16$ mN/m, continuous phase density $\rho_o = 900$ kg/m^3 and restriction diameter 0.1 m (=1 cm). Calculations were carried out for several $nK/2$ values and flow velocities from 1 to 4 m/s. The maximum water bubble diameters μm in an oil-continuous emulsion, are shown in Figure 7.15. The calculations were carried out for energy dissipation factors of 10^3, 10^4 and 10^5. Clearly, the bubble diameter decreases with flow velocity due to increased shear forces. It decreases also with the energy dissipation coefficient; the higher the energy dissipation, the smaller the bubble diameter.

The bubble diameters in Figure 7.15, appear to be similar to the water bubble diameters in real crude oils, shown in two figures below, and in *Appendix X2 – Emulsions of Crude Oil and Brine*. The present results may not be accurate, but they illustrate the physical effects that dominate for water-in-oil emulsions in the *absence of surfactants*. The diameters in Figure 7.15 are uniform in size and can be considered as *maximum* bubble diameters.

The particle (droplet) size distribution of oil-in-water emulsion has been reported by Sjöblom *et al.* (2003). Experiments were carried out to investigate the effect of mixing on the particle size distribution. A water-in-oil emulsion containing 30% vol. water, was passed through first one choke valve and immediately downstream, through another choke valve. The total pressure drop across the two consecutive choke valves was 20 bar. The droplet size distributions are shown in Figure 7.16. The droplet size distribution was smaller for two choke valves compared to one choke valve. The greater

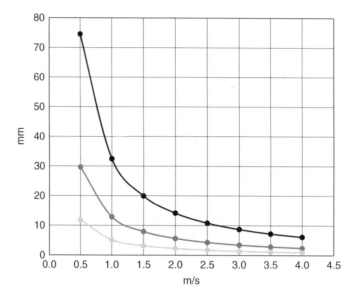

Figure 7.15 Maximum and uniform water bubble diameter in mm versus flow velocity m/s for energy dissipation coefficient *nK/2* values of 10^3, 10^4 and 10^5 (from top to bottom).

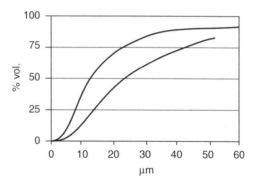

Figure 7.16 Droplet size distribution for 30% vol. water-in-oil emulsion after 20 bar total pressure drop over one choke valve and over two choke valves. Droplet sizes decreased from one to two chokes (Sjöblom *et al.*, 2003).

the number of mixing points (flow restrictions), the smaller the water droplets in an oil-continuous petroleum emulsion (stock-tank oils were used). The vertical axis is cumulative volume; the horizontal axis is droplet size in microns (=μm).

The effect of watercut (WC) was also investigated by Sjöblom *et al.* (2013). The total pressure drop was the same as above, 20 bar. The droplet size distributions are shown in Figure 7.17, for emulsions having 10% vol., 30% vol. and 70% vol. water fractions. The droplet sizes increased with increased watercut. The energy dissipation due to the pressure drop, will be distributed throughout the emulsion volume.

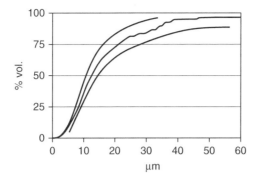

Figure 7.17 Droplet size distribution for 10% vol., 30% vol. and 70% vol. water-in-oil emulsion after 20 bar total pressure drop over one choke valve. Droplet sizes increase with watercut (Sjöblom *et al.*, 2003).

The emulsion viscosity will affect (lower) the energy dissipation rate. The lower the energy dissipation rate, the less breakup of water droplets in oil. The crude oil(s) used by Sjöblom *et al.* (2003) in the droplet size measurements, were not specified. However, they were presumably among the many listed in the paper. The North Sea crudes in the paper, are the same as shown in Table 3.1 and Figure 3.3, taken from Aske (2002).

The droplet size distributions in Figures 7.16 and 7.17 are in reasonable agreement with the distributions shown for a tight emulsion, a medium emulsion and a loose emulsion in Figure X2.3. The figures represent the range of water droplet sizes found in crude oil emulsions.

7.9 FIELD CHALLENGES

An oil field operator wants to maximize the daily production volume and to deliver high-quality processed crude for transport (pipeline or shuttle tanker). To achieve this goal the operator will meet several technical challenges and constraints. Important constraints are the flow assurance solids. Naphthenates form along the whole length of the flow steam. The effects of temperature, pressure and pH were discussed above. The main additional constraints are shear forces (emulsion making) and shut-down problems. None of the challenges/constraints, of course, should affect the safety of daily operations. Use of acid injection and/or inhibitor chemical dosage can alleviate naphthene surfactant (NaN) and deposition (CaN) problems.

Naphthene deposits may come as a surprise in both new and mature field developments. The reason being that reliable methods are not available to assess the risks due to naphthenate deposition. Neither in the design of new facilities in anticipation of naphthene problems, nor in adjusting on-going operations to avoid technical hazards and excessive costs. Crude oil analysis, including TAN-value and API gravity, cannot give a clear indication of whether naphthenate problems may arise. The front-end designers and field operators have a blind-zone for naphthenate deposits.

Table 7.8 Publicly available assay values for two crude oils, Field G blend (couple of sources) in the North Sea and Field H (single source) in the Norwegian Sea. Also, the water depth and reservoir depth (from seabed) and temperature.

Assay Property	Field G	Field H
°API	23.7	25.4
Specific gravity	0.9110	0.9014
Sulphur % mass	0.73	0.52
Pour point °C	−20	−65
TAN mg KOH/g	1.9	3.1
Nickel ppm	3.8	2.2
Vanadium ppm	11.8	10.2
Dynamic viscosity* cSt	76	42.0
Water depth m	128	350
Reservoir depth m	N/A	2300
Temperature °C	67	85

Dynamic viscosity μ/ρ at 20°C. 1 cSt $= 10^{-6}$ m²/s.

The assay values for two oil fields are shown in Table 7.8. Field G produces a blend described as heavy, high TAN crude. The API gravity places the crude in the medium category. Field H crude is described as low sulphur, heavy, high TAN crude. The API gravity places the crude in the medium category (see Table 7.1). The crudes are described as heavy by the operator but as medium based on API gravity. Field H has experienced naphthene deposition problems while Field G has not. High acid crudes are stipulated to have TAN > 1.0 and acid crude TAN > 0.5 (see *Section 7.1 – Acidity of Crude Oil*). According to this demarcation, both of the crudes are high acid crudes. It turns out that the lighter crude in Field H has naphthene problems while the not as light crude in Field G has not, apparently. A cursory glance at Table 7.8 could not distinguish which of the two fields suffers from major naphthenate problems.

Added to the assay values in Table 7.8, are the water depth, reservoir depth referenced to the seabed and a calculated reservoir temperature, assuming a geothermal gradient of 39°C/km (Ramm and Bjørlykke, 1994). The reservoir data are added here to suggest that because naphthenate deposit formation depends on both the crude oil and formation brine, both need to be known. The composition of formation brine depends not only on the subsurface mineralogy, but importantly also on the local (*in situ*) temperature and pressure.

The classical way of making an emulsion in industry is to stir two immiscible fluids strongly. Oil and water are two immiscible fluids. Emulsions in oil and gas production and processing can be classified as follows: (a) loose emulsions that separate in a few minutes (b) medium emulsions that separate in tens of minutes and (c) tight emulsions that separate in hours. Centrifugal pumping, downhole and in surface (top-side offshore and on-land) facilities. The used of positive displacement pumps may lead to less shear forces and therefore loose-to-medium emulsions instead of tight emulsions. The choke valve is where the largest local pressure drop occurs in oil production. The bubble size downstream a choke depends mainly on the power dissipation, as illustrated above. The coalescence of water droplets in oil is beneficial in oil-water separation. Surface active molecules and particles are known to hinder the coalescence.

Field operators experience difficult CaN deposits when separators and associated equipment are shut-down (planned and unplanned) and opened for maintenance. The deposits are soft at the very start of shut-down but harden with time due to exposure to air. Naphthenic acids are commonly used in oil paints to harden paint upon application. To prevent the autoxidation of CaN deposits, separators and associated equipment should perhaps be nitrogen blanketed. The technology would be similar to nitrogen injected into oil storage tanks.

High pressure and low temperature aid in preventing/reducing the formation of naphthenate (Figure 7.8). However, pH control seems to be superior to temperature and pressure adjustments in reducing solids precipitation and deposition. It is not the carbon dioxide reduction that affect the formation of solid deposits. It is the pH of the co-produced formation brine. A sharp bend is shown in the pH-curve in Figure 7.8. A prudent approach to controlling naphthenate formation would be to find this balancing point (sharp bend) for a particular crude oil and formation brine. The balancing point could then be used to determine the injection volume rate of mineral acid to reduce the pH. An important implication of acid injection is that it will balance/buffer all the other organic acids in a particular crude oil. The acid injection can easily be excessive, resulting in chemical costs and detrimental corrosion. The operator needs to find the right balance to optimize the oil volume produced and the operating and capital costs, all within the mandatory health, safety and environmental (HSE) regulations.

Naphthenate deposition control using acids and inhibitors was presented by Kelland (2009). The book chapter is a short and useful presentation of the naphthenate problem in industry, with many relevant references. The main acids reported used include the following: (1) mineral acids such as phosphoric acid and hydrochloric acid (for short-term use to remove deposits), (2) organic acids such as acetic acid and glycolic acid and (3) surfactant acids. What acids are most commonly used differ from publication to publication. A useful discussion of inhibition of naphthenate deposition is offered by Ahmed (2010), who tested eight proprietary compounds. The proprietary nature of specialized chemicals make it difficult to carry out research and development outside the commercial sphere. To find the best inhibitor is still based on the trial-and-error approach.

The quality of processed crude oil can be affected by CaN and NaN deposits. Separator performance depends greatly on the absence/presence of surface active agents. Naphthenate deposits are such agents. Poor separator performance can lead to excessive water content of the processed crude. BS&W (Basic Sediment and Water) of crude oil is important for custody transfers. The measurement includes the free water content and solid sediments and should not exceed about 0.5% by volume. Unfortunately, the BS&W of poorly separated crude oil may contain NORM (Naturally Occurring Radioactive Material). Poor separator performance may also lead to an increased oil content in separated water.

The rising velocity of an oil bubble in water and the sinking velocity of a water bubble in oil depends on *Stokes's Law*

$$u = \frac{gd^2}{18}\left(\frac{\Delta\rho}{\mu}\right)$$

where g is the gravitational constant, d the bubble diameter, $\Delta\rho$ the density difference between oil and water and the μ the viscosity of the continuous phase. In general, the viscosity of liquids depends exponentially on temperature

$$\mu = a\exp(-bT)$$

where a and b are liquid depending coefficients. The viscosity decreases with increasing temperature. The viscosity is in the denominator of *Stokes's Law*. It means that the rising and sinking velocities of oil-in-water and water-in-oil, increase with decreasing viscosity.

The formation of naphthenate increases with temperature as shown in Figure 7.8. The increase may follow an Arrhenius-type equation, meaning increasing exponentially with temperature

$$k = a\exp\left(\frac{-b}{T}\right)$$

where a and b are rate of formation coefficients. The rate of formation coefficients are not the same as the liquid dependent coefficients above. The formation rate increases with increases temperature.

The ease of crude oil-water separation increases with temperature because the viscosity of the continuous phase decreases with temperature. The concentration of CaN and NaN in crude oil increases with temperature. The naphthenates being surface active agents, the greater their concentration, the poorer the oil-water separation. Surface active agents are widely reported to decrease the rising and sinking velocities of droplets. Poor separation leads to higher concentration of BS&W in processed crude oil. Schematically, the BS&W concentration decreases and reaches a minimum, thereafter to increases because of the formation of CaN and NaN, as illustrated in Figure 7.18. The broken line shown the BS&W concentration with increasing temperature in the *absence* of surface active naphthenates.

The old dictum that the complexity of crude oil increases with density, can be paraphrased for naphthenate solids. The complexity of flow assurance solids increases with the co-precipitation of organic and inorganic solids. Mixed solids are more prevalent in the oil field than apparent at first sight (Frenier and Ziauddin, 2010). Naphthenate solids are sticky and act as a lodestone for other solids carried with crude oil and various precipitates. They facilitate the aggregation of solids. Naphthenates have been described as a binding agent for other solids (Turner and Smith, 2005). Heteroatoms may be included in solid deposits in the oil field. The atoms make solid deposits polar and are thus sensitive to electrostatic forces. The electrostatic coalescence process is reported to suffer from naphthene deposits. The polar nature of naphthenes facilitates their agglomeration.

The following anecdotal statements illustrate how complex the phenomena of naphthenate deposits are faced by operators of affected wells, flowlines and processing facilities: (1) deposits are more likely in fields with low TDS brines, (2) not all crude oils containing naphthenic acids exhibit operational problems and (3) although both CaN and NaN can occur in the same oil field, it is more common that one dominates in a particular field.

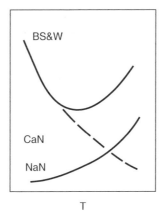

Figure 7.18 Schematic BS&W in separation of crude oil and water, improving with temperature and decreasing with increased naphthenate formation.

7.10 CONCLUDING REMARKS

Great progress has been made in recent years to identify the carboxylic acids that lead to the precipitation and deposition of naphthenates in oil and gas facilities. Medium molecular weight acids have been shown to be responsible for NaN deposits and high molecular weight acids for CaN deposits. The sodium deposits are more common and act as surfactants that gravely affect the separation of water from crude oil. The calcium deposits are considered to be responsible for the troublesome naphthenate deposits in processing facilities.

Field data show that the naphthenic acids content of crude oils is proportional with the TAN-value. The TAN-value represents *all* the acids in crude oil of which naphthenic acids are only a small fraction. Field data show also that the logarithm of TAN-values is proportional with the API gravity of crude oil. The gravity is inversely proportional with crude oil density. Despite these relationships, there are no clear relationships between TAN-values and naphthenate deposition. The inconclusive results may be related to the fact that it takes two-to-tango: crude oil and formation brine. Reservoir data and brine compositions should be included in the analysis of naphthene deposition. Oilfield geothermometry may also offer a useful insight.

An important consideration is the emulsion behaviour of water-in-oil and oil-in-water. The whole gamut of heat, mass and momentum transfer, plus chemical reactions, combine to make the problem complex, not to mention multiphase flow. Whether naphthenate deposition is more likely in oil continuous or water continuous emulsions remains unclear. A bold suggestion has been made in the present context, suggesting that naphthenate precipitation and deposition are more severe in oil continuous systems. Because the watercut increases with time, an emulsion may exhibit inversion from oil-continuous to water-continuous, potentially reducing the risk of naphthenate deposition.

The organic chemistry of the carboxylic functional group plays an important role in the deposition of naphthenates. Two postulates were suggested: (1) monovalent NaN deposits are more easily formed than divalent CaN deposits, and that the latter have a tighter structure and are denser and (2) an analogy exists between carbonic acid and naphthenic acid, explaining why naphthenates deposit at pH-values above six. Chemical data on the solubility product constant of naphthenic acid in water, and more importantly the equilibrium constants of its species, would contribute greatly to a better understanding of naphthenate deposition.

Risk assessment of naphthenate precipitation and deposition in new and operating oil production facilities, is still based on what is likely, and what is not likely. Research and development work needs to progress in the direction of quantifying the risks involved.

References

Abdel-Waly, A.A. (1997): New Correlation Estimates Viscosity of Paraffinic Stocks, Oil & Gas Journal, June 16, 61–65.

Abdolahi, F., Mesbah, A., Boozajomehry, R.B. & Svrcek, W.Y. (2007): The Effect of Major Parameters on Simulation Results of Gas Pipelines, International Journal of Mechanical Sciences, Vol. 49, 989–1000.

Ahmed, M.M. (2010): Characterization, Modelling, Prediction and Inhibition of Naphthenate Deposition in Oilfield Production, Ph.D. Thesis, Petroleum Engineering, Heriot-Watt University, 251 pp.

Akbarzadeh, K., Alboudwarej, H., Svercek, W.Y & Yarranton, H.W. (2005): A Generalized Regular Solution Model for Asphaltene Precipitation from n-Alkane diluted heavy oils and bitumens, Fluid Phase Equilibria, 232, 159–170.

Akbarzadeh, K., Hammami, A., Kharrat, A. & Zhang, D. (2007): Asphaltenes – Problematic but Rich in Potential, Oilfield Review, Summer, 22–43, (8 additional authors were listed), 22–43.

Altschuller, A.P. (1953): The Dipole Moments of Hydrocarbons, J. Phys. Chem., Vol. 57, No. 5, 538–540.

Al-Zahrani, S.M. & Al-Fariss, T.F. (1998): A General Model for the Viscosity of Waxy Oils, Chemical Engineering and Processing, 37, 433–437.

Andersson, V. (1999): Flow Properties of Natural Gas Hydrate Slurries – An Experimental Study, Dr.Ing. Thesis, Department of Petroleum and Applied Geophysics, Norwegian University of Science and Technology, Trondheim, 156 pp.

Andersson, V. & Gudmundsson, J.S. (1999): Flow Experiments on Concentrated Hydrate Slurries, SPE Annual Technical Conference and Exhibition, Paper SPE-56567, 3–6 October, Houston, 6 pp.

Angle, C.W. (2001): Chemical Demulsification of Stable Crude Oil and Bitumen Emulsions in Petroleum Recovery. Chapter 24, Encyclopedic Handbook of Emulsion Technology, Sjöblom, J. (editor), Marcel Dekker, (reference from Silset, 2008).

Angle, C.W., Long, Y., Hamza, H. & Lue, L. (2006): Precipitation of aspahltenes from solvent-diluted Heavy Oil and Thermodynamic Properties of Solvent-Diluted Heavy Oil solutions, Fuel, Vol. 85, 492–506.

Aske, N. (2002): Characterization of Crude Oil Components, Asphaltene Aggregation and Emulsion Stability by Means of Near Infrared Spectroscopy and Multivariate Analysis, Ph.D. Thesis, Department of Chemical Engineering, Norwegian University of Science and Technology, Trondheim, 51 pp. (excluding published papers).

Aske, N., Kallevik, H., Johnsen, E.E. & Sjöblom J. (2002): Asphaltene Aggregation from Crude Oils and Model Systems Studied by High-Pressure NIR Spectroscopy, Energy and Fuels, 16 (5), 1287–1295.

Aslan, S. & Firoozabadi, A. (2014): Effect of Water on Deposition, Aggregate Size, and Viscosity of Asphaltenes, Langmuir, Vol. 30, 3658–3664.

Aslan, S. & Firoozabadi, A. (2014): Deposition, Stabilization, and Removal of Asphaltenes in Flow-Lines, SPE Paper 172049, International Petroleum Engineering and Conference, Abu Dhabi, 10–13 November, 17 pp.

Atkinson, G. & Raju, K. (1991): The Thermodynamics of Scale Prediction, SPE Paper 21021, International Symposium on Oilfield Chemistry, Anaheim, CA, February, 209–215.

Avramov, I. (2007): Viscosity Activation Energy, Physics and Chemistry of Glasses, European Journal of Glass Science and Technology, Part B, Vol. 48, No. 1, February, 61–63.

Bacon, M.M., Romero-Zerón, L.B. & Chong, K.K. (2009): Using Crosss-Polarized Microscopy to Optimize Wax-Treatment Methods, SPE Annual Conference and Exhibition, New Orleans, 4–7 October, SPE 124799, 8 pp.

Baldi, S., Hann, D. & Yianneskis, M. (2002): On the Measurement of Turbulence Energy Dissipation in Stirred Vessels with PIV Techniques, 11th International Symposium on Applications of Laser Technology in Fluid Mechanics, Lisboa, 8–11 July, 12 pp.

Ball, P. (2015): Forging Patterns and Making Waves From Geology to Geology – A Commentary on Turing (1952) "The Chemical Basis of Morphogenesis, Phil. Trans. R. Soc. B, Vol. 370, 10 pp.

Balou, N.H.A. (2014): Destabilization and Aggregation Kinetics of Asphaltenes, Ph.D. Thesis, Department of Chemical Engineering, University of Michigan, xv + 175 pp.

Barros, J.M., Alves, D.P.P., Barroso, A.L., Souza, R.O. & Azevedo, L.F.A. (2005): Experimental Validation of Models for Predicting Wax Removal Forces in Pipgging Operations, Proc. 18th International Congress of Mechanical Engineering, Ouro Preto, November 6–11, 8 pp.

Barth, T., Høiland, S., Fotland, P., Askvik, K.M., Pedersen, B.S. & Borgund, A.E. (2004): Acid Compounds in Biodegraded Petroleum, Organic Geochemistry, 35, 1513–1525.

Baugh, T.D., Grande, K.V., Mediaas, H. & Vindstad, J.E. (2005): The Discovery of High-Molecule-Weight Naphthenic Acids (ARN Acid) Responsible for Calcium Naphthenate Deposits, SPE Paper 93011, SPE International Symposium on Oilfield Scale, Aberdeen, 11–12 May, 7 pp.

Baumann, W. & Rehme, K. (1975): Friction Correlations for Rectangular Roughness, Int. J. Heat Mass Transfer, Vol. 18, 1189–1197.

Berecz, E. & Balla-Achs, M. (1983): Gas Hydrates, Elsevier, 343 pp.

Birchwood, R., Dai, J., Shelander, D. & Ten Additional Authors (2010): Developments in Gas Hydrates, Oilfield Review, Vol. 22, No. 1 (Spring), 18–33.

Blevins, R.D. (2003): Applied Fluid Dynamics Handbook, Krieger Publishing Company, 558 pp.

Botne, K.K. (2012): Modeling Wax Thickness in Single-Phase Turbulent Flow, M.Sc. Thesis, Department of Petroleum Engineering and Applied Geophysics, Norwegian University of Science and Technology, Trondheim, v+33 pp.

Botros, K.K. & Golshan, H. (2010): Field Validation of a Dynamic Pig Model for an MFL ILI Tool in Gas Pipelines, Proc. 8th International Pipeline Conference, Calgary, Alberta, September 27 – October 1, 12 pp.

Bott, T.R. (1995): Fouling of Heat Exchangers, Elsevier Science & Technology, 548 pp.

Bott, T.R. & Gudmundsson, J.S. (1976): Deposition of Paraffin Wax from Kerosene in Cooled Heat Exchanger Tubes, 16th National Heat Transfer Conference, St. Louis, August 8–11, Paper 76-CSME/CSChE-21, 8 pp.

Bott, T.R. & Gudmundsson, J.S. (1978): Rippled Silica Deposits in Heat Exchanger Tubes, Proc., 6th International Heat Transfer Conference, Vol. 4, 373–378.

Brocart, B. & Hurtevent, C. (2008): Flow Assurance Issues and Control With Naphthenic Oils, Journal of Dispersion Science and Technology, 29, 1496–1504.

Bromley, L.A., Singh, D., Ray, P., Sridhar, S. & Read, S.M. (1974): Thermodynamic Properties of Sea Salt Solutions, AIChE Journal, Vol. 20, No. 2, 326–335.

Bryers, R.W. & Cole, S.S. (1982): Fouling of Heat Transfer Surfaces, Proc. Engineering Foundation Conference, October 31 to November 5, White Haven, Pennsylvania, 825 pp.

Buckley, J.S. & Fan, T. (2007): Crude Oil/Brine Interfacial Tensions, Petrophysics, Vol. 48, No. 3, June, 175–185.

Buckley, J.S., Hirasaki, G.J., Liu, Y., Von Drasek, S., Wang, J.X. & Gill, B.S (1998): Asphaltene Precipitation and Solvent Properties of Crude Oils, Petroleum Science and Technology, Vol. 16, No. 3–4, 251–285.

Burger, E.D., Perkins, T.K. & Striegler, J.H. (1981): Study of Wax Deposition in the Trans Alaska Pipeline, Journal Petroleum Technology, June, 1075–1086.

Calles, J.A., Dufour, J., Marugán, J., Peña, J.L., Giménez-Aguirre, R. & Merino-García, D. (2008): Properties of Asphaltenes Precipitated with Different n-Alkanes. A Study to Assess the Most Representative Species for Modeling, Energy & Fuels, Vol. 22, No. 2, 763–769.

Carroll, J.J. (2003): Natural Gas Hydrates – A Guide for Engineers, Gulf Professional Publishing, Elsevier Science, 270 pp.

Carroll, J.J. (1999): Henry's Law Revisited, Chem. Eng. Progress, Vol. 95, No. 1, 49–56.

Chapoy, A. (2004): Phase Behaviour in Water/Hydrocarbon Mixtures Involved in Gas Production Systems, Doctoral Thesis, Ecole des Mines de Paris, Paris, 258 pp.

Chen, C.Y. (1993): Measurement of Fine Particle Deposition From Flowing Suspensions, M.Sc. Thesis, Department of Chemical Engineering, University of British Columbia, 132 pp.

Chen, W. & Zhao, Z. (2006): Thermodynamic Modeling of Wax Precipitation in Crude Oils, Chinese J. Chem. Eng., Vol. 14, No. 5, 685–689.

Chisholm, D. (1967): A Theoretical Basis for the Lockhart-Martinelli Correlation for Two Phase Flow, Int. J. Heat Mass Transfer, Vol. 10, 1767–1778.

Colebrook, C.F. (1939): Turbulent Flow in Pipes With Particular Reference to the Transition Region Between the Smooth and Rough Pipe Laws, Proc. Institution Civil Engrs, 12, 393–422.

Colebrook, C.F. & White, C.M. (1937): Experiments With Fluid-Friction in Roughened Pipes, Proc. Royal Soc. London, 161, 367–381.

Coto, B., Martos, C., Peña, J.L., Espada, J.J. & Robustillo, M.D. (2008a): A New Method for the Determination of Wax Precipitation from Non-Diluted Crude Oils by Fractional Precipitation, Fuel, Vol. 87, 2090–2094.

Coto, B., Martos, C., Espada, J.J., Robustillo, M.D., Peña, J.L. & Gómez, S. (2008b): Assessment of a Thermodynamic Model to Describe Wax Precipitation in Flow Assurance Problems, Energy & Fuel, 5 pp.

Coto, B., Martos, C., Espada, J.J., Robustillo, M.D. & Peña, J.L. (2010): Analysis of Paraffin Precipitation from Petroleum Mixtures by Means of DSC: Iterative Procedure Considering Solid-Liquid Equilibrium Equations, Fuel, Vol. 89, 1087–1094.

Coutinho, J.A.P., Edmonds, B., Moorwood, T., Szczepanski, R. & Zhang, X. (2006): Reliable Wax Prediction for Flow Assurance, Energy & Fuels, Vol. 20, No. 3, 1081–1088.

Crabtree, M., Eslinger, D., Fletcher, P., Miller, M., Johnson, A. & King, G. (1999): Fighting Scale – Removal and Prevention., Oilfield Review, Autumn, 30–41.

Cragoe, C.S. (1929): Thermophysical Properties of Petroleum Products, Publication No. 97, U.S. Burau of Standards, 48 pp.

D'Amore, F. & Arnórsson, S. (2000): Chapter 10 – Geotermometry, Isotopic and Chemical Techniques in Geothermal Exploration, Development and Use, Arnórsson, S. (editor), International Atomic Energy Agency, Vienna, 152–199.

De Boar, R.B., Leelooyer, K., Eigner, M.R.P. & van Bergen, A.R.D. (1995): Screening of Crude Oils for Asphalt Precipitation – Theory, Practice, and the Selection of Inhibitors, SPE Production and Facilities, February, 55–61.

Davies, J.T. (1972): Turbulence Phenomena, Academic Press, 412 pp.

Davies, J.T. (1986): High Local Energy Dissipation Rates in Fine-Clearance Equipment, 9th Australian Fluid Mechanics Conference, Auckland, 8–12 December, 2 pp.

Davidson, R. (2002): An Introduction to Pipeline Pigging, Pigging Products and Services Association, Aberdeen, 12 pp.

Davison, W. (1991): The Solubility of Iron Sulphides in Synthetic and Natural Waters at Ambient Temperature, Aquatic Sciences, Volume 53, Issue 4, 309–329.

Debye, P. & Hükel, E. (1923): Zur Theorie der Elektrolyte, Physikalische Zeitschrift, 24, 185–206.

Dietzel, M. (2000): Dissolution of Silicates and the Stability of Polysilicic Acid, Geochimica et Cosmochimica Acta, Vol. 64, No. 19, 3275–3281.

Dong, L. & Gudmundsson, J.S. (1993): Model for Sound Speed in Multiphase Mixtures, 3rd Lerkendal Petroleum Engineering Workshop, Trondheim, January 20–21, 12 pp.

Drever, J.I. (1997): The Geochemistry of Natural Waters, Prentice Hall, 436 pp.

Duan, Z. & Mao, S. (2006): A thermodynamic Model for Calculating Methane Solubility, Density and Gas Phase Composition of Methane-Bearing Aqueous Fluids from 273 to 523 K and from 1 to 2000 bar, Ceochimica et Cosmochimica Acta, Vol. 70, 3369–3386.

Dyer, S.J. & Graham, G.M. (2002): The Effect of Temperature and Pressure on Oilfield Scale Formation, Journal of Petroleum Science and Engineering, Vol. 35, Issue 1–2, July, 95–107.

Dyer, S.J., Graham, G.M. & Arnott, C. (2003): Naphthenate Scale Formation – Examination of Molecular Controls in Idealized Systems, SPE Paper 80395, SPE 5th International Symposium on Oilfield Scale, Aberdeen, 29–30 January, 14 pp.

Edmonds, B., Moorwood, R.A.S., Szczepanski, R., Zhang, X., Hayward, M. & Hurle, R. (1999): Measurement and Prediction of Asphaltene Precipitation from Live Oils, Third International Symposium on Colloid Chemistry in Oil Production (ISCOP '99), Asphaltenes and Wax Deposition, 14–17 November, Huatulco, Mexico, 16 pp.

Edmonds, B., Moorwood, T., Szczepanski, R. & Zhang, X. (2008): Simulating Wax Depostion in Pipelines for Flow Assurance, Energy & Fuels, Vol. 22, No. 2, 729–741.

Einstein, A. (1906): Eine neue Bestimmung der Moleküldimensjonen, Ann. Phys., Vol. 19, No. 2, 289–306.

Ellis, A.J. & Golding, R.M. (1963): The solubility of carbon dioxide above 100°C in water and in sodium chloride solutions, American Journal of Science, 261, 47–60.

Elnour, M.M., Ahmed, I.M. & Ibrahim, M.T. (2014): Study the Effects of Naphthenic Acid in Crude Oil Equipment Corrosion, Journal of Applied and Industrial Sciences, Vol. 2, No. 6, 255–260.

Eyring, H. (1935): The Activated Complex in Chemical Reactions, J. Chem. Phys., Vol. 3, No. 2, 107–115.

Falcone, G., Hewitt, G.F. & Alimonti, C. (2010): Multiphase Flow Metering, Elsevier, 329 pp.

Falk, K. (1999): Pressure Pulse Propagation in Gas-Liquid Pipe Flow, Doctoral Thesis, Department of Petroleum Engineering and Applied Geophysics, NTNU, Trondheim, 210 pp.

Fan, C., Kan, A.T., Zhang, P., Lu, H., Work, S., Yu, J. & Tomson, M.B. (2012): Scale Prediction and Inhibition for Oil and Gas Production at High Temperature/High Pressure, SPE Journal, June, 379–392.

Fan, T., Wang, J. & Buckley, J.S. (2002): Evaluating Crude Oils by SARA Analysis, SPE/DOE Improved Oil Recovery Symposium, Tulsa, Oklahoma, 13–17 April, SPE 75228, 7 pp.

Fahre-Skau, A., O'Donoghue, A., Rønningsen, H.P. & Rognved, K. (2013): Heimdal Brae Dewaxing Operation, Seminar Presentation, PPSA (Pigging Products & Services Association), Aberdeen, November 20, 41 pp.

Farshad, F.F. & Rieke, H.H. (2005): Technology Innovation for Determining Surface Roughness in Pipes, Journal Petroleum Technology, October, 82–86.

Fernández-Prini, R., Alvarez, J.L. & Harvey, A.H. (2003): Henry's Constants and Vapor-Liquid Distribution Constants for Gaseous Solutes in H_2O and D_2O at High Temperatures, J. Phys. Chem. Ref. Data, Vol. 32, No. 2, 903–916.

Firoozabadi, A. (2016): Thermodynamics and Applications in Hydrocarbon Energy Production, McGraw Hill, 467 pp.

Firoozabadi, A. & Ramey, H.J. (1988): Surface Tension of Water-Hydrocarbon Systems at Reservoir Conditions, J. Can. Pet. Technol., 27, May–June, 41–48.

Flaten, E.M., Watterud, G., Andreassen, J.P. & Seiersten, M. (2008): Prediction of Iron and Calcium Carbonate in Pipelines at Varying MEG Contents, SPE Paper 114089, SPE International Oilfield Scale Conference, Aberdeen, 28–29 May, 9 pp.

Flory, P.J. (1942): Thermodynamics of High Polymer Solutions, J. Chem. Phys., Vol. 10, No. 1, 51–61.

Fournier, R.O. & Rowe, J.J. (1977): The Solubility of Amorphous Silica in Water at High Temperature and High Pressure, American Mineralogist, Volume 62, 1052–1056.

Fossum, T. (2007): Temperature Control in Multiphase Pipelines, Guest Lecture, TPG 4136, Processing of Petroleum, March 19, Department of Petroleum and Applied Geophysics, Norwegian University of Science and Technology, 26 pp.

Frenier, W.W. & Ziauddin, M. (2008): Formation, Removal and Inhibition of Inorganic Scale in the Oilfield Environment, Society of Petroleum Engineers, 240 pp.

Galta, T. (2014): Bypass Pigging of Subsea Pipelines Suffering Wax Deposition, M.Sc. Thesis, Department of Petroleum Engineering and Applied Geophysics, NTNU, June, xiv+92 pp.

Gas Processors Suppliers Association (1998): GPSA Engineering Data Book, Volume II, Section 20, Dehydration, SI Version, 11th Edition, Tulsa, 45 pp.

George, A.K. & Singh, R.N. (2015): Correlation of Refractive Index and Density of Crude Oil and Liquid Hydrocarbon, International Journal of Chemical, Environmental & Biological Sciences, Vol. 3, No. 5, 420–422.

Glasstone, S. & Lewis, D. (1960): Elements of Physical Chemistry, 2nd Edition, Reprint 1966, Macmillan, 759 pp.

Glover, P.W., Walker, E. & Jackson, M.E. (2012): Steaming-Potential of Reservoir Rock – A Theoretical Model, Geophysics, Vol. 77, No. 2, 17–43.

Goldszal, A., Hurtevent, C. & Rousseau, G. (2002): Scale and Naphthenate Inhibition in Deep-Offshore Fields, SPE Paper 74661, SPE Oilfield Scale Symposium, Aberdeen, 30–31 January, 11 pp.

Goual, L. & Firoozabadi, A. (2002): Measuring Asphaltenes and Resins, and Dipole Moment in Petroleum Fluids, AIChE Journal, Vol. 48, No. 11, 2646–2663.

Goual, L. & Firoozabadi, A. (2004): Effect of Resins and DBSA on Asphaltene Precipitation from Petroleum Fluids, AIChE Journal, Vol. 50, No. 2, 470–479.

Granados, E.E. & Gudmundsson, J.S. (1985): Production Testing in Miravalles Geothermal Field, Cost Rica, Geothermics, Vol. 14, No. 4, 517–524.

Griesbaum, K., Behr, A., Biedenkapp, D., Voges, H.W., Garbe, D., Paetz, C., Collin, G., Mayer, D. & Höke, H. (2012): Hydrocarbons, Vol. 18, Ullmann's Encyclopedia of Industrial Chemistry, Wiley, 133–189.

Groenzin, H. & Mullins, O.C. (2000): Molecular Size and Structure of Asphaltenes from Various Sources, Energy & Fuels, Vol. 14, No. 3, 677–684.

Gudmundsson, J.S. (1975): Paraffin Wax Deposition and Fouling, Draft M.Sc. Thesis, Department of Chemical Engineering, University of Birmingham, 122 pp.

Gudmundsson, J.S. (1977): Fouling of Surfaces, Ph.D. Thesis, Department of Chemical Engineering, University of Birmingham, 196 pp.

Gudmundsson, J.S. (1990): Method and Equipment for Production of Gas Hydrates, Norwegian Patent Number 172080, 11 pp.

Gudmundsson, J.S. & Bott, T.R. (1977): Deposition – The Geothermal Constraint, I. Chem. E. Symp. Ser., No. 48, 27.1–27.20.

Gudmundsson, J.S. & Celius, H.K. (1999): Gas-Liquid Metering Using Pressure-Pulse Technology, SPE Paper 56584, Annual Technical Conference and Exhibition, Houston, 3–6 October, 10 pp.

Gudmundsson, J.S., Hveding, F. & Børrehaug, A. (1995): Transport of Natural Gas as Frozen Hydrate, Proceedings, Fifth International Offshore and Polar Engineering Conference, The Hague, The Netherlands, June 11–16, Vol. I, 282–288.

Gudmundsson, J.S. & Parlaktuna, M. (1992): Storage of Natural Gas Hydrate at Refrigerated Conditions, AIChE Spring National Meeting, New Orleans, March 29–April 2, 26 pp.

Gudmundsson, J.S., Parlaktuna, M. & Khokhar, A.A. (1994): Storage of Natural Gas as Frozen Hydrate, SPE Production Engineering, February, 69–73.

Gudmundsson, J.S., Mork, M. & Graff, O.F. (2002): Hydrate Non-Pipeline Technology, 4th International Conference on Gas Hydrates, May 19–23, Yokohama, 6 pp.

Gudmundsson, J.S. & Bott, T.R. (1977): Solubility of Paraffin Wax in Kerosene, Fuel, Vol. 56, 15–16.

Gudmundsson, J.S. (1981): Particulate Fouling, Fouling of Heat Transfer Surfaces, Somerscales, E.F.C. and Knudsen J.G. (eds.), Hemisphere Publishing Corporation, Washington, 357–387.

Gudmundsson, J.S. & Bott, T.R. (1977): Particle Diffusivity in Turbulent Pipe Flow, J. Aerosol Sci., Vol. 8, 317–319.

Gudmundsson, J.S. & Thomas, D.M. (1988): Deposition of Solids in Geothermal Systems, Proc. International Workshop, 16–19 August, Reykjavik, Iceland (published in Geothermics 1989).

Gudmundsson, J.S. (1992): Molecular Diffusion, SPOR Monograph, Recent Advances in Improved Oil Recovery Methods for North Sea Sandstone Reservoirs, S.M. Skjæveland & J. Kleppe (editors), Norwegian Petroleum Directorate, 20–24.

Gudmundsson, J.S., Durgut, I., Celius, H.K. & Korsan, K. (2001): Detection and Monitoring of Deposits in Multiphase Flow Pipelines Using Pressure Pulse Technology, 12th International Oil Field Chemistry Symposium, Gailo, Norway, April 1–4, 2001, 13 pp.

Gudmundsson, J.S. (2002): Cold Flow Hydrate Technology, 4th International Conference on Gas Hydrates, May 19–23, Yokohama, 5 pp.

Guha, A. (2008): Transport and Deposition of Particles in Turbulent and Laminar Flow, Ann. Rev. Fluid Mech., 40, 311–341.

Gunnarsson, I. & Arnórsson, S. (2003): Silica Scaling – The Main Obstacle in Efficient Use of High-Temperature Geothermal Fluids, International Geothermal Conference, Reykjavik, Paper 118, 30–36.

Haaland, S. (1983): Simple and Explicit Formulas for the Friction Factor in Turbulent Pipe Flow, Journal Fluid Mechanics, 105, 89–90.

Hammerschmidt, E.G. (1934): Formation of Gas Hydrates in Natural Gas Transmission Lines, Ind. Eng. Chem., Vol. 26, No. 8, 851–855, (taken from Oellrich 2004).

Hammerschmidt, E.G. (1939): Oil and Gas Journal, May, 66–72.

Hammerschmidt, E.G. (1969): Possible Technical Control of Hydrate Formation in Natural Gas Pipelines, Brennstoff-Chemie, Vol. 50, 117–123, (taken from Pedersen et al., 1989).

Han, S., Huang, Z., Senra, M., Hoffmann, R. & Fogler, H.S. (2010): Method to Determine the Wax Solubility Curve in Crude Oil from Centrifugation and High Temperature Gas Chromatography Measurements, Energy & Fuels, 9 pp.

Hanneseth, A.M.D. (2009): An Experimental Study of Tetrameric Naphthenic Acids at W/O Interfaces, Ph.D. Thesis, Chemical Engineering, Norwegian University of Science and Technology, January, 59 pp. (published papers not included).

Hansen, A.B., Larsen, E., Pedersen, W.B., Nielsen, A.B. & Ronningsen, H.P. (1991): Wax Precipitation from North Sea Crude Oils. 3. Precipitation and Dissolution of Wax Studied by Differential Scanning Calorimetry, Energy & Fuels, Vol. 5, 914–923.

Hansen, A.B., Larsen, E., Pedersen, W.B., Nielsen, A.B., Roenningsen, H.P. (1991): Wax Precipitation from North Sea Crude Oils. 3. Precipitation and Dissolution of Wax Studied by Differential Scanning Calorimetry, Energy & Fuels, Vol. 5, 914–923.

Hansen, C.M. (2007): Hansen Solubility Parameters – An Introduction, Hansen Solubility Parameters, User's Handbook, Hansen, C.M. (editor), CRC Press, 2nd edition, 1–26.

Hashmi, S.M., Quintiliano, L.A. & Firoozabadi, A. (2010): Polymeric Dispersants Delay Sedimentation in Colloidal Asphaltene Suspensions, Langmuir, Vol. 26, 8021–8029 pp.

Hashmi, S.M., Loewenberg, M. & Firoozabadi, A. (2015): Colloidal Asphaltene Deposition in Laminar Pipe Flow – Flow Rate and Parametric Effects, Physics of Fluids, Vol. 27, 22 pp.

Hashmi, S.M. & Firoozabadi, A. (2015): Effective Removal of Asphaltene Deposition in Metal-Capillary Tubes, SPE Journal, February, 9 pp.

Håvåg, S. (2006): Qualitative and Quantitative Determination of Naphthenic Acids in Heidrun Crude Oil, Cand. Sci. Thesis, Department of Chemistry, University of Oslo, 85 pp.

Havre, T.E. (2002): Formation of Calcium Napthenate in Water/Oil Systems, Naphthenic Acid Chemistry and Emulsion Stability, Ph.D Thesis, Department of Chemical Engineering, Norwegian University of Science and Technology, 44 pp., (plus 5 scientific papers).

Havre, T.E., Sjöblom, J. & Vindstad, J.E. (2003): Oil/Water-Partitioning and Interfacial Behavior of Naphthenic Acids, Journal of Dispersion Science and Technology, Vol. 24, No. 6, 789–801.

Hayduk, W. & Minhas, B.S. (1982): Correlation for Prediction of Molecular Diffusivities in Liquids, Canadian J. Chem. Eng., Vol. 60, No. 2, 295–299.

Henry, W. (1803): Experiments on the Quality of Gases Absorbed by Water at Different Temperatures, and Under Different Pressures, Phil. Trans. R. Soc., Vol. 93, 29–274.

Hernandez, O.C. (2002): Investigation of Single-Phase Paraffin Deposition Characteristics, M.Sc. Thesis, University of Tulsa, 172 pp.

Hess, G.H. (1840): Recherches sur les quantités de chaleur dégaées dans les combinaisons chimiques, Comptes rendus de l'Académie des Sciences, Académie des sciences, vol. 10, 759–763.

Hewitt, G.F., Shires, G.L. & Bott, T.R. (1994): Process Heat Transfer, CRC Press, 1042 pp.

Hildebrand, J. & Scott, R.L. (1962): Regular Solutions, Prentice-Hall, 180 pp.

Hirschberg, A., deJong, L.N.J., Schipper, B.A. & Meijer, J.G. (1984): Influence of Temperature and Pressure on Asphaltene Flocculation, SPE Journal, June, 283–293.

Hoffmann, R. & Amundsen, L. (2010): Single-Phase Wax Deposition Experiments, Energy & Fuels, Vol. 24, 1069–1080.

Holder, G.D., Zetts, S.P. & Pradhan, N. (1988): Phase Behavior in Systems Containing Clathrate Hydrates, Reviews in Chemical Engineering, Vol. 5, Issue 1–4, 1–70.

Hossain, M.E., Ketata, C. & Islam, M.R. (2009): Experimental Study of Physical and Mechanical Properties of Natural and Synthetic Waxes Using Uniaxial Compressive Strength, Proc. Third International Conference on Modeling, Simulation and Applied Optimization, Sharjah, U.A.E., January 20–22, 5 pp.

Hovden, L., Rønningsen, H.P., Xu, Z.G., Labes-Carrier, C. & Rydahl, A. (2004): Pipeline Wax Deposition Models and Model for Removal of Wax by Pigging: Comparison Between Model Predictions and Operational Experience, 4th North American Conference on Multiphase Technology, Banff, 3–4 June, 20 pp.

Huggins, M.L. (1942): Some Properties of Solutions of Long-Chain Compounds, J. Phys. Chem, Vol. 46, No. 1, 151–158.

Hunter, K.A. (1998): Acid-Base Chemistry of Aquatic Systems, Department of Chemistry, University of Otago, New Zealand, 163 pp.

Igwebueze, C., Oduola, L., Smith, O., Vijn, P., Delden, B.V. & Shepherd, A.G. (2013): Calcium Naphthenate Solid Deposits Identification and Control in Offshore Nigerian Fields, SPE 164055, International Symposium of Oilfield Chemistry, Woodlands, Texas, 8–10 April, 13 pp.

Ilahi, M. (2006): Evaluation of Cold Flow Concepts, M.Sc. Thesis, Department of Petroleum Engineering and Applied Geophysics, Norwegian University of Science and Technology, Trondheim, 123 pp., (wrong year on front page).

Ilahi, M. (2005): Cold Flow Concepts Compared, B.Sc. Thesis, Department of Petroleum Engineering and Applied Geophysics, Norwegian University of Science and Technology, Trondheim, 108 pp.

Ilia Anisa, A.N., Nour, A.H. & Nour, A.H. (2010). Catastrophic and Transitional Phase Inversion of Water-in-Oil Emulsion for Heavy and Light Crude Oil, Journal of Applied Sciences, Vol. 10, 3076–3083.

Ishkov, O., Mackay, E.J. & Vazquez, O. (2015): Reservoir Simulation and Near-Well Bore Modelling to Aid Scale Management in a Low Temperature Development with Multilateral Wells, SPE Paper 173739, SPE International Symposium on Oilfield Chemistry, Woodlands, Texas, 13–15 April, 22 pp.

Jeffrey, G.A. (1972): Pentagonal Dodecahedral Water Structure in Crystalline Hydrates, Mat. Res. Bull., Vol. 7, 1259–1270.

Jones, D.M., Watson, J.S., Meredith, W., Chen, M. & Bennett, B. (2001): Determination of naphthenic Acids in Crude Oils Using Non-Aqueous Ion Exchange Solid-Phase Extraction, Analytical Chemistry, Vol. 73, No. 3, 703–707.

Juyal, P., Mapolelo, M.M., Yen, A., Rogers, R.P. & Allenson, S.J. (2015): Identification of Naphthenate Deposition in South American Oil Fields, Energy & Fuels, Vol. 29, No. 4, 2342–2350.

Kaasa, B. (1998): Prediction of pH, Mineral Precipitation and Multiphase Equilibria During Oil Recover, Doctoral Thesis, Department of Inorganic Chemistry, Norwegian University of Science and Technology, Trondheim, 219 pp.

Kaasa, B., Sandengen, K. & Østvold, T. (2005): Thermodynamic Predictions of Scale Potential, pH and Gas Solubility in Glycol Containing Systems, Paper SPE 95075, 7th International Symposium on Oilfield Scale, Aberdeen, 11–12 May, 14 pp.

Kaasa, B. & Østvold, T. (1997): Alkalinity in Oil Field Waters. What Alkalinity is and How it is Measured, Paper PE 37277, International Symposium on Oilfield Chemistry, Houston, 18–21 February, 657–660.

Kaftori, D., Hetsroni, G. & Banerjee, S. (1995): Particle Behavior in the Turbulent Boundary Layer. I. Motion, Deposition and Entrainment, Phys. Fluids, Vol. 7, No. 5, May, 1095–1106.

Kaminsky, R. & Radke, C.J. (1997): Asphaltenes, Water Films, and Wettability Reversals, SPE Journal, December, 485–493.

Kan, A.T. & Tomson, M.B (2010): Scale Prediction for Oil and Gas Production, International Oil and Gas Conference and Exhibition, Paper SPE 132237, Beijing 8–10 June.

Karan, K., Hammami, A., Flannery, M. & Stankiewicz, B.A. (2003): Evaluation of Asphaltene Instability and a Chemical Control During Production of Live Oils, Petroleum Science and Technology, Vol. 21, No. 3 & 4, 629–645.

Katz, D.L. & Lee, R.L. (1990): Natural Gas Engineering, McGraw Hill, 760 pp.

Katz, B.J. & Robinson, V.D. (2006): Oil Quality in Deep-Water Settings – Concerns, Perceptions, Observations, and Reality, AAPG Bulletin, Vol. 90, No. 6, 909–920.

Kay, J.M. & Nedderman, R.M. (1985): Fluid Mechanics and Transfer Processes, Cambridge University Press, 602 pp.

Keetch, M. (2005): Activity and Centration, Lecture 7, Geochemistry, Geology 480, San Francisco State University, 12 pp.

Kelland, M.A. (2014): Production Chemicals for the Oil and Gas Industry, Second Edition, CRC Press, 454 pp.

Kelland, M.A. (2009): Production Chemicals for the Oil and Gas Industry, CRC Press, 437 pp.

Kelland, M.A., Svartaas, T.M. & Dybvik, L. (1995): A New Generation of Gas Hydrate Inhibitors, SPE Annual Technical Conference, Paper SPE 30695, Dallas, 22–25 October, 529–537.

Khaleel, A., Abutaqiya, M., Tavakkoli, M., Melendez-Alvarez, A.A. & Vargas, F.M. (2015): One the Prediction, Prevention and Remediation of Asphaltene Deposition, International Petroleum Exhibition and Conference, Abu Dhabi, 9–12 November, 25 pp.

Kim, S.M. & Mudawar, I. (2012): Universal Approach to Predicting Two-Phase Frictional Pressure Drop for Adiabatic and Condensing Mini/Micro-Channel Flows, International Journal of Heat and Mass Transfer, Vol. 55, 3246–3261.

Kinker, B.G. (2000): Fluid Viscosity and Viscosity Classification, Chapter 5, Handbook of Hydraulic Fluid Technology, Totten, G.E. (editor), Marcel Dekker, 305–338.

Kløv, T. (2000): High-Velocity Flow in Fractures, Doctoral Thesis, Department of Petroleum Engineering and Applied Geophysics, NTNU, xxii + 181 pp.

Koh, C.A. (2002): Towards a Fundamental Understanding of Natural Gas Hydrates, Chem. Soc. Review, Vol. 31, 157–167.

Kohler, N., Courbin, G. & Ropital, F. (2001): Static and Dynamic Evaluation of Calcium Carbonate Scale Formation and Inhibition, SPE Paper 68963, SPE European Formation Damage Conference, The Hague, 21–22 May, 11 pp.

Kokal, S.L. (2007): Crude Oil Emulsions, Chapter 12, Vol. I, SPE Petroleum Engineering Handbook, Fanchi, J.R. (editor), 533–570.

Kumar, B. (2012): Effect of Salinity on the Interfacial Tension of Model and Crude Oil Systems, M.Sc. Thesis, University of Calgary, September, 146 pp.

Kyrkjeeide, J. (1992): Experiments Concerning Wax Precipitation in Oil Production, Diploma Thesis, Department of Industrial Chemistry, Norwegian Institute of Technology, 96 pp. (in Norwegian).

Langelandsvik, L.I. (2008): Modeling of Natural Gas Transport and Friction Factor for Large-Scale Pipelines. Ph.D. Thesis, Department of Energy and Process Engineering, Norwegian University of Science and Technology, Trondheim, 165 pp. (additionally several papers).

Larsen, R., Lund, A., Hjarbo, K.W. & Wolden, M. (2009): Robustness Testing of Cold Flow, 20th International Oil Field Chemistry Symposium, 22–25 March, Geilo, Norway, 16 pp.

Lee, S.Y. & Holder, G.D. (2001): Methane Hydrates Potential as a Future Energy Source, Fuel Processing Technology, Vol. 71, 181–186.

Lennie, A.R., Redfern, S.A.T., Champness, P.E., Stoddardt, C.P., Schofield, P.F. & Vaughan, D.J. (1997): Transformation of Mackinawite to Greigite, American Mineralogist, Volume 82, 302–309.

Leontaritis, K.J. & Mansoori, G.A. (1988): Asphaltene Deposition – A Survey of Field Experiences and Research Approaches, Journal of Petroleum Science and Engineering, Vol. 1, 229–239.

Levik, O.I., Monfort, J.-P. & Gudmundsson, J.S. (2002): Effect of the Driving Force on the Composition of Natural Gas Hydrates, 4th International Conference on Gas Hydrates, May 19–23, Yokohama, 5 pp.

Lewis, G.N. (1901): The Law of Physico-Chemical Change, Proc. American Academy of Arts and Sciences, 37 (4), 49–69.

Lewis, G.N. & Randall, M. (1921): The Activity Coefficient of Strong Electrolytes, J. Am. Chem. Soc., 43 (5), 1112–1154.

Lide, D.R. (2004): CRC Handbook of Chemistry and Physics, Editor-in-Chief, 84th Edition, CRC Press, 3475 pp.

Ling, K. & He, J. (2012): A New Correlation to Calculate Oil-Water Interfacial Tension, SPE Paper 163328, Kuwait International Petroleum Conference and Exhibition, Kuwait, 10–12 December, 9 pp.

Lira-Galeana, C., Firoozabadi, A. & Prausnitz, J.M. (1996): Thermodynamics of Wax Precipitation in Petroleum Mixtures, A.I.Ch.E. Journal, Vol. 42, No. 1, 239–248.

Liu, B., Sun, W., Liu, C. & Guo, L. (2015): The Thermodynamic Model on Paraffin Wax Deposition Prediction, Engineering, Vol. 7, 827–832.

Lockhart, R.W. & Martinelli, R.C. (1949): Proposed Correlation of Data for Isothermal Two Phase Flow, Two Component Flow in Pipes, Chem. Eng. Prog., Vol. 45, 39–48.

Long, R.B. (1981): The Concept of Asphaltenes, Chemistry of Asphaltenes, (editors, Bunger, J.W. & Li, N.C.), Advances in Chemistry Series 195, American Chemical Society, 17–27.

Lovell, D. & Pakulski, M. (2002): Hydrate Inhibition in Gas Wells Treated With Two Low-Dosage Hydrate Inhibitors, SPE Gas Technology Symposium, Paper SPE 75668, Calgary, Alberta, 30 April–2 May, 7 pp.

Lund, H.J. (1998): Investigation of Paraffin Deposition During Single Phase Flow in Pipelines, M.Sc. Thesis, University of Tulsa, 158 pp.

Lund, A., Hjarbo, K.W., Wolden, M., Høiland, S., Skjetne, P. Straume, E.O. & Larsen, R. (2010): Cold Flow – Towards a Fundamental Understanding, 21st International Oil Field Chemistry Symposium, 15–17 March, Geilo, Norway, 20 pp.

Lund, A., Lysne, D., Larsen, R. & Hjarbo, K.W. (1998): Method and System for Transporting a Flow of Fluid Hydrocarbons Containing Water, Norwegian Patent NO311854.

Lutnaes, B.F., Brandal, Ø., Sjöblom, J. & Krane J. (2006): Archeal C80 Isoprennoid Tetraacids Responsible for Naphthenate Deposition in Crude Oil Processing, Org. Biomol. Chem., Vol. 4, No. 4, 616–620.

Mackay, E.J. (2008): Oilfield Scale – A New Integrated Approach to Tackle an Old Foe, SPE Distinguished Lecture, SPE 87459, 47 pp.

Mak, T.C.W. & McMullan (1965): Polyhedral Clathrate Hydrates. X. Structure of the Double Hydrate of Tetrahydrofuran and Hydrogen Sulfide, J. Chem. Phys., Vol. 42, 2732–2737 (reference from Lee & Holder, 2001).

Makogon, Y.F. (1981): Hydrates of Natural Gas, PennWell Books, Tulsa, Oklahoma, 237 pp.

Makogon, Y.F. (1997): Hydrates of Hydrocarbons, PennWell Books, Tulsa, Oklahoma, 482 pp.

Makogon, Y.F. (1988): Natural Gas Hydrates – The State of Study in the USSR and Perspectives for its Using, 3rd Chemical Congress of North America, June 5–10, (reference taken from Katz & Lee 1990).

Martínez, I. (2016): Properties of Liquids, ETSIA, Universidad Politécnica de Madrid, 3 pp.

Mapolelo, M.M., Stanford, L.A., Rodgers, R.R., Yen, A.T., Debord, J.D., Asomaning, S. & Marshall, A.G. (2009): Chemical Speciation of Calcium and Sodium Naphthenate Deposits by Electrospray Ionization FT-ICR Mass Spectrometry, Energy & Fuels, 23, 349–355.

Maqbool, T., Balgoa, A.T. & Fogler, H.S. (2009): Revisiting Asphaltene Precipitation from Crude Oils: A Case of Neglected Kinetics Effects, Energy & Fuel, Vol. 23, 3681–3686.

Marshall, W.L. (2008): Dielectric Constant of Liquids (Fluids) Shown to be Simple Fundamental Relation of Density over Extreme Ranges from −50° to +600°C, Believed Universal, Nature Proceedings, Oak Ridge National Laboratory, 22 pp.

Martos, C., Coto, B., Espada, J.J., Robustillo, M.D., Gómez, S. & Peña, J.L. (2008): Experimental Determination and Characterization of Wax Fractions Precipitated as a Function of Temprature, Energy & Fuels, Vol. 22, No. 2, 708–714.

Martos, C., Cato, B., Espada, J.J., Robustillo, M.D., Peña, J.L. & Merino-Garcia, D. (2009): Characterization of Brazilian Crude Oil Sample to Improve the Prediction of Wax Precipitation in Flow Assurance Problems, Energy & Fuels, 6 pp.

Marugán, J., Calles, J.A., Dufour, J., Giménez-Aguirre, R., Peñja, J.L. & Merino-García, D. (2009): Characterization of the Asphaltene Onset Region by Focused-Beam Laser Reflectance: A Tool for Additives Screening, Energy & Fuels, Vol. 23, 1155–1161.

Matzain, A., Apte, M.S., Zhang H., Volk, M., Redus, C.L. Brill, J.P. & Creek, J.L. (2001): Multiphase Flow Wax Deposition Modeling, Proc. ETCE, Paper ETCE2001-17114, Houston, (reference taken from Rosvold 2008).

McAdams, W.H., Woods, W.K. & Heroman, L.C. (1942): Vaporization Inside Horizontal Tubes, II – Benzene-Oil Mixture, ASME 64, 193–200.

McCabe, W.L., Smith, J.C. & Harriott, P. (1993): Unit Operations of Chemical Engineering, McGraw-Hill, 1130 pp.

McCartney, R.A., Duppenbecker, S. & Cone, R. (2014): How Accurate are Scale Prediction Models? An Example from a Gas Condensate Well, 25th International Oil Field Chemistry Symposium, 23–26 March, Geilo, TEKNA, 21 pp.

McGovern, J. (2011): Friction Factor Diagrams for Pipe Flow, Technical Note, Dublin Institute of Technology, 15 pp.

Meredith, W., Kelland, S.J. & Jones, D.M. (2000): Influence of Biodegradation on Crude Oil Acidity and Carboxylic Acid Composition, Organic Geochemistry, 31, 1059–1073.

Meyerhoff, M.E. (2003): A More Detailed Look at Chemical Equilibria, Lecture Note 11, Biomedical Engineering, University of Michigan, 10 pp.

Mohammadi, A.H. & Richon, D. (2007a): Use of Boiling Point Elevation Data of Aqueous Solutions for Estimating Hydrate Stability Zone, Ind. Eng. Chem. Res., Vol. 47, 987–989.

Mohammadi, A.H. & Richon, D. (2007b): A Monodisperse Thermodynamic Model for Estimating Asphaltene Precipitation, AIChE Journal, Vol. 53, No. 11, 2940–2946.

Mohammadi, A.H., Samieyan, V. & Tohidi, B. (2005): Estimation of Water Content in Sour Gases, Paper SPE 94133, SPE Europec/EAGE Annual Conference, Madrid, 13–16 June, 10 pp.

Moreira A.P.D. & Teixeira A.M.R.F (2009): An Investigation of the Formation of Calcium Naphthenate From Commercial Naphthenic Acid Solutions by Thermogravimetric Analysis, Brazilian Journal of Petroleum and Gas, Vol. 3, No. 2, 051–056.

Mork, M. & Gudmundsson, J.S. (2001): Rate of Hydrate Formation in Subsea Pipelines – Correlation Based on Reactor Experiments, 12th International Oil Field Chemistry Symposium, Geilo, Norway, April 1–4, 14 pp.

Mullins, O.C., Sabbah, H., Eyssautier, J., Pomerantz, A.E., Barré, Andrewst, A.B., Ruiz-Morales, Y., Mostofwfi, R., McFarlanne, R., Goual, L., Lepkowicz, R., Cooper, T., Orbulescu, J., Leblanc, R.M., Edwards, J. & Zare, R.N. (2012): Advances in Asphaltene Science and the Yen-Mullins Model, Energy & Fuels, 26 (7), 3986–4003.

Muzychka, Y.S. & Awad, M.M. (2010): Asymptotic Generalizations of the Lockhart-Martinelli Method for Two Phase Flow, Journal of Fluids Engineering, Vol. 132, 12 pp.

Nalwaya, V., Tantayakom, V., Piumsomboon, P. & Fougler, S. (1999): Studies on Asphaltenes through Analysis of Polar Fractions, Ind. Eng. Chem. Res., Vol. 38, No. 3, 964–972.

Nasr-El-Din, H.A. & Al-Humaidan, A.Y. (2001): Iron Sulfide Scale – Formation, Removal and Prevention, SPE Paper 68315, SPE International Symposium on Oilfield Scale, Aberdeen, 30–31 January, 13 pp.

Nieckele, A.O., Braga, A.M.B. & Azevedo, L.F.A. (2001): Transient Pig Motion Through Gas and Liquid Pipelines. ASME, Journal of Energy Resources Technology, Vol. 123, No. 4, 260–269.

Nielsen, R.B. & Bucklin, R.W. (1983): Why Not Use Methanol for Hydrate Control?, Hydro. Proc., Vol. 55, No. 7, 71–75, (taken from Carroll 2003).

Ning, J., Zheng, Y., Brown, B., Young, D. & Nesic, S. (2015): Construction and Verification of Pourbaix Diagrams for Hydrogen Sulfide Corrosion of Mild Steel, Paper 5507, Corrosion 2015, NACE International Conference & Expo, Dallas, TX, 15–19 March, 19 pp.

Nikuradse, J. (1933): Stromnungsgesetze in Rauhen Rohren. Forschungsheft 361, Vol. B, VDI Verlag, Berlin (translated in NACA Technical Memorandum Nr. 1292, 1950).

Nordgård, E.L. (2009): Model Compounds for Heavy Crude Oil Components and Tetrameric Acids, Ph.D. Thesis, Department of Chemical Engineering, Norwegian University of Science and Technology, September, 69 pp. (excluding technical papers).

NORSOK (1997): Process Design, Standard P-001, Rev. 3, 27 pp.

O'Donoghue, A. (2004): Pigging as a Flow Assurance Solution – Estimating Pigging Frequency for Dewaxing, Pipeline Pigging and Integrity Management Conference, Amsterdam, May 17–18, 10 pp.

Oellrich, L.R. (2004): Natural Gas Hydrates and Their Potential for Future Energy Supply, ASME Heat and Mass Transfer Conference, IGCAR, Kalpakkam, 70–78.

Ohen, H.S., Williams, L.E., Lynn, J.S. & Ali, L. (2004): Assessment and Diagnosis of Inorganic-Scaling Potential Using Near-Infrared Technology of Effective Treatment, SPE Production and Facilities, November, 245–252.

Olesik, S.V. & Woddruff, J.L. (1991): Liquid Mass-Transport Theories Applied to Molecular Diffusion in Binary and Ternary Supercritical Fluid Mixtures, Analytical Chemistry, Vol. 63, No. 7, April 1, 670–676.

Oliveira, M.C.K., Rosário, F.F., Bertelli, J.N., Pereira, R.C.L., Albuquerque, F.C. & Marques, L.C.C. (2013): Flow Assurance Solutions to Mitigate Naphthenates Problems in Crude Oil Production, SPE Paper 166235, SPE Annual Technical Conference and Exhibition, New Orleans, 30 September to 2 October, 10 pp.

Ostergaard, K.K., Tohidi, B., Danesh, A., Todd, A.C. & Burgass, R.W. (2000): A General Correlation for Predicting the Hydrate-Free Zone of Reservoir Fluids, SPE Prod. & Facilities, Vol. 14, No. 4, November, 228–233.

Østergaard, K.K., Tohidi, B., Danesh, A., Todd, A.C. & Burgass, R.W. (2000): A General Correlation for Predicting the Hydrate-Free Zone of Reservoir Fluids, SPE Prod. & Facilities, Vol. 14, No. 4, November, 228–232.

Østvold, T. & Randhol, P. (2002): Prediction and Kinetics of Carbonate Scaling from Oil Field Waters, Paper 02317, NACE Corrosion 2002, 15 pp.

Østvold, T., Mackay, E.J., McCartney, R.A., Davis, I. & Aune, E. (2010): Re-Development of the Frøy Field: Selection of the Injection Water, SPE Paper 130567, SPE International Conference on Oilfield Scale, 26–27 May, Aberdeen, 23 pp.

Panuganti, S.R. (2013): Asphaltene Behavior in Crude Oil Systems, Ph.D. Thesis, Department of Chemical and Biomolecular Engineering, Rice University, Texas, xv + 156 pp.

Passade-Boupat, N., Gonzalez, M.R., Brocart, B., Hurtevent, C. & Palermo, T. (2012): Risk Assessment of Calcium Naphtenates and Separation Mechanisms of Acidic Crude Oil, SPE 155229, International Conference and Exhibition on Oilfield Scale, Aberdeen, 30–31 May, 12 pp.

Pedersen, K.S., Fredenslund, Aa. & Thomassen, P. (1989): Properties of Oils and Natural Gases, Gulf Publishing Company, 252 pp.

Pedersen, K.S., Christensen, P.L. & Shaikh, J.A. (2014): Phase Behavior of Petroleum Fluids, 2nd Edition, CRC Press, 465 pp.

Pedersen, K.S., Skovborg, P. & Roenningsen, H.P. (1991): Wax Precipitation from North Sea Crude Oils. 4. Thermodynamic Modeling, Energy & Fuels, Vol. 5, 924–932.

Pedersen, W.B., Hansen, A.B., Larsen, E., Nielsen, A.B. & Roenningsen, H.P. (1991): Wax Precipitation from North Sea Crude Oils. 2. Solid-Phase Content as Function of Temperature Determined by Pulsed NMR, Energy & Fuels, Vol. 5, 908–913.

Paso, K., Senra, M., Yi, Y., Sastry, A.M. & Fogler, H.S. (2005): Paraffin Polydispersity Facilitates Mechanical Gelation, Ind. Eng. Chem. Res., Vol. 44, No. 18, 7242–7254.

Paso, K., Kompalla, T., Oschmann, H.J. & Sjöblom J. (2009): Rheological Degradation of Model Wax-Oil Gels, J. Dispersion Science Technology, Vol. 30, 472–480.

Perry, R.H., Green, D.W. & Maloney, J.O. (1998): Perry's Chemical Engineer' Handbook, 7th Edition, McGraw Hill, 2.6 kg.

Pierre, B. (2016): Private communication.

Pitzer, K.S. (1973): Thermodynamics of Electrolytes, Journal of Physical Chemistry, 77, 268–277.

Pohl, H.A. (1962): Solubility of Iron Sulfides, J. Chem. Eng. Data, 7(2), 295–306.

Ramey, H.J. (1973): Correlations of Surface and Interfacial Tensions of Reservoir Fluids, SPE Paper 4429, 27 pp.

Ramm, M. & Bjørlykke, K. (1994): Porosity/Depth Trends in Reservoir Sandstones – Assessing the Qualitative Effects of Varying Pore-Pressure, Temperature History and Mineralogy, Norwegian Shelf Data, Clay Minerals, 29, 475–490.

Reeves, M.J. (2009): Aqueous Geochemistry, GEOE 475 Advanced Hydrogeology, Lecture Notes, Department of Civil and Geological Engineering, University of Saskatchewan, 40 pp.

Reistle, C.E. (1932): Paraffin and Congealing Oil Problems, Bull. USBM, 348.

Riazi, M.R. (2005): Characterization and Properties of Petroleum Fractions, ASTM, xxii + 405 pp.

Rittirong, A. (2014): Paraffin Deposition Under Two-Phase Gas-Oil Slug Flow in Horizontal Pipes, Ph.D. Thesis, University of Tulsa, Oklahoma, 374 pp.

Rittirong, A., Panacharoensawad, E. & Sarica, C. (2017): Experimental Study of Paraffin Deposition Under Two-Phase Gas/Oil Slug Flow in Horizontal Pipes, SPE Production & Operations, February, 99–117.

Roenningsen, H.P., Bjoerndal, B., Hansen, A.H. & Pedersen, W.B. (1991): Wax Precipitation from North Sea Crude Oils: 1. Crystallization and Dissolution Temperatures, and Newtonian and Non-Newtonian Flow Properties, Energy & Fuels, Vol. 5, 895–908.

Rogel, E. & Carbognani, L. (2003): Density Estimation of Asphaltenes Using Molecular Dynamics Simulationa, Energy & Fuels, Vol. 17, 378–386.

Rogers, G.F.C. & Mayhew, Y.R. (1980): Thermodynamics and Transport Properties of Fluids, 4th Edition, Blackwell Publishers, 25 pp.

Rojey, A., Jaffret, C., Carnot-Gandolphe, S., Durand, B., Jullian, S. & Valais, M. (1997): Natural Gas – Production, Processing, Transport, Éditions Technip, Paris, 429 pp.

Ronander, M.E. (2015): Magnetic Nanoparticles for Extraction of Naphthenic Acids from Model Oil, MSc Thesis, Chemical Engineering and Biotechnology, Norwegian University of Science and Technology, June, 94 pp.

Rosvold, K. (2008): Wax Deposition Models, M.Sc. Thesis, Department of Petroleum Engineering and Applied Geophysics, Norwegian University of Science and Technology, 104 pp.

Rousseau, G., Zhou, H. & Hurtevent, C. (2001): Calcium Carbonate and Naphthenate Mixed Scale in Deep-Offshore Fields, SPE Paper 68307, 3rd International Conference on Oilfield Scale, Aberdeen, 30–31 January, 8 pp.

Rygg, O.B., Rydahl, A.K. & Rønningsen, H.P. (1998): Wax Deposition in Offshore Pipeline Systems, 1st North American Conference, Multiphase Technology, Banff, 9–11 June, 14 pp.

Saffman, P.G. (1965): The Lift on a Small Sphere in a Slow Shear Flow, J. Fluid Mech., Vol. 22, No. 2, 385–400.

Sandengen, K. (2006): Prediction of Mineral Scale Formation in Wet Gas Condensate Pipelines and in MEG (Mono Ethylene Glycol) Regeneration Plants, Doctoral Thesis, Department of Materials Science and Engineering, Norwegian University of Science and Technology, Trondheim, 209 pp.

Sandengen, K. (2012): Scale – Precipitation of Salts Creating Problems, Guest Lecture, TPG 4140 Natural Gas, Department of Petroleum Engineering and Applied Geophysics, Norwegian University of Science and Technology, Trondheim, 25 pp.

Sarica, C. & Volk, M. (2004): Tulsa University Deposition Projects, Final Technical Report, University of Tulsa, Tulsa, 114 pp.

Schabron, J.F., Rovani, J.F. & Sanderson, M.M. (2010): Optimized Asphaltene and Wax Separations, Presentation, Pavement Performance and Prediction Symposium, Western Research Institute, July 15, 82 pp.

Schabron, J.F., Pauli, A. T. & Rovani, J.F. (1999): Petroleum Residua Solubility Parameter/ Polarity Map – Stability Studies of Residua Pyrolysis, Report WRI-99-R004, Western Research Institute, 24 pp.

Schramm, L.L. (1992): Emulsions – Fundamentals and Applications in the Petroleum Industry, Advances in Chemistry Series, No. 231, ACS, 428 pp.

Schmitt, D.R. (2005): Heavy and Bituminous Oils – Can Alberta Save the World, Australian Society of Exploration Geophysicists, Preview/Issue 118, 22–29.

Shafiee, N.S. (2014): Carboxylic Acid Composition and Acidity in Crude Oil and Bitumens, Ph.D. Thesis, School of Civil Engineering and Geosciences, Newcastle University, 185 pp.

Shafizadeh, A., McAteer, G. & Sigmon, J. (2003): High Acid Crudes, Presentation, Crude Oil Quality Group Meeting, New Orleans, January 30, 50 pp.

Sharqawy, M.H., Lienhard, J.H. & Zubair, S.M. (2010): Thermophysical Properties of Seawater: A Review of Existing Correlations and Data, Desalination and Water Treatment, Vol. 16, 354–380.

Shell Global Solutions International B.V. (2007): Gas/Liquid Separators – Type Selection and Design Rules, Design and Engineering Practice, DEP 31.22.05.11, 100 pp.

Shepherd, A.G. (2008): A Mechanistic Analysis of Naphthenate and Carboxylate Soap-Forming Systems in Oilfield Exploration and Production, Ph.D. Thesis, School of Engineering and Physical Sciences, Heriot-Watt University, 337 pp.

Shepherd, A.G., Thomson, G.B., Westacott, R., Sorbie, K.S., Turner, M. & Smith, P.C. (2006): Analysis of Organic Field Deposits – New Types of Calcium Naphthenate Scale or the Effect of Chemical Treatment, SPE 100517, International Oilfield Scale Symposium, Aberdeen, 30 May–1 June, 16 pp.

Shock, D.A., Sudbury, J.D., & Crockett, J.J. (1955): Studies of the Mechanism of Paraffin Deposition and its Control, J. Pet. Tech. (Sept.), 23–28.

Siljuberg, M.K. (2012): Modelling of Paraffin Wax in Oil Pipelines, M.Sc. Thesis, Department of Petroleum Engineering and Applied Geophysics, Norwegian University of Science and Technology, Trondheim, xi+68 pp.

Silset, A. (2008): Emulsions (W/O and O/W) if Heavy Crude Oils – Characterization, Stabilization, Destabilization and Produced Water Quality, Ph.D. Thesis, Department of Chemical Engineering, Norwegian University of Science and Technology, Trondheim, 88 pp. (excluding published papers).

Simon, S., Reisen, C., Bersås, A. & Sjöblom, J. (2012): Reaction Between Tetrameric Acids and Ca^{2+} in Oil/Water Systems, Ind. Eng. Chem., 51, 5669–5676.

Simon, S. & Sjöblom, J. (2013): Fundamental Chemistry of Heavy Crude Oils, TEKNA Presentation, Stavanger, September 25, 28 pp.

Singh, A., Lee, H., Singh, P. & Sarica, C. (2011): Validation of Wax Deposition Models Using Field Data from a Subsea Pipeline, Offshore Technology Conference, OTC 21641, May 2–5, Houston, Texas, 19 pp.

Singh, P., Venkatesan, R., Fogler, H.S. & Nagarajan, N.R. (2000): Formation and Aging of Incipient Thin Film Wax-Oil Gels, AIChE J., Vol. 46, No. 5, (reference taken from Rosvold 2008).

Singh, P., Venkatesan, R., Fogler, H.S. & Nagarajan, N.R. (2000): Morphological Evaluation of Thick Wax Deposits During Aging, AIChE J., Vol. 47, No. 1, (reference taken from Rosvold 2008).

Sippola, M.R. (2002): Particle Deposition in Ventilation Ducts, Ph.D. Thesis, Department of Civil and Environmental Engineering, University of California, Berkeley, 483 pp.

Sippola, M.R. & Nazaroff, W.W. (2002): Particle Deposition from Turbulent Flow: Review of Published Research and its Applicability to Ventilation Ducts in Commercial Buildings, Lawrence Berkeley National Laboratory Report, LBNL-51432, 145 pp.

Sjöblom, J., Simon, S. & Xu, Z. (2014): The Chemistry of Tetrameric Acids in Petroleum, Advances in Colloid and Interface Science, 205, 319–338.

Sjöblom, J., Aske, N., Auflem, I.H., Brandal, Ø., Havre, T.E., Sæther, Ø., Westvik, A., Johnsen, E.E. & Kallevik, H. (2003): Our Current Understanding of Water-in-Crude Oil Emulsions, Advances in Colloid and Interfacial Science, 100–102, 399–473.

Skjetne, E. (1995): High-Velocity Flow in Porous Media – Analytical, Numerical and Experimental Studies, Doctoral Thesis, Department of Petroleum Engineering and Applied Geophysics, NTNU, vi + 180 pp.

Sletfjerding, E. (1999): Friction Factor in Smooth and Rough Gas Pipelines, Dr.Ing. Thesis, Department of Petroleum Engineering and Applied Geophysics, Norwegian University of Science and Technology, Trondheim (includes several published papers).

Sletfjerding, E. & Gudmundsson, J.S. (2003): Friction Factor Directly from Roughness Measurements, Journal of Energy Resources Technology, Transactions of the ASME, Vol. 125, No. 2, 126–130.

Sloan, E.D., Koh, C.A., Sum, A.K., Ballard, A.L., Shoup, G.J., McMullen, N., Creek, J.L. & Palermo, T. (2009): Hydrates: State of the Art Inside and Outside Flowlines, Journal of Petroleum Technology, December, 89–94.

Sloan, E.D. (2000): Hydrate Engineering, SPE Monograph, Vol. 21, Society of Petroleum Engineers, Richardson, Texas, 89 pp.

Sloan, E.D. (1998): Clathrate Hydrates of Natural Gases, Marcel Dekker, 705 pp.

Smith, B.E., Sutton, P.A., Lewis, C.A., Dunsmore, B., Fowler, G., Krane, J., Lutnaes, B.F., Brandal, Ø., Sjöblom, J. & Rowland, S.J. (2007): Analysis of "ARN" Naphthenic Acids by High Temperature Gas Chromatography and High Performance Liquid Chromatography, Journal of Separation Science, 30 (3), 375–380.

Smith, F.L. & Harvey, A.H. (2007): Avoid Common Pitfalls When Using Henry's Law, Chemical Engineering Progress, September, 33–39.

Smith, H.V. & Arnold, K.E. (1987): Crude Oil Emulsions, Chapter 19, SPE Petroleum Engineering Handbook, Bradley, H.B. (editor), 34 pp.

Smith, J.M., Van Ness, H.C. & Abbott, M.M. (1996): Introduction to Chemical Engineering Thermodynamics, Fifth Edition, McGraw Hill, 763 pp.

Somerscales, E.F.C. & Knudsen, J.G. (1981): Fouling of Heat Transfer Equipment, Hemisphere Publishing Corporation 743 pp.

Spalding, D.B. (1961): A Single Formula for the Law of the Wall, J. Appl. Mech. Vol. 28, 455–457 (taken from White 1991).

Speight, J.G. (1980): Chemistry and Technology of Petroleum, Marcel-Dekker, New York (reference taken from Thou et al., 2002).

Speight, J.G. & Moschopedis, S.E. (1981): On the Molecular Nature of Petroleum Asphaltenes, Chemistry of Asphaltenes, (editors Bunger, J.W. & Li, N.C.), Advances in Chemistry Series, No. 195, American Chemical Society, 1–15.

Spiecker, P.M., Gawrys, K.L., Trail, C.B. & Kilpatrick, P.K. (2003): Effect of Petroleum Resins on Asphaltene Aggregation and Water-in-Oil Emulsion Formation, Colloids and Surfaces A, Physicochemical Aspects, 200, 9–27.

Stevenson, C.J., Davies, S.R., Gasanov, I., Hawkins, P., Demiroglu, M. & Marwood, A.P. (2015): Development and Execution of a Wax Remediation Pigging Program for a Subsea Oil Export Pipeline, Paper OTC-25889, Offshore Technology Conference, 4–7 May, Houston, 17 pp.

Sumestry, M. & Tedjawidjaja, H. (2013): Case Study of Calcium Carbonate Scale Inhibitor Performance Degradation Because of H_2S Scavenger Injection in Semoga Field, SPE Paper 150705, Oil and Gas Facilities, Society of Petroleum Engineers, February, 40–45.

Sun, M. & Firoozabadi, A. (2014): New Hydrate Anti-Agglomerant Formulation for Offshore Flow Assurance and Oil Capture, Paper OTC-25439, Offshore Technology Conference, 05–08 May, 8 pp.

Sundman, O., Nordgård, E.L., Grimes, B. & Sjöblom, J. (2010): Potentiometric Titrations of Five Synthetic Tetraacids as Models for Indigenous C_{80} Tetraacids, Langmujir, Vol. 26, No. 3, 1619–1629.

Taiwo, E., Otolorin, J. & Afolabi, T. (2012): Crude Oil Transportation – Nigerian Niger Delta Waxy Crude, Chapter 8, Crude Oil Exploration in the World, Younes, M. (editor), InTech, 135–154.

Tardos, T.F. (2013): Emulsion Formation, Stability and Rheology, Chapter 1, Emulsion Formation and Stability, Tardos, T.F. (editor), Wiley-VCH, 272 pp.

Thomas, D.G. (1965): Transport Characteristics of Suspensions, J. Coll. Sci., 20(3), 267–277.

Thorley, A.R.D. (1991): Fluid Transients in Pipeline Systems, D. & L. George Ltd., 266 pp.

Thou, S., Ruthammer, G. & Potsch, K. (2002): Detection of Asphaltenes Flocculation Onset in a Gas Condensate System, SPE 13th European Petroleum Conference, Aberdeen, 29–31 October, SPE 78321.

Tiwary, D. & Mehrotra, A.K. (2004): Phase Transformation and Rheological Behaviour of Highly Paraffinic "Waxy" Mixtures, Canadian J. Chem. Eng., Vol. 82, February, 162–174.

Turin, A. (1952): The Chemical Basis of Morphogenesis, Phil. Trans. R. Soc. B, Vol. 237, 37–72.

Turner, M.S. & Smith, P.C. (2005): Controls on Soap Scale Formation, Including Naphthenate Soaps – Drivers and Mitigation, SPE Paper 94339, SPE International Symposium on Oilfield Scale, Aberdeen, 11–12 May, 14 pp.

Tvedt, V. (2006): Evaluation of Different Commercialisation Strategies for Cold Flow Technology, M.Sc. Thesis, Department of Industrial Economics and Technology Management, Norwegian University of Science and Technology, Trondheim, 142 pp.

Tvedt, V. (2005): Transportation of Petroleum in Subsea Pipelines – A Comparative Analysis, B.Sc. Thesis, Department of Industrial Economics and Technology Management, Norwegian University of Science and Technology, Trondheim, 107 pp.

Umesi, N.O. & Danner R.P. (1981): Predicting Diffusion Coefficients in Nonpolar Solvents, Ing. Eng. Chem. Process Des. Dev., Vol. 20, No. 4, 662–665.

Van Driest, E.R. (1956): On Turbulent Flow Near a Wall, Journal Aeronautical Sciences, Vol. 23, No. 11, 1007–1011 & 1036.

Venkatesan, R. (2004): The Deposition and Rheology of Organic Gels, Doctoral Thesis in Chemical Engineering, University of Michigan, 225 pp.

Venkatesan, R. & Creek, J.L. (2007): Wax Deposition During Production Operations, OTC 18798, Offshore Technology Conference, Houston, 30 April–3 May, 5 pp.

Venkatesan, R. & Fogler, H.S. (2004): Comments on Analogies for Correlated Heat and Mass Transfer in Turbulent Flow, AIChE Journal, Vol. 50, No. 7, July, 1623–1626.

Venkatesan, R. & Creek, J. (2010): Wax Deposition and Rheology: Progress and Problems from an Operator's View, Paper OTC 20668, Offshore Technology Conference, Houston, 3–6 May, 9 pp.

Verdier, S. (2006): Experimental Study and Modelling of Asphaltene Precipitation Caused by Gas Injection, Ph.D. Thesis, Department of Chemical Engineering, Technical University of Denmark, xviii + 266 pp.

Vindstad, J.E., Bye, A.S., Grande, K.V., Hustad, B.M., Hustvedt, E. & Nergård, B. (2003): Fighting Naphthenate Deposition at the Heidrun Field, SPE Paper 80375, SPE 5th International Sumposium on Oilfield Scale, Aberdeen, 29–30 January, 7 pp.

Waage, P. & Guldberg, C.M. (1864): Studies Concerning Affinity, Proceeding of the Norwegian Academy of Science and Letters, 35–45, (in Norwegian).

Walther, J.V. (2009): Essentials of Geochemistry, Jones and Bartlett Publishers, Boston, 797 pp.

Wang, J.X., Creek, J.L. & Buckley, J.S. (2006): Screening for Potential Asphaltene Problems, SPE Annual Technical Conference and Exhibition, 24–27 September, San Antonio, Texas, SPE 103137, 6 pp.

Wang, Q., Sarica, C. & Chen, T.X. (2005): An Experimental Study on Mechanics of Wax Removal in Pipeline, Trans. ASME, Vol. 127, December, 302–309.

Warren, K.W. (2007): Emulsion Treating, Chapter 3, Vol. III, SPE Petroleum Engineering Handbook, Arnold, K.E. (editor), 61–122.

White, F.M. (1991): Viscous Fluid Flow, McGraw Hill, 614 pp.

Wilke, C.R. & Chang, P. (1955): Correlation of Diffusion Coefficient in Dilute Solutions, A.I.Ch.E. Journal, Vol. 1, No. 2, 264–270.

Wilkinson, J.J. (2001): Fluid Inclusions in Hydrothermal Ore Deposits, Lithos, 55 (1–4), 229–272.

Wolden, M., Lund, A., Oza, N., Makogon, T., Argo, C.B. & Larsen, R. (2005): Cold Flow Black Oil Slurry Transport of Suspended Hydrate and Wax Solids, Proc. 5th International Conference Gas Hydrates, June 12–16, Trondheim, Norway, 6 pp.

Won, K.W. (1989): Thermodynamic Calculation of Cloud Point Temperatures and Wax Phase Compositions of Refined Hydrocarbon Mixtures, Fluid Phase Equilibria, Vol. 53, 377–379 (reference taken from Cato et al., 2010).

Wood, A.B. (1941): A Textbook of Sound, G. Bell & Sons, xvi + 578 pp.

Yarranton, H.W. & Masliyah, J.H. (1996): Molar Mass Distribution and Solubility Modeling of Asphaltenes, AIChE Journal, Vol. 42, No. 12, 3533–3543.

Yarraton, H.W. (1997): Asphaltene Solubility and Asphaltene Stabilized Water-in-Oil Emulsions, Ph.D. Thesis, Department of Chemical and Materials Engineering, University of Alberta, 23+365 pp.

Yaws, C.L. (2010): Thermophysical Properties of Chemicals and Hydrocarbons, Knovel (Electronic Edition).

Yaws, C.L. (1999): Chemical Properties Handbook, McGraw-Hill.

Zhang, A., Ma, Q., Wang, K., Tang, Y. & Goddard, W.A. (2005): Improved Process to Remove Naphthenic Acids, Technical Report, California Institute of Technology, Pasadena, 96 pp.

Zhang, X., Queimada, A., Szczepanski, R. & Moorwood, A. (2014): Modeling the Shearing Effect of Flowing Fluid and Wax Aging on Wax Deposition in Pipelines, Paper OTC 24797, Offshore Technology Conference Asia, Kuala Lumpur, 25–28 March, 22 pp.

Zheng, X. & Silber-Li, Z. (2009): The Influence of Saffman Lift Force on Nanoparticle Concentration Distribution Near a Wall, Applied Physics Letters, Vol. 95, 3 pp.

Zuo, J.Y., Elshahawi, H., Mullins, O.C., Dong, C., Zang, D., Jia, N. & Zhao, H. (2012): Asphaltene Gradients and Tar Mat Formation in Reservoirs Under Active Gas Charging, Fluid Phase Equilibria, 315, 91–98.

Zuo, J.Y., Dumont, H., Mullins, O.C., Dong, C., Elshahwai, H. & Seifert, D.J. (2013): Integration of Downhole Fluid Analysis and the Flory-Huggins Zuo EOS for Asphaltene Gradient and Advanced Formation Evaluation, SPE Paper 166385, SPE Annual Technical Conference and Exhibition, New Orleans, 30 September–2 October, 11 pp.

Temperature in pipelines

Two situations are envisioned: (1) steady-state flow and (2) transient cooling on shut-in. In steady-state flow, the fluid temperature along a pipeline will be constant at any location, typically decreasing with distance. Upon shut-in, the fluid temperature at any location along a pipeline, will typically decrease exponentially with time. It will typically decrease from the steady-state flow temperature and approach the ambient temperature. The above holds true for subsea pipelines carrying predominantly liquid phases. In some natural gas pipelines, the steady-state temperature may be below the ambient temperature, due to expansion cooling.

STEADY-STATE FLOW

The flowing fluid is cooled from the outside. Joule-Thomson cooling due to gas expansion is not included in the derivations below. A subsea pipeline is assumed surrounded by cold seawater at constant temperature. For an on-land pipeline, the surrounding constant temperature will be the air temperature. The following variables are used:

$m =$ mass flow rate inside pipeline, kg/s
$T_o =$ constant outside temperature, K
$T_1 =$ pipeline inlet (upstream) fluid temperature, K
$T_2 =$ pipeline outlet (downstream) fluid temperature, K
$L =$ pipeline length, m
$d =$ pipeline internal diameter, m

The pipeline is envisioned as a long heat exchanger tube with cooling from the outside

$$q = UA\Delta T_{\mathrm{LMTD}}$$

where q W is the heat transferred, U W/m$^2 \cdot$ K the overall heat transfer coefficient, A m^2 the heat transfer area and ΔT_{LMTD} K the logarithmic mean temperature difference.

The cooling of the flowing fluid inside a pipeline can be expressed by the equation

$$q = mC_p(T_1 - T_2)$$

where C_p J/kg $\cdot$ K is the heat capacity of the flowing fluid (fluid mixture).

Based on the definition of the logarithmic mean temperature difference

$$\Delta T_{\text{LMTD}} = \frac{(T_1 - T_o) - (T_2 - T_o)}{\ln \dfrac{T_1 - T_o}{T_2 - T_o}}$$

and setting in for constant outside temperature, the following simplification results

$$\Delta T_{\text{LMTD}} = \frac{T_1 - T_2}{\ln \dfrac{T_1 - T_o}{T_2 - T_o}}$$

The cooling from outside the pipeline corresponds to (equal to) the cooling of the fluid flowing inside the pipeline, resulting in the relationship

$$mC_p(T_1 - T_2) = U\pi d(L)\frac{T_1 - T_2}{\ln \dfrac{T_1 - T_o}{T_2 - T_o}}$$

where the heat transfer area is expressed by $A = \pi d(L)$. The bracket around L is used to avoid confusion with usual differential, like dL. The relationship can be rewritten as

$$\ln \frac{T_1 - T_o}{T_2 - T_o} = \frac{U\pi d(L)}{mC_p}$$

and

$$\ln \frac{T_2 - T_o}{T_1 - T_o} = \frac{-U\pi d(L)}{mC_p}$$

The result can be written

$$T_2 = T_o + (T_1 - T_o)\exp\left[\frac{-U\pi d}{mC_p}L\right]$$

The temperature T_2 at some downstream location L can be for a pipeline segment or a whole pipeline with upstream inlet temperature T_1. Length segments can be used if the pipeline has different outside temperatures or different overall heat transfer coefficients.

The main uncertainty in calculations concerns the value of the overall heat transfer coefficient U. It can be calculated theoretically or be based on experimental values. Subsea pipelines without insulation have U-values in the range 15–25 W/m² · K, while insulated subsea pipelines have U-values in the range 2–4 W/m² · K, based on measurements carried out on pipelines in operation. The U-value for subsea pipelines without insulation (carbon steel with external coating) may be as high as 80 W/m² · K. However, this high value applies to a free-standing pipeline, meaning not partly buried in seabed sediments. Non-insulated flexible rises have U-values of about 10 W/m² · K.

Flexible rises have several layers of metal and composites, so the U-value will vary with the design and construction.

The mass flowrate m kg/s is that of the total flow. Assuming homogeneous flow (oil, gas and water well mixed) the heat capacity can be estimated from the relationship

$$C_{pM} = x_o C_{po} + x_g C_{pg} + x_w C_{pw}$$

where the x-values are the mass fractions of the relevant phases (oil, gas, water). Heat capacity depends on temperature but is practically independent of pressure. The accuracy of pipeline temperatures, where the temperature change is large, can be improved by carrying out the calculations for pipeline segments.

COOLING ON SHUT-IN

Time dependent (transient) heat transfer from outside a pipeline (cooling) can be expressed by the relationship

$$q(t) = UA[T(t) - T_o]$$

The cooling applies to some volume V m^3, for example a pipeline segment of an appropriate length. The cooling of the fluid volume can be expressed as

$$q(t) = -\rho V C_p \frac{dT}{dt}$$

The cooling from outside the pipeline corresponds to the cooling of the fluid volume (in the pipeline segment) inside the pipeline resulting in the relationship

$$UA[T(t) - T_o] = -\rho V C_p \frac{dT}{dt}$$

which can be written for integration as

$$\frac{-UA}{\rho V C_p} \int_{t_1}^{t_2} dt = \int_{T_1}^{T_2} \frac{1}{[T(t) - T_o]} dT$$

giving

$$\frac{-UA}{\rho V C_p} (t_2 - t_1) = \ln \frac{T_2 - T_o}{T_1 - T_o}$$

The resulting equation is the following

$$T_2 = T_o + (T_1 - T_o) \exp \left[\frac{-UA}{\rho V C_p} \Delta t \right]$$

because

$$A = \pi d(L)$$

and

$$V = \left(\frac{\pi d^2}{4}\right) L$$

The exponential term can also be written as

$$\left[\frac{-UA}{\rho V C_p}\Delta t\right] = \left[\frac{-4U}{\rho C_p d}\Delta t\right]$$

In the transient temperature equation, the T_1 and T_2 temperatures are *not* the inlet and outlet temperatures (as in steady-state flow), but the local segment temperatures. An alternative way to express these temperatures would be to use a double subscript, for example $T_{x,t}$ where x stands for location and t for time.

Appendix B

Pipeline wall heat transfer

Heat flow through a pipe wall from a high temperature to a low temperature is given by *Fourier's Law*

$$q = -kA\frac{dT}{dr}$$

where the units are q W, k W/m·K, A m², T K and r m. At steady-state conditions, the heat flow is constant. The thermal conductivity of the pipe material is constant.

The heat transfer area of a length of pipe is given by the equation

$$A = 2\pi rL$$

and is thus a function of pipe radius. *Fourier's Law* can now be written as

$$\frac{1}{r}dr = \frac{-k2\pi L}{q}dT$$

Integration from inside (i) to outside (o) gives

$$\ln\left(\frac{r_o}{r_i}\right) = \frac{-k2\pi L}{q}(T_o - T_i)$$

The minus sign can be removed by switching the places of the temperatures. The steady-state heat flow can thus be written as

$$q = \frac{k2\pi L}{\ln\left(\dfrac{r_o}{r_i}\right)}(T_i - T_o)$$

In heat exchanger *tubes*. it is usual to use the outside heat transfer area as a reference. In pipeline engineering, the outside diameter (OD) is quoted in international standards. In deposition studies (flow assurance solids), however, it seems more natural to use the inside heat transfer area as a reference. It is important to know whether an overall heat transfer coefficient is based on the ID or the OD. For the same heat transfer situation, $U_{ID} > U_{OD}$. The steady-state heat flow equation can be written as

$$\frac{q}{A_i} = \frac{k2\pi L}{A_i \ln\left(\dfrac{r_o}{r_i}\right)}(T_i - T_o)$$

such that

$$\frac{q}{A_i} = \frac{k2\pi L}{2\pi r_i L \ln\left(\dfrac{r_o}{r_i}\right)}(T_i - T_o)$$

and

$$\frac{q}{A_i} = \frac{(T_i - T_o)}{\dfrac{r_i}{k} \ln\left(\dfrac{r_o}{r_i}\right)}$$

The general heat transfer equation is

$$q = UA(T_i - T_o)$$

where U W/m$^2 \cdot$ K is the overall heat transfer coefficient. The equation is sometimes called *Newton's Law of Cooling* and can be written as

$$\frac{q}{A} = \frac{(T_i - T_o)}{R}$$

where R m$^2 \cdot$ K/W is the overall heat transfer *resistance*.

For a circular pipe, referenced to the inside heat transfer area, the overall heat transfer resistance can be written

$$R_i = \frac{r_i}{k} \ln\left(\frac{r_o}{r_i}\right)$$

The overall equation for the total heat transfer coefficient for a circular pipe (thick-walled pipe), is given by the well-known expression

$$\frac{1}{U_i} = \frac{1}{h_i} + R_{fi} + \frac{r_i}{k} \ln\left(\frac{r_o}{r_i}\right) + \left(\frac{1}{h_o} + R_{fo}\right)\frac{r_i}{r_o}$$

The overall heat transfer equation is referenced to the pipe inner radius. The individual heat transfer coefficients, h_i and h_o are referenced to the inside and outside heat transfer areas, respectively. Hence the correction term r_i/r_o. The terms R_{fi} and R_{fo} are the inside and outside fouling resistances, respectively.

The term

$$\frac{r_i}{k} \ln\left(\frac{r_o}{r_i}\right)$$

Table B.1 Assumed diameter and fluid properties.

Property	Value
d	0.3048 m
k	0.1 W/m · K
ρ	750 kg/m^3
u	2 m/s
μ	0.5 mPa · s
C_p	2300 J/kg · K

can be used to represent the heat transfer resistance of any *circular layer* of a pipeline, for example thermal insulation. Note that care must be taken to use the correct radius values.

Consider a 12 inch (=304.8 mm) ID (inside *diameter*) steel pipeline transporting liquid hydrocarbon subsea. The inside *radius* is therefore 152.4 mm. The pipe wall thickness is assumed 12 mm, based on rule-of-thumb (wall thickness in mm is assumed to have the same numerical value as the ID in inches). For simplicity, the thermal conductivity of duplex steel can be assumed 20 W/m · K. The pipe wall thermal resistance R_w will therefore amount to

$$R_w = \frac{152.4 \cdot 10^{-3}}{20} \ln\left(\frac{152.4 + 12}{152.4}\right) = 0.578 \cdot 10^{-3} \, \text{m}^2 \cdot \text{K/W}$$

The inside individual heat transfer coefficient can be estimated from the classical Dittus-Boelter equation for pipe flow cooling (for heating the exponent on the Prandtl number should be 0.4)

$$\text{Nu} = 0.023 \text{Re}^{0.8} \text{Pr}^{0.3}$$

where the dimensionless numbers of the fluid (not the steel wall) and fluid flow are given by

$$\text{Nu} = \frac{hd}{k}$$

$$\text{Re} = \frac{\rho u d}{\mu}$$

$$\text{Pr} = \frac{C_p \mu}{k}$$

Assumed fluid properties needed to calculate the dimensionless numbers are shown in Table B.1. The calculated values of the dimensionless numbers and the inside individual heat transfer coefficient and resistance are shown in Table B.2.

The overall heat transfer coefficient U, for bare (no insulation) subsea pipelines is usually in the range 15–25 W/m^2 · K. Assuming an average value of 20 W/m^2 · K the individual outside heat transfer coefficient and resistance can be

Table B.2 Dimensionless numbers and calculated
heat transfer coefficient and resistance
(referenced to inside pipe diameter).

Number	Value
Pr	11.5
Re	914,400
Nu	2811
h_i	922 W/m$^2 \cdot$ K
R_i	$1.08 \cdot 10^{-3}$ m$^2 \cdot$ K/W

calculated. Assuming no fouling resistance the outside heat transfer resistance
was calculated $R_o = 0.0483$ m$^2 \cdot$ K/W, representing 97% of the total resistance of
$1/20 = 0.05$ m$^2 \cdot$ K/W. The steel wall resistance contributes 1.2% to the total resistance
and the individual inside resistance contributes 0.2%.

The heat flux for a pipeline carrying fluid at 50°C with a 5°C outside temperature
(seawater) will be

$$\frac{q}{A} = 20 \cdot (50 - 5) = 900 \; \frac{W}{m^2}$$

The inside wall temperature can be calculated from

$$900 = 926(50 - T_{wi})$$

resulting in 49°C. The outside wall temperature can be calculated from

$$900 = \frac{1}{0.578 \cdot 10^{-3}}(49 - T_{wo})$$

resulting in 48.5°C.

Assuming that the subsea pipeline is surrounded (in reality, such pipelines have
concrete coating) by stationary water-saturated mud, having a thermal conductivity
of 0.6 W/m $\cdot$ K, which is similar to that of water, the thickness of the mud assuming
circular form can be calculated from

$$0.0485 = \frac{r_i}{k} \ln\left(\frac{r_o}{r_i}\right) = \frac{0.1664}{0.6} \ln\left(\frac{r_o}{0.1644}\right)$$

resulting in $r_o = 0.196$ m. Note that r_i for this calculation is given by 0.152 m plus
0.012 m. The thickness of the water-saturated mud is therefore 32 mm which seems
reasonable. Subsea pipelines have a reinforced concrete coating 1–2 inches (=25–
50 mm) in thickness, to give weight to prevent floating during laying. Concrete coating
prevents also displacement due to *in-situ* sea currents. There is a difference in coating
thickness between flowlines and export pipelines. The thermal conductivity of pipeline
concrete is typically 2.5 W/m $\cdot$ K. Knowing the thickness of the coating makes it possible

to calculate its outside temperature and the mud thickness estimated above, will be less than calculated.

The overall conclusion of the above is that most of the temperature drop from inside to outside, is across the material outside the steel of a subsea pipeline. The temperature drop is across the concrete coating and the surrounding water-saturated mud. The above calculations are based on a clean pipe wall. With paraffin wax deposit, for example, the metal pipe wall temperature will be lower and the deposit surface temperature will be closer to the bulk temperature. Lower pipe wall temperature will facilitate aging of a paraffin wax deposit (see *Section 4.6 Nature of Deposits*). Higher deposit surface temperature will decrease the driving force of deposition.

Appendix C

Boundary layer temperature profile

One example of the temperature law-of-the-wall is that by Kay and Nedderman (1985) as reported by Sippola (2002). In the present text, the term boundary layer is used for the viscous sub-layer *and* the buffer layer (other distinctions may be used in other texts). In the viscous sub-layer where $y^+ < 5$, the temperature profile is given by the equation

$$T^+ = T_w^+ + (\text{Pr})y^+$$

and in the buffer layer where $5 < y^+ < 30$, the temperature profile is given by the equation

$$T^+ = T_w^+[+/-][5\,\text{Pr} + 5\ln(0.2(\text{Pr})y^+ + 1 - \text{Pr})]$$

In the turbulent core where $y^+ > 30$, the dimensionless temperature is given by the equation

$$T^+ = T_w^+[+/-][5\,\text{Pr} + 5\ln(1 + 5\text{Pr})][+/-]2.5\ln\left(\frac{y^+}{30}\right)$$

In two of the above equations, the term $[+/-]$ means that $[+]$ is to be used in cooling and $[-]$ in heating. In subsea pipelines cooled from the outside (ambient temperature lower than the fluid bulk temperature) the $[+]$ should be used.

The dimensionless distance is given by

$$y^+ = \frac{\rho u^* y}{\mu}$$

where u^* is the friction velocity (also called the shear stress velocity)

$$u^* = \sqrt{\frac{\tau_w}{\rho}} = \bar{u}\sqrt{\frac{f}{8}}$$

where τ_w is the wall shear stress, $\bar{u}$ is the average flow velocity and f the friction factor.

Table C.1 Assumed pipe diameter and fluid properties.

Property	Value
d	0.3048 m
k	0.1 W/m · K
ρ	750 kg/m^3
u	2 m/s
μ	0.5 mPa · s
C_p	2300 J/kg · K
Pr	11.5
f	0.015
q/A	900 W/m^2
T_w	20°C

The dimensionless wall temperature is given by

$$T_w^+ = T \left(\frac{\rho C_p u^*}{q/A} \right)$$

and the subscript w stands for wall. Fluid density and fluid heat capacity have the usual symbols. Heat flow q W divided by heat transfer area A m^2 gives heat flux.

The dimensionless Prandtl number contains the thermal properties of fluids

$$\Pr = \frac{C_p \mu}{k}$$

expressed in terms of heat capacity and viscosity in the numerator and thermal conductivity in the denominator.

Noted with curiosity is that the dimensionless temperature equations reported by Sippola (2002), are not the same as given in Kay and Nedderman (1985). Perhaps the equations were up-dated in subsequent editions of the Kay and Nedderman book, not available in the present work. Other expressions of the temperature law-of-the-wall are found in the literature.

Calculations were carried out using the assumed properties shown in Table C.1. The properties are the same as used in *Appendix B – Pipe Wall Heat Transfer*, with a few modifications. While the same heat flux of 900 W/m^2 was used, the wall temperature chosen was 20°C instead of 50°C, for no other reason than to avoid repeating already carried out calculations.

Applying the temperature profile equations using the assumed properties, the plot shown in Figure C.1 was constructed. The temperature increases from 20°C at the wall to 20.5°C at about 1.5 mm from the wall. This small increase is primarily due to the low heat flux used. The radius of the pipe is 152 mm. The figure illustrates how close to the wall, within the viscous sub-layer, most of the temperature change occurs. This is typical for most wall-based transfer processes (heat, mass, momentum).

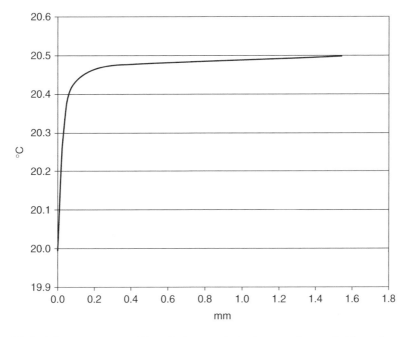

Figure C.1 Radial temperature profile in fluid in a typical subsea pipeline cooled from the outside.

The heat flux of 900 W/m² was based on an inside pipe temperature of 50°C and an outside temperature of 5°C and an overall heat transfer coefficient (for subsea pipelines) of 20 W/m² · K which is quite low compared to industrial heat exchangers. Typical U-values for water-kerosene heat exchangers are in the range 450–500 W/m² · K while for water-water exchangers typical values are in the range 1000–1200 W/m² · K.

Darcy-Weisbach equation

Consider an element of length dL subjected to a pressure drop dp in a pipeline. The pressure drop acts on the flow area πr^2 while the wall shear stress τ (often written as τ_w) acts on the wall area $2\pi r dL$ (circumference times length). The wall shear stress has the unit N/m^2.

A force balance can be written as follows

$$dF_p = dp\pi r^2$$

and

$$dF_\tau = 2\pi r dL \tau$$

such that

$$dp\pi r^2 = 2\pi r dL \tau$$

The wall shear stress can then be expressed by

$$\tau = \frac{r}{2}\frac{dp}{dL}$$

According to *tradition*, the wall shear stress in pipes is related to kinetic energy by the empirical equation

$$\tau = \frac{1}{8}f\rho u^2$$

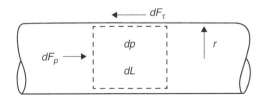

Figure D.1 Force balance for pressure drop in a pipeline.

where f is the friction factor, the Darcy = Moody friction factor. Shear stress and kinetic energy per volume have the same unit of $N/m^2 = J/m^3$ such that

$$\frac{r}{2}\frac{dp}{dL} = \frac{1}{8}f\rho u^2$$

Solving for pressure drop, gives the well know Darcy-Weisbach equation for pressure drop in pipes

$$\Delta p_f = \frac{f}{2}\frac{L}{d}\rho u^2$$

The pressure drop is given the subscript f to indicated frictional pressure drop and L is the pipe length. The equation can be used for both laminar flow and turbulent flow in pipes.

The Darcy-Weisbach equation is used for non-compressible flow, such as oil and water. However, it can be used to estimate the pressure drop in natural gas pipe flow (compressible flow), provided an average friction factor and an average density are used.

$$\Delta p_f = \frac{\bar{f}}{2}\frac{L}{d}\bar{\rho} u^2$$

Transfer equations

The conservation of energy, mass and momentum are fundamental in engineering science and technology. The transfer of energy (heat), mass and momentum can be described by fundamental laws on the molecular scale in fluids and solids. However, the focus here are fluids in laminar *and* turbulent flow. The fundamental laws can be used for laminar flow but semi-empirical correlations are needed for turbulent flow situations.

The fundamental law for the transfer of thermal energy is expressed by the equation

$$q = -kA\frac{dT}{dx}$$

which is *Fourier's Law*. Energy in the form of heat q W depends on the thermal conductivity k W/m·K, the area A m^2 and the temperature gradient dT/dx K/m. The minus sign is used because heat flows from high temperature to low temperature, by convention. Dividing the heat flow by area gives heat flux W/m^2 = J/s·m^2.

The fundamental law for the transfer of mass is expressed by the equation

$$J = -D\frac{dc}{dx}$$

which *Fick's Law*. Molar flowrate or molar flux J mol/s·m^2 depends on the molecular diffusivity coefficient D m^2/s and the concentration gradient dc/dx mol/m^3. Again, the minus sign is used to express mass diffuses from high concentration to low concentration.

The fundamental law for the transfer of momentum is expressed by the equation

$$\tau = \mu\frac{du}{dx}$$

Which is *Newton's Law of Viscosity* (Newtonian fluids). Wall shear stress τ N/m^2 depends on fluid viscosity μ Pa·s and velocity gradient du/dx s^{-1}. The minus sign is not used because wall shear stress increases with velocity gradient.

The conservation of momentum is expressed by *Newtons Second Law* by the product of mass kg and velocity m/s. The product of mass kg and acceleration m/s^2 has the unit of force N. For fluid flow in pipes, the mass can be replaced by mass rate kg/s such that product of mass and velocity has the unit (kg/s) × (m/s), which is the same as force N.

The fundamental law for *filtration velocity* in petroleum engineering is expressed by the equation

$$u = -\frac{k}{\mu}\frac{dp}{dx}$$

which is *Darcy's Law*. The filtration velocity u m/s depends on intrinsic permeability k m^2 divided by fluid viscosity μ Pa · s and the pressure gradient dp/dx Pa/m.

Darcy's Law describes fluid flow in porous media. The filtration velocity is analogous to superficial velocity in pipes, because it refers to the *whole* cross-sectional area, not the porous space available for fluid flow. The minus sign is used because fluids flow from high pressure to low pressure.

The fundamental law for cross-sectional *fluid velocity* in a circular pipe is expressed by the equation

$$u = \frac{1}{8\mu}(R^2 - r^2)\frac{dp}{dx}$$

which is *Poiseuille's Law*. The velocity should perhaps be expressed as $u(r)$ to emphasize that it is the cross-sectional velocity, from a centreline to a solid wall. At a solid wall, $r = R$ and the velocity $u = 0$. The pressure gradient dp/dx is in the longitudinal direction (along the pipe). The law applies to incompressible laminar flow in pipes of uniform cross-section. The friction factor is given by the relationship

$$f = \frac{64}{\text{Re}}$$

The fundamental laws of heat, mass and momentum, can be extended to turbulent flow by the use of semi-empirical correlations. *Darcy's Law* and *Poiseuille's Law* are only applicable for low velocity flow in porous media and laminar flow in pipes. High-velocity flows in porous media and fractures have been studied by Skjetne (1995) and Kløv (2000), respectively.

Based on the concept of *mixing length*, the wall shear stress in turbulent flow can be expressed by the equation

$$\tau = (\mu + \rho\varepsilon)\frac{du}{dx} = (\mu + \eta)\frac{du}{dx}$$

where ε m^2/s is eddy diffusivity and η Pa · s is eddy viscosity. The equation is based on the movement (transfer) of turbulent eddies. While eddy diffusivity is a property analogous to molecular diffusivity, eddy viscosity is not analogous to dynamic viscosity, because it depends on fluid density, mixing length and velocity gradient. A semi-empirical expression of eddy diffusivity (and hence eddy viscosity) is used to obtain the universal velocity profile (see *Appendix H – Universal Velocity Profile*).

Based on the concept of *turbulent diffusivity*, the transfer of mass in turbulent flow can be expressed by the equation

$$J = -(D + D_\varepsilon)\frac{dc}{dx}$$

where D_ε m²/s is turbulent diffusivity. The turbulent diffusivity depends on the nature of the turbulent flow causing mass transfer and must be derived from experiments. In pipe flow, the Sherwood number is commonly used to correlate turbulent mass transfer

$$Sh = \frac{hd}{D}$$

where h m/s is the convective *mass transfer* coefficient, d m is pipe diameter and D m²/s the molecular diffusivity. The turbulent diffusivity in pipe flow can therefore be expressed by

$$D_\varepsilon = hd = DSh$$

Based on the concept of *heat transfer* coefficient, the transfer of heat in turbulent flow can be expressed by the equation

$$q = -(k + k_\varepsilon)A\frac{dT}{dx}$$

where k_ε W/m · K is the turbulent thermal conductivity. Analogous to mass transfer in turbulent pipe flow, the Nusselt number

$$Nu = \frac{hd}{k}$$

is used in heat transfer. The symbol h W/m² · K represents the convective heat transfer coefficient (same symbol as the mass transfer coefficient). The following can be written

$$k_\varepsilon = hd = kNu$$

The three *dimensionless* numbers that are used to correlate heat transfer, mass transfer and momentum transfer are respectively the Nusselt number, the Sherwood number and the friction factor, as in the Darcy-Weisbach Equation (*Appendix D*). The friction factor is proportional to the pressure loss coefficient K (see *Appendix X3 – Energy Dissipation and Bubble Diameter*). The friction factor is also proportional to the Euler number, given by the ratio of pressure forces over inertial forces.

$$Eu = \frac{\Delta \rho}{\rho u^2}$$

Friction factor of structured deposits

The friction factor in pipes depends on the Reynolds number and the relative roughness. The structure of the roughness is also an important consideration. For example, the friction factor for uniformly distributed roughness will be different from a structured deposit such as transverse ripples.

Pressure drop in pipes can be calculated from the *Darcy-Weisbach Equation*

$$\Delta p_f = \frac{f}{2}\frac{L}{d}\rho u^2$$

where f is the friction factor for laminar and turbulent flow. In turbulent flow, the friction factor is based on experimental data. Nikuradse (1933) carried out experiments on sand grain roughness in pipes and proposed the following equation for the roughness function

$$B = \sqrt{\frac{8}{f}} + 2.5\ln\frac{k}{r} + 3.75$$

where k is the roughness height and r the pipe radius. Several correlations have been developed based on Nikuradse's experiments and a wide range of other experiments and tests on pipelines. The implicit Colebrook-White correlation is probably the most widely used. A more convenient correlation is the explicit Haaland (1983) correlation

$$\frac{1}{\sqrt{f}} = -\frac{1.8}{n}\log\left[\left(\frac{6.9}{\mathrm{Re}}\right)^n + \left(\frac{k}{3.75d}\right)^{1.11n}\right]$$

where Re is the Reynolds number, d the pipe diameter and n an exponent $= 1$ for liquids and $= 3$ for gases. The friction factor in liquid pipelines is not exactly the same as in gas pipelines.

Sletfjerding (1999) carried out test on pipelines with a wide range of sand grain roughness and high Reynolds number, much higher than used as basis for the Colebrook-White correlation. The tests confirmed the general applicability of the above relationships (Haaland correlation).

Sand grain roughness is a convenient representation of the real roughness of a pipe wall surface. A measured average roughness of $35\,\mu m$ is not necessarily equivalent to

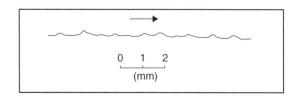

Figure F.1 Cross-section of a transverse rippled geothermal silica deposit. Arrow shows direction of flow.

35 μm sand grain roughness. In practice, therefore, each pipe surface is assigned a sand grain roughness that will give the same pressure drop as the real average roughness.

The work of Nikuradse (1933) and the friction factor correlations found in the literature, including the Colebrook-White and Haaland correlations, are based on pipe wall surfaces with uniform roughness. But structured deposits are also found in pipes. Bott and Gudmundsson (1978) reported rippled deposits in pipes carrying geothermal water laden with silica. Similar results for other systems were reported by Gudmundsson (1977). The cross-section of the rippled silica deposit is shown in Figure F.1.

The Fanning friction factor against Reynolds number was plotted by Bott and Gudmundsson (1978), shown in Figure F.2. The solid line shows the friction factor for a smooth commercial pipe surface. The dashed line shows the friction factor in the same kind of stainless steel pipe with a rippled deposit (geothermal silica). The Fanning friction factor is not to be confused with the friction factor used in the Darcy-Weisbach Equation (*Appendix D*). They are different by a factor of four

$$f_{Darcy} = f_{Moody} = 4f_{Fanning}$$

$$f_D = f_M = 4f_F$$

The common friction factor in pipes is also called the Moody friction factor, originating from an early compilation of pressure drop data in pipelines, presented in the well-known Moody diagram.

The roughness function B (also called friction function) plotted in Figure F.3 along with the roughness function for the sand grain roughness (Nikuradse, 1933), against dimensionless roughness height

$$k^+ = \frac{\rho u^* k}{\mu}$$

where u^* is the friction velocity and k the roughness height. The friction velocity is also called the shear stress velocity or wall friction velocity (see *Appendix H – Universal Velocity Profile*). The dimensionless height has the same kind of variables as the Reynolds number. The roughness function for rippled roughness was equal to about $B = 3.0$ while for sand grain roughness was equal to about $B = 8.5$. These values are for fully rough flow where the friction factor is independent of Reynolds number.

The average ripple height was measured 0.123 mm (Bott and Gudmundsson, 1978). Assuming sand grain roughness, the friction factor was calculated 0.0416 while

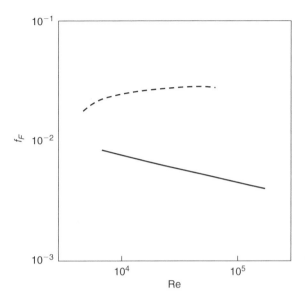

Figure F.2 Fanning friction factor for a clean stainless steel tube (solid line) and the same kind of tube with a rippled silica deposit (dashed line).

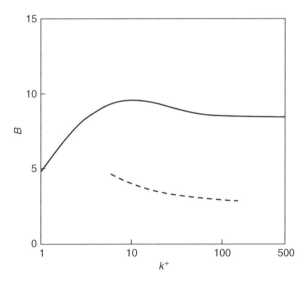

Figure F.3 Roughness function B for sand grain roughness (solid line) and rippled deposit (dashed line) against dimensionless roughness height k^+.

the measured friction factor was 0.112. In other words, the rippled surface friction factor was 2.7-times larger. The Fanning friction was used in the original paper and the figures above, being one-quarter as large as the Moody friction factor used in the Darcy-Weisbach equation. The two friction factor values immediately above, have

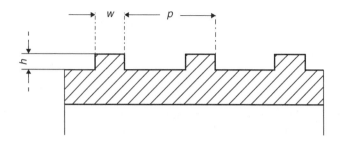

Figure F.4 Geometry of rectangular transverse roughness elements. Height *h*, width *w* and pitch *p*.

been converted from Fanning to Moody (same as Darcy-Weisbach friction factor). To generalize, the friction factor of rippled surfaces can be 2–3 times higher than the friction factor of sand grain surfaces. The pressure drop in a pipeline is directly proportional to the friction factor. The structure of rippled surfaces is shaped by natural forces, akin to ripples on the beach and in the desert.

Baumann and Rehme (1975) studied the roughness function for rectangular transverse roughness elements. Figure F.4 shows the geometry studied.

Rectangular roughness can be used to illustrate how the height *h*, width *w* and pitch *p* of transverse structured deposits affect the friction factor in pipeline flow. Baumann and Rehme (1975), based on a wide range of experimental data, suggested the following correlation for the roughness function

$$B = a_1 \left(\frac{p}{h} \right)^{a_2} + a_3 \left(\frac{p}{h} \right)^{a_4}$$

$$a_1 = 18.5 \left(\frac{h}{w} \right)^{-0.9475}$$

$$a_2 = -1.143 \left(\frac{h}{w} \right)^{-0.147}$$

$$a_3 = 0.33 \left(\frac{h}{w} \right)^{0.1483}$$

$$a_4 = 0.758 \left(\frac{h}{w} \right)^{-0.11}$$

The correlation is valid for the following ranges of the parameters

$$1 \le \frac{p}{h} \le 40$$

$$0.3 \le \frac{h}{w} \le 8$$

$$B \le 10$$

In the Bott and Gudmundsson (1978) experiments, the average ripple height was $h = 0.123$ mm, the average ripple pitch was $p = 0.87$ mm, both with a standard deviation of about 33%. If the width of the ripples is assumed one-half the height $w = h/2$, the following values will apply

$$\frac{p}{h} = 7$$

$$\frac{h}{w} = 2$$

which is within the limiting values of Baumann and Rehme (1975). Based on these values the roughness function was calculated to be $B = 2.7$. This value is not far off from the experimental value of $B = 3.0$. Therefore, it appears that the rectangular transverse roughness correlation can be used to model rippled deposits in pipelines.

Pressure drop in gas pipelines and wells

The *momentum equation* for steady-state, isothermal and one-dimensional flow (compressible and incompressible) in pipes and pipelines can be written as

$$\frac{d(\rho u^2)}{dx} = -\frac{dp}{dx} - \frac{f\rho u^2}{2d} - \rho g \sin \alpha$$

The term on the left-hand-side represents the pressure drop due to acceleration, the first term on the right-hand-side represents the total pressure drop, the second term the pressure drop due to wall friction and the third term the pressure drop due to hydrostatic head. The pressure decreases with distance, hence the minus terms. In general, compressible flow means *gas flow* and incompressible flow means *liquid flow*. The angel α is referenced to horizontal.

From differential calculus, the acceleration term can be written

$$d(\rho u^2) = d(\rho u \cdot u) = \rho u du + u d(\rho u)$$

Because the mass flux ρu is constant (conserved) the following applies

$$d(\rho u^2) = \rho u du$$

and the momentum equation can be written

$$\rho u \frac{du}{dx} = -\frac{dp}{dx} - \frac{f\rho u^2}{2d} - \rho g \sin \alpha$$

COMPRESSIBLE FLOW

Based on the real gas law, the density of natural gas can be calculated from

$$\rho = \frac{pM}{zRT}$$

Velocity is given by

$$u = \frac{q}{A} = \frac{m}{\rho A}$$

Setting in for ρ and ρu in the momentum equation gives

$$\frac{m}{A}\frac{du}{dx} = -\frac{dp}{dx} - \frac{fmu}{2Ad} - \frac{pM}{zRT}g\sin\alpha$$

The velocity can also be expressed by

$$u = \frac{m}{A}\frac{zRT}{pM}$$

such that

$$du = -\frac{m}{A}\frac{zRT}{M}\frac{1}{p^2}dp$$

Setting in for the velocity results in

$$-\frac{m^2zRT}{A^2Mp^2}\frac{dp}{dx} = -\frac{dp}{dx} - \frac{fm^2}{2A^2d}\frac{zRT}{pM} - \frac{pM}{zRT}g\sin\alpha$$

and

$$\frac{dp}{dx} - \frac{m^2zRT}{A^2Mp^2}\frac{dp}{dx} = \frac{dp}{dx}\left(1 - \frac{m^2zRT}{A^2Mp^2}\right) = -\frac{fm^2}{2A^2d}\frac{zRT}{pM} - \frac{pM}{zRT}g\sin\alpha$$

The differential equation can be reorganized to give

$$dx = \left[\frac{1 - \dfrac{m^2zRT}{A^2Mp^2}}{-\dfrac{fm^2zRT}{2A^2d \cdot pM} - \dfrac{pMg\sin\alpha}{zRT}}\right]dp$$

This equation contains all the parts necessary to calculated pressure drop in gas pipelines and wells, taking into account wall friction, hydrostatic head and acceleration. The equation can be integrated (see below) but there is a discontinuity (singularity) in the solution for horizontal flow. However, the differential form does not have that problem.

HORIZONTAL GAS PIPELINE

In a horizontal pipeline $\sin\alpha = 0$ such that the differential form of the equation can be written

$$dx = \left[\frac{1 - \dfrac{m^2zRT}{A^2Mp^2}}{-\dfrac{fm^2zRT}{2A^2d \cdot pM}}\right]dp$$

To get a simpler view of the integration, the equation can be written

$$dx = \left[\frac{1 - \dfrac{a}{p^2}}{-\dfrac{b}{p}} \right] dp$$

or

$$dx = \left[\frac{a}{b}\frac{1}{p} - \frac{p}{b} \right] dp$$

In terms of integrals

$$\frac{a}{b}\int \frac{1}{p}dp - \frac{1}{b}\int pdp = \int dx$$

such that

$$\frac{a}{b}\left[\ln \frac{p_2}{p_1} \right] - \frac{1}{2b}[p_2^2 - p_1^2] = L$$

where L is the length of the pipeline or pipeline segment. Setting in the real variables gives

$$\frac{2d}{f}\left[\ln \frac{p_2}{p_1} \right] - \frac{A^2 d}{fm^2}\frac{M}{zRT}[p_2^2 - p_1^2] - L = 0$$

or

$$\frac{dA^2 M}{fm^2 zRT}(p_2^2 - p_1^2) - \frac{d}{f}\ln\left(\frac{p_2^2}{p_1^2} \right) + L = 0$$

Inlet pressure is p_1 and outlet pressure is p_2. The first term is usually much larger than the second term. Therefore, the following approximation may give an accurate enough result

$$\frac{dA^2 M}{fm^2 zRT}(p_2^2 - p_1^2) + L \cong 0$$

The logarithmic term can be written

$$\ln\left(\frac{p_2^2}{p_1^2} \right) = \ln\left(\frac{p_1}{p_2} \right)^{-2} = -2\ln\left(\frac{p_1}{p_2} \right)$$

AVERAGE GAS DENSITY EQUATION

The terms with the pressures squared above can be written

$$(p_2^2 - p_1^2) = (p_2 + p_1)(p_2 - p_1)$$

The density of gas is given by the equation

$$\rho = \frac{pM}{zRT}$$

The *average* density of gas is given by the equation

$$\overline{\rho} = \frac{\overline{p}M}{\overline{z}R\overline{T}}$$

where the average pressure is given by

$$\overline{p} = \frac{p_1 + p_2}{2}$$

By algebra, it can be shown when using average gas density, that the pressure drop can be calculated from the equation

$$p_1 - p_2 = \frac{\overline{f}}{2} \frac{L}{d} \overline{\rho} u^2$$

which is the *Darcy-Weisbach Equation* for incompressible fluid flow in pipes. An average friction factor and an average gas density are used.

PRESSURE IN STATIC GAS WELL

The differential form of the equation derived above, for pressure drop in natural gas pipelines and wells, can be simplified to express the pressure in a static gas well (non-flowing well or vertical risers). By setting the mass flowrate $m = 0$ the equation can be simplified to give

$$dx = \left[\frac{-1}{\dfrac{pMg \sin \alpha}{zRT}} \right] dp$$

Integration gives

$$p_2 = p_1 \exp\left[\frac{-Mg \sin \alpha}{\overline{z}R\overline{T}} L \right]$$

Average values for z and T need to be used. The inlet static pressure is p_1 and the outlet static pressure is p_2. For an uphill pipeline, $p_2 < p_1$ and for a downhill pipeline,

$p_2 > p_1$. For a natural gas wellbore (production tubing), p_1 is the wellhead pressure and p_2 the downhole pressure.

FRICTIONAL AND HYDROSTATIC PRESSURE DROP

In the general *momentum equation* given at the start of this appendix

$$\rho u \frac{du}{dx} = -\frac{dp}{dx} - \frac{f \rho u^2}{2d} - \rho g \sin \alpha$$

the gas density and average velocity can be expressed by

$$\rho = \frac{pM}{zRT}$$

$$u = \frac{m}{\rho A}$$

Setting in for ρ and ρu gives

$$\frac{m}{A} \frac{du}{dx} = -\frac{dp}{dx} - \frac{fmu}{2Ad} - \frac{pM}{zRT} g \sin \alpha$$

In natural gas wells, the frictional and hydrostatic pressure drops dominate. Therefore, pressure drop due to acceleration can be ignored. The pressure drop due to acceleration in horizontal pipelines is also negligible, compared to frictional pressure drop. Simplification gives

$$\frac{dp}{dx} = -\frac{fmu}{2Ad} - \frac{pM}{zRT} g \sin \alpha$$

Defining

$$a_1 = \frac{M}{zRT}$$

$$b_1 = \frac{fm^2}{2A^2 d}$$

means that

$$\rho = a_1 p$$

and

$$u = \frac{m}{a_1 pA}$$

The pressure drop equation (frictional and hydrostatic) can thus be written as

$$\frac{dp}{dx} = -\frac{b_1}{a_1}\frac{1}{p} - (a_1 g \sin \alpha)p$$

and

$$p\,dp = -\left[\frac{b_1}{a_1} + (a_1 g \sin \alpha)p^2\right]dx$$

Defining two new variables, a_2 and b_2, gives

$$\int_{p_1}^{p_2} \frac{p}{(a_2 + b_2 p^2)}\,dp = -\int_0^L dx$$

Integration and rearrangement gives

$$\left[\frac{1}{2b_2}\ln(a_2 + b_2 p^2)\right]_{p_1}^{p_2} = -L$$

$$\frac{1}{2b_2}\left[\ln(a_2 + b_2 p_2^2) - \ln(a_2 + b_2 p_1^2)\right] = -L$$

$$\ln\left[\frac{a_2 + b_2 p_2^2}{a_2 + b_2 p_1^2}\right] = -2b_2 L$$

$$\left[\frac{a_2 + b_2 p_2^2}{a_2 + b_2 p_1^2}\right] = \exp(-2b_2 L)$$

$$b_2 p_2^2 = (a_2 + b_2 p_1^2)\exp(-2b_2 L) - a_2$$

$$p_2^2 = \frac{a_2}{b_2}\exp(-2b_2 L) + p_1^2 \exp(-2b_2 L) - \frac{a_2}{b_2}$$

$$p_2^2 = p_1^2 \exp(-2b_2 L) - \frac{a_2}{b_2}\left[1 - \exp(-2b_2 L)\right]$$

The variables are

$$b_2 = a_1 g \sin \alpha$$

$$a_2 = \frac{b_1}{a_1}$$

Using the variable without subscript, the equation can be generalized and written as follows

$$p_2^2 = p_1^2 \exp(-2ag \sin \alpha L) - \frac{b}{a^2 g \sin \alpha}[1 - \exp(-2ag \sin \alpha L)]$$

where

$$a = \frac{M}{zRT}$$

$$b = \frac{fm^2}{2A^2d}$$

The equation expresses pressure drop in gas pipelines and wells due to wall friction and hydrostatic head. The variable a includes gas properties while the variable b includes fluid flow properties.

The equation can be tested by assuming a static gas well where $m = 0$ which gives

$$p_2^2 = p_1^2 \exp(-2ag \sin \alpha L)$$

and thus

$$p_2 = p_1 \exp(-ag \sin \alpha L)$$

which is the same equation that expresses hydrostatic pressure drop with depth in a gas well (see above). The angel α is measured from the horizontal on the natural trigonometric scale. Meaning that $\alpha = +90$ for a vertical well such that p_1 is the bottomhole pressure and p_2 the wellhead pressure. In the variable a, the average values of z and T from wellhead to depth L need to be used.

Appendix H

Universal velocity profile

The shear stress in fluid flow has the unit force per unit area N/m^2. In laminar flow the shear stress is given by *Newton's Law of Viscosity*

$$\tau_\mu = \mu \frac{du}{dy}$$

where μ Pa $\cdot$ s $=$ kg/s $\cdot$ m is the dynamic viscosity and du/dy the velocity gradient away from the wall. In laminar flow, the wall shear stress is equal to the shear stress.

The turbulence model derived below, is the classical *mixing-length* model. The model gives the velocity gradient away from a pipe wall in turbulent flow. This perpendicular velocity gradient determines the longitudinal pressure drop in turbulent pipe flow. It relates directly to momentum transport (see *Appendix E – Transfer Equation*). The *analytical* mixing-length model has been extended/replaced by the *numerical* k-ε model, which is widely used in computational fluid dynamics (CFD).

In turbulent flow, there is continuous mixing of fluid eddies. Assuming two-dimensional flow where U represents the instantaneous velocity in the longitudinal direction, x and V the instantaneous velocity in the transverse direction y, normal to the wall, the following can be written

$$U = u + u'$$

$$V = v + v'$$

The lower-case letters u and v represent the time averaged velocities and the corresponding primed letters represent the instantaneous velocity fluctuations. The fluctuations can be relatively large, typically 10% of the averages.

In turbulent flow, the random movement of eddies causes a continuous exchange of momentum within the fluid. The momentum exchange gives rise to turbulent shear stress, also called Reynolds stress. The turbulent shear stress in the fluid is given by the expression

$$\tau_\varepsilon = -\rho \overline{u'v'}$$

derived from the fundamental *Navier-Stokes Equations* with Reynolds averaging.

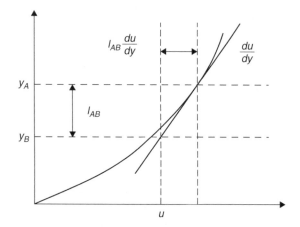

Figure H.1 Mixing length in turbulent flow, illustrated by velocity transverse to the longitudinal flow direction.

In turbulent pipe flow, the fully turbulent core occupies more than 99% of the pipe cross-sectional area. At the pipe wall, there is a very thin viscous sub-layer and a buffer-layer (between the viscous sub-layer and the fully turbulent core). The wall shear stress in turbulent flow depends on both the shear stress in the viscous sub-layer and the shear stress in the turbulent core.

The linear viscosity law (see above) can be used in calculations for viscous shear stress, but the turbulent shear stress expression needs some reformulation for practical calculations. The *mixing length theory* developed by Ludwig Prandtl (1875–1953), can be used for this purpose. The mixing length is analogous to the mean-free-path in the kinetic theory of gas molecules.

Imagine a fluid eddy, at some distance y_A from the pipe wall, that moves to some distance y_B closer to the wall. The distance y_A-y_B is defined as the mixing length l_{AB}. The longitudinal velocity at location A is greater than the longitudinal velocity at location B. The momentum (higher fluid velocity) carried from location A will increase the instantaneous longitudinal (x-direction) velocity at location B and the following can be written

$$u' = u_A - u_B = l_{AB}\frac{du}{dy}$$

At the new location, the momentum exchange results in the turbulent eddy slowing down and the longitudinal velocity increasing. The velocity fluctuation acts in the longitudinal direction x, the longitudinal velocity gradient is expressed in the transverse direction y. The mixing length concept is illustrated in Figure H.1.

The velocity fluctuation in the transverse direction

$$v' = v_A - v_B = l_{AB}\frac{du}{dy}$$

is assumed equal in magnitude and is given by the mixing length and the velocity gradient in the transverse direction also. Whether an eddy moves towards or away from the wall the mixing length, is assumed to have a positive value in both cases. The turbulent shear stress equation can thus be written as

$$\tau_\varepsilon = -\rho\overline{u'v'} = \rho l_{AB}^2 \left(\frac{du}{dy}\right)^2$$

Defining an eddy diffusivity as

$$\varepsilon = l_{AB}^2 \frac{du}{dy}$$

the turbulent shear equation can be written

$$\tau_\varepsilon = \rho\varepsilon \frac{du}{dy}$$

The eddy diffusivity ε m^2/s has the same unit as molecular diffusivity.

The total wall shear stress in turbulent flow, due to the viscous sub-layer and the turbulent core, can be expressed by adding the two equations

$$\tau_w = (\mu + \rho\varepsilon)\frac{du}{dy}$$

The product of fluid density and eddy diffusivity can be expressed in terms of an eddy viscosity η Pa $\cdot$ s $=$ kg/s $\cdot$ m such that

$$\tau_w = (\mu + \eta)\frac{du}{dy}$$

The eddy viscosity is not a fluid property such as dynamic viscosity because it depends on the fluid density, the mixing length and the velocity gradient. Experiments are required to determine the eddy viscosity. For turbulent flow in pipes, such experiments have been carried out in terms of pressure drop and velocity profile with distance from the wall. The universal velocity profile is one of the results derived from such experiments.

The *mixing length theory* by Ludwig Prandtl (1875–1953) can be found in numerous books on fluid mechanics. The relationship between the mixing length and the velocity profile in turbulent flow, was suggested by Theodore von Kármán (1881–1963) to follow the differential equation

$$l_{AB} = \kappa \frac{du/dy}{d^2u/dy^2}$$

where κ is a universal constant regardless of boundary conditions (wall bounded flow and free jet flow). The value of the von Kármán constant is about 0.4 for turbulent flow in pipes, based on experiments.

Consider flow over a smooth surface such as a pipe wall. Assume a thin film/layer at the wall with thickness $y = \delta$, where the fluid dynamic viscosity predominates, meaning that $\eta \ll \mu$. The wall shear stress equation can be integrated to give

$$\frac{\tau_w}{\rho} = \frac{\mu}{\rho}\frac{u}{y}$$

because the velocity is zero at the wall. The equation holds for $y \leq \delta$. Defining the *friction velocity* (also called the shear stress velocity or wall friction velocity)

$$u^* = \sqrt{\frac{\tau_w}{\rho}}$$

the following can be written

$$\frac{u}{u^*} = \frac{\rho u^* y}{\mu}$$

or

$$u^+ = y^+$$

This is a dimensionless equation and shows a linear relationship between u and y. It represents the viscous sub-layer closes to the wall in pipe flow. The dimensionless distance has the same/similar variable as the Reynolds number.

Away from the wall, where $y > \delta$ it can be assumed that the eddy viscosity predominates, meaning that $\eta \gg \mu$. The mixing length l_{AB} has the dimension of length; the distance from the wall y has the dimension of length. These are the two significant linear dimensions such that the following can be assumed

$$l = \kappa y$$

The subscript AB is dropped, because the assumption suggests generality. Substitution into the wall shear stress equation gives

$$\sqrt{\frac{\tau_w}{\rho}} = \kappa y \frac{du}{dy}$$

and therefore

$$\frac{du}{u^*} = \frac{1}{\kappa}\frac{dy}{y}$$

Integration gives

$$\frac{u}{u^*} = \frac{1}{\kappa}\ln y + C_1$$

To relate the dimensional velocity to the dimensionless distance, assume that the constant C_1 can be expressed by

$$C_1 = C_2 - \frac{1}{\kappa} \ln \frac{\mu}{\rho u^*}$$

where C_2 is a second constant. Therefore, the dimensionless velocity will be given by the expression

$$\frac{u}{u^*} = \frac{1}{\kappa} \ln \frac{\rho u^* y}{\mu} + C_2$$

or

$$u^+ = \frac{1}{\kappa} \ln y^+ + C_2$$

Based on experimental data where u^+ is plotted against $\ln(y^+)$ the gradient gives κ and the intercept C_2. In the fluid mechanics literature, the symbol B is commonly used for the constant C_2. As already mentioned, for turbulent flow in general, the von Kármán constant $\kappa = 0.4$, while the constant in smooth pipe flow assumes a value of 5.5 giving

$$u^+ = 2.5 \ln y^+ + 5.5$$

The equation expresses the velocity profile in the core of turbulent flow in a smooth pipe (hydraulically smooth). The values $\kappa = 0.41$ and $B = 5.0$ are now considered more correct. The constant B will have a different value for turbulent flow with a rough wall. It is called the *roughness function* is typically used to find the correct value for different wall roughness (see *Appendix F – Friction Factor of Structured Deposits*).

The mixing length theory was derived for two-dimensional flow, for example between two flat plates. Flow in pipes is three-dimensional. However, through experimental work the constants for fully developed turbulent pipe flow can be found.

Between the laminar sub-layer and the turbulent core there is a buffer layer. The velocity profile in the buffer layer is commonly expressed by the same functional form as the turbulent core, but with different constants

$$u^+ = 5.00 \ln y^+ - 3.05$$

This equation for the buffer layer is an approximation. The dimensionless velocity profile equations are found in most books on fluid mechanics and related works (McCabe *et al.*, 1993). The equations have the following ranges of applicability:

Viscous sub-layer	$y^+ < 5$
Buffer layer	$5 < y^+ < 30$
Turbulent core	$y^+ > 30$

A single correlation by Spalding (1961) covers the entire range from the wall to fully turbulent flow

$$y^+ = u^+ + \exp(-\kappa B)\left[\exp(\kappa B) - 1 - \kappa u^+ - \frac{(\kappa u^+)^2}{2} - \frac{(\kappa u^+)^3}{6}\right]$$

$$y^+ = y^+ + 0.1108\left[\exp(0.4u^*) - 1 - 0.4u^* - \frac{(0.4u^*)^2}{2} - \frac{(0.4u^*)^3}{6} - \cdots\right]$$

from the wall to about $y^+ \simeq 100$. The fourth-degree term is not shown in the above equation. Other correlations are to be found in the literature, one of which is that of van Driest (1956).

Considering the analogy between momentum, heat and mass transfer, the following can be written for convective (turbulent) heat flux

$$Q_\varepsilon = -\rho C_p \overline{v'T'}$$

and

$$Q = -(k + k_\varepsilon)\frac{dT}{dy}$$

Similarly, for convective mass transfer the following can be written

$$J_\varepsilon = -\overline{v'c'} = -D_\varepsilon\frac{dc}{dy}$$

such that

$$J = -(D + D_\varepsilon)\frac{dc}{dy}$$

Particle velocity and stopping distance

The gravity force acting on a particle that accelerates through a fluid is

$$F_g = M_p g$$

where M_p stands for particle mass, and g represents the gravitational constant. The opposite buoyancy force between the particle and the surrounding fluid is

$$F_b = M_f g$$

where M_f stands for fluid mass. The frictional drag force on the accelerating particle is

$$F_f = \frac{1}{2} f_D A_p \rho_f u_p^2$$

where f_D is a drag coefficient (friction coefficient), A_p is the aerodynamic area and ρ_f the fluid density (not the particle density).

The force balance on an accelerating particle can be written

$$M_p \frac{du_p}{dt} = F_g - F_b - F_f$$

The term on the left-hand-side is the inertial force.

Assuming a spherical particle of diameter d_p

$$V_p = \frac{\pi d_p^3}{6}$$

$$A_p = \frac{\pi d_p^2}{4}$$

$$M_p = \rho_p V_p$$

$$M_f = \rho_f V_p$$

The drag coefficient f_D is not the same as the friction factor in laminar pipe flow. At low velocities where *Stoke's Law* applies

$$f_D = \frac{24}{Re_p}$$

where Re_p is the *particle* Reynolds number

$$Re_p = \frac{\rho_f u_p d_p}{\mu_f}$$

based on fluid density (not particle density) and fluid viscosity μ_f.

Rearranging the force balance equation leads to

$$\frac{du_p}{dt} + \frac{18\mu_f}{\rho_p d_p^2} u_p = \left(\frac{\rho_p - \rho_f}{\rho_p}\right) g$$

The above is a first order linear ordinary differential equation, with a characteristic time, τ, called particle relaxation time

$$\tau \equiv \frac{\rho_p d_p^2}{18\mu_f}$$

The general solution of the above equation is

$$u_p(t) = C \exp\left(\frac{-t}{\tau}\right) + \left(\frac{\rho_p - \rho_f}{\rho_p}\right) \tau g$$

where C is a constant that can be evaluated based on initial conditions. The right hand term is the *terminal settling velocity* of particles due to gravity (constant for a given particle-fluid system), obtained when $t \gg \tau$, commonly written

$$u_{TS} = \frac{gd_p^2}{18\mu_f} (\rho_p - \rho_f)$$

A *first* practical case can be considered where a particle is initially at rest and accelerates due to gravity. The solution to the above first order linear ordinary differential equation considering the particle initially at rest ($u_p(t=0)=0$) leads to

$$u_p(t) = u_{TS} \left(1 - \exp\left(\frac{-t}{\tau}\right)\right)$$

giving the particle velocity u_p m/s at any time t s. At large times the exponential term approaches zero and the particle velocity becomes the terminal settling velocity. In practical engineering situations, a particle will reach its terminal settling velocity rapidly, as in gas-liquid separation calculations.

A *second* practical case can be considered where a particle decelerates from an initial velocity u_i to a final velocity, u_∞, due to frictional drag forces. In such a case the gravitational and buoyancy forces can be considered negligible. The first order linear ordinary differential equation becomes

$$\frac{du_p(t)}{dt} + \frac{1}{\tau} u_p(t) \simeq 0$$

which has the following general solution

$$u(t) = u_i - (u_i - u_\infty)\left[1 - \exp\left(\frac{-t}{\tau}\right)\right]$$

The general solution can also be written

$$u(t) = u_\infty + (u_i - u_\infty)\exp\left(\frac{-t}{\tau}\right)$$

Writing the velocity in terms of distance with time gives

$$\frac{dx(t)}{dt} = u_\infty - (u_\infty - u_i)\exp\left(\frac{-t}{\tau}\right)$$

and

$$\int_0^x dx = \int_0^t u_\infty dt - \int_0^t (u_\infty - u_i)\exp\left(\frac{-t}{\tau}\right) dt$$

Integration gives

$$x(t) = u_\infty t - (u_\infty - u_i)(-\tau)\left[\exp\left(\frac{-t}{\tau}\right) - 1\right]$$

When $t \gg \tau$ the final velocity $u_\infty \simeq 0$, the exponential term becomes negligible and the result becomes

$$x = \lambda = \tau u_i$$

where λ is the particle stopping distance.

Diffusion coefficient

The diffusion coefficient is used in *Fick's Law*. The coefficient (also called diffusivity) is needed to calculate molar flux (see *Appendix E – Transfer Equations*) from the equation

$$J = -D\frac{dc}{dx}$$

The units are flux J mol/s $\cdot$ m^2, diffusivity D m^2/s, concentration c mol/m^3 and distance x m. The concentration can also be expressed in terms of kg/m^3 such that the mole flux becomes mass flux kg/s $\cdot$ m^2. The diffusion coefficient has commonly a subscript D_{AB} to indicate the diffusion of solute molecule A in solvent B.

A number of correlations are found in the literature for the diffusion coefficient of molecules in hydrocarbon mixtures. Some of these have been reviewed by Gudmundsson (1992). Typical values of the diffusion coefficient (D m^2/s) of hydrocarbons are in the range

$$0.1 \cdot 10^{-9} < D < 10 \cdot 10^{-9}$$

A fundamental equation for the diffusivity of solute molecule in solvent is the *Stokes-Einstein Equation*

$$D = \frac{k_B T}{3\pi \mu d}$$

where d m is the solute hydrodynamic diameter, μ the solvent viscosity and T the absolute temperature. The equation contains the Boltzmann constant (Ludwig Boltzmann, 1844–1906), with the value $1.38 \cdot 10^{-23}$ J/K. It has been reported by Olesik and Woodruff (1991) that diffusion of solutes in liquid hydrocarbons is *not* well described by the *Stokes-Einstein Equation*. The equation appears to work better for high viscosity solvents than low viscosity solvents.

Several correlations for the prediction of diffusion coefficient in nonpolar solvents were studied by Umesi and Danner (1981). Based on functional form of the

Stokes-Einstein Equation and values from the literature, they proposed the semi-empirical correlation

$$D = 2.75 \cdot 10^{-12} \frac{T}{\mu} \left[\frac{r_B}{r_A^{2/3}} \right]$$

The units used were D m²/s, T K and μ mPa·s. The symbols r_A Å and r_B Å stand for the radius of gyration of the solute and solvent molecules, respectively. The radius of gyration of molecules can be found in the literature, for example Yaws (2010). It expresses how the mass of a molecule is distributed around its centre of rotation. Methane has gyration radius of 1.1 Å and hexane 1.8 Å.

For illustration purposes, assume that the solvent has a radius of gyration of 10 Å and the solute a radius of gyration twice that value; that is, 20 Å. Solvent viscosity 0.5 mPa·s (=cp) and temperature 20°C (=293 K). From the above correlation, the diffusivity coefficient becomes $2.2 \cdot 10^{-9}$ m²/s. This value is within the range expected for hydrocarbons. Similarly, for a 2 Å solvent and a 4 Å solute, the diffusivity coefficient is $1.3 \cdot 10^{-9}$ m²/s.

> The symbol Å stands for the unit Ångström, equivalent to 10^{-10} m. As often in science and technology, fundamental units are defined to be close to unity, at the scale of investigation; the diameter of common molecules is of-the-order-of one Ångström (Anders J. Ångström, 1814–1874). The diameter of light hydrocarbons that can form hydrates must be less than 6–7 Å, which is the size of clathrate cavities. Methane 3.9 Å in diameter and hexane 5.9 Å in diameter are hydrate formers.

Applying the *Stokes-Einstein Equation* to methane (diameter 3.9 Å) and the same liquid viscosity and temperature as above, the theoretical diffusivity coefficient becomes $2.2 \cdot 10^{-9}$ m²/s. This value is coincidentally the same as that obtained from the semi-empirical correlation above for solvent and solute having a gyration radius of 10 Å and 20 Å, respectively.

While the *Stokes-Einstein Equation* is based on particles in gases, its functional form can be used for large molecules and particles in liquids systems by writing

$$D = C_1 \frac{T}{\mu}$$

where C_1 is an experimental constant. The diffusivity of paraffin wax in crude oil at ambient temperature has been measured and reported by Burger *et al.* (1981) as

$$D = 2.4 \cdot 10^{-9} \frac{1}{\mu}$$

in terms of the units D m²/s and μ mPa·s. (In the original paper, the units used were different). Note that the absolute temperature is omitted in this semi-empirical relationship. For an oil having viscosity of 0.5 mPa·s, the diffusion coefficient becomes $4.8 \cdot 10^{-9}$ m²/s. Again, this value is within the range expected for hydrocarbons.

Diffusivity values often used in the petroleum industry, include the correlations by Wilke and Chang (1955) and Hayduk and Minhas (1982).

Deposition-release models

EXPONENTIAL MODEL

The increase in deposit thickness with time can be described by the equation

$$\frac{dx}{dt} = k_1 - k_2 x$$

where k_1 and k_2 are constants coefficients or functions. Rearranging to

$$\frac{1}{(k_1 - k_2 x)} dx = dt$$

results in the integrals

$$\int_0^x \frac{1}{(k_1 - k_2 x)} dx = \int_0^t dt$$

Integration gives

$$\frac{-\ln(k_1 - k_2 x) - \ln(k_1)}{k_2} = t$$

and

$$k_1 - k_2 x = \exp[\ln k_1 - k_2 t]$$

In terms of deposit thickness

$$x = \frac{1}{k_2}[k_1 - \exp(\ln k_1 - k_2 t)]$$

and

$$x = \frac{1}{k_2}\left[k_1 - \frac{\exp(\ln k_1)}{\exp(k_2 t)}\right]$$

The main result becomes

$$x = \frac{k_1}{k_2}[1 - \exp(-k_2 t)]$$

The asymptotic value is given by the limit

$$\lim_{t \to \infty} x = \frac{k_1}{k_2}$$

LOGARITHMIC MODEL

The increase in deposit thickness with time can be described by the equation

$$\frac{dx}{dt} = k_1 k_2^{-x}$$

where k_1 and k_2 are constants, coefficients or functions. Rearrangement gives

$$\frac{k_2^x dx}{k_1} = dt$$

and the integrals

$$\int_0^x \frac{k_2^x dx}{k_1} = \int_0^t dt$$

Integration gives

$$\frac{1}{k_1}\left[\frac{k_2^x}{\ln k_2}\right]_0^x = [t]_0^t$$

and

$$\frac{k_2^x}{\ln k_2} - \frac{1}{\ln k_2} = k_1 t$$

Rewriting gives

$$k_2^x - 1 = k_1 \ln k_2 t$$

and

$$k_2^x = 1 + k_1 \ln k_2 t$$

Applying natural logarithm

$$\ln k_2^x = \ln[1 + (k_1 \ln k_2)t]$$

results in

$$x \ln k_2 = \ln[1 + (k_1 \ln k_2)t]$$

The main result becomes

$$x = \frac{1}{\ln k_2} \ln[1 + (k_1 \ln k_2)t]$$

Based on the first equation

$$\frac{dx}{dt_{x=0}} = k_1$$

which is the same as in the exponential deposition-release model.

Dipole moment of hydrocarbons

The polarity of hydrocarbon molecules is directly related to their dipole moment. Large polarity contributes to the aggregation/flocculation of asphaltene molecules.

The *Onsager Equation* can be used to calculate the dipole moment of liquids

$$\mu^2 = \frac{9k_B T}{4\pi N_A} \frac{M}{\rho} \frac{(\varepsilon - n_\infty^2)(2\varepsilon + n_\infty^2)}{\varepsilon(n_\infty^2 + 2)^2}$$

where k_B and N_A are the Boltzmann constant and Avagadro's number, respectively. The symbols ε and n_∞ stand for the dielectric constant and refractive index at infinite wavelength, respectively. The other symbols have the usual meaning. The dipole moment μ D ($=$ Debye) represents the physics of charge times displacement (charge $\times$ distance).

Lars Onsager, 1903–1976, graduated in physics in 1925 from NTH (called NTNU from 1996). He received the Nobel Prize in chemistry in 1968.

For hydrocarbons at ambient temperature, according to Altschullere (1953), the *Onsager Equation* can be simplified to the approximate expression

$$\mu = (\varepsilon - n^2)^{0.5} \cdot 10^{-18}$$

The dielectric constant of hydrocarbons (and other liquids) can be estimated from an equation of the form

$$\log(\varepsilon - 1) = A + B \log \rho$$

where A and B are constants for each particular liquid (Marshall, 2008). For example, for n-pentane $A = 0.1789$ and $B = 1.229$ when density ρ is expressed in g/cm^3.

Based on literature values of dielectric constant and refractive index, the dipole moments of n-pentane, benzene and toluene were calculated and are shown in Table L1.1. The dipole moments are shown to increase linearly with molecular weight. Smaller molecules than pentane (methane, ethane, propane and butane, gases at ambient conditions) are practically non-polar (dipole moments of zero).

Table L1.1 Calculated dipole moment of three hydrocarbons based on literature values at 25°C.

	M g/mol	ρ g/cm^3	ε	n	μ D
Pentane	72.15	0.626	1.844	1.3575	0.0346
Benzene	78.11	0.877	2.274	1.5011	0.1439
Toluene	92.14	0.867	2.379	1.4969	0.3719

Regular solution model

The solubility of heavy molecules (for example, asphaltene) in crude oils, depends on the difference in the solubility parameters of the solute and the solvent. The solubility parameters are related to the change in *Gibbs Free Energy*.

The dissolution and precipitation of asphaltenes can be expressed in terms of thermodynamics. Precipitation is the opposite process to dissolution. In general, *Gibbs Free Energy* (see *Appendix S*) can be written as

$$\Delta G = \Delta H - \Delta(TS) = \Delta H - \Delta TS - T\Delta S$$

For a constant temperature process $\Delta T = 0$ and the equation becomes

$$\Delta G = \Delta H - T\Delta S$$

Any process is considered spontaneous (occurs) when $\Delta G < 0$ (negative), as illustrated in Figure 6.8. A process occurs to achieve minimum energy at equilibrium, meaning that energy must be released (exothermic process). The entropy S increases in a closed constant internal energy process, such that $\Delta S > 0$ (positive). A dissolution/precipitation process is an approximation of such a process. The spontaneity of a process depends on the sign (negative or positive) and value of the enthalpy change ΔH.

Applying *Hess's Law*, the enthalpy term has three parts for a dissolution process

$$\Delta H = \Delta H_1 + \Delta H_2 + \Delta H_3$$

The *first* part is the energy added (positive) to break intermolecular forces between solvent molecules, the *second* part is the energy added (positive) to break intermolecular forces between the solute molecules and the *third* part the energy released (negative) from the attraction between solvent and solute molecules. If the numerical value of the process is exothermic (gives energy), and because ΔS is positive, the process will occur (dissolution).

$$|\Delta H_3| > |\Delta H_1 + \Delta H_2|$$

In terms of enthalpy, it means that the negative third part is larger than the sum of the positive first and second parts. In terms of entropy, the positive ΔS turns to negative in the *Gibbs Free Energy* equation.

However, if the numerical value of

$$|\Delta H_3| < |\Delta H_1 + \Delta H_2|$$

the process is endothermic (takes energy).

A process will occur if $|\Delta H| < |T\Delta S|$. Large entropy change and high temperature favour the spontaneity of a thermodynamic process. The enthalpy and entropy symbols above should perhaps have the subscript M to signify mixing of solute in solvent, meaning the same as dissolution of asphaltene.

The solubility parameter δ is an important quantity in predicting solubility relations. An introduction to solubility parameter has been presented by Hansen (2007). The *Hildebrand Solubility Parameter* is defined as the square-root of the cohesive energy density

$$\delta = \left(\frac{E}{v}\right)^{\frac{1}{2}}$$

where v is the molar volume of a pure solvent and E is the measurable energy of vaporisation. Expressed more commonly

$$\delta = \left(\frac{\Delta H_{LV} - RT}{v}\right)^{\frac{1}{2}}$$

where ΔH_{LV} represents the heat of vaporization from liquid (fluid) to vapour (gas). The subscript fg (fluid to gas) is commonly used instead of LV. RT is the gas constant multiplied by the absolute temperature. Interestingly, the gas constant has the same unit as heat capacity. Note also the equivalence

$$R = k_B N_A$$

is used above in the *Onsager Equation* (see *Appendix L1 – Dipole Moment of Hydrocarbons*). Asphaltenes exist as molecules in solution in crude oil at reservoir conditions, but precipitate as solids due to reduced pressure from reservoir conditions to surface conditions. The *Hildebrand Solubility Parameter* has been used with success in predicting the solubility/dissolution of asphaltenes. The parameter is used in a wide range of processes, frequently involving polymer solutions.

According to Hansen (2007), the energy term in the *Hildebrand Equation* was originally (Hildebrand and Scott, 1962) expressed as

$$\Delta H = \phi_1 \phi_2 V (\delta_1 - \delta_2)^2$$

where the volume fractions ϕ are solvent (1) and solute (2) and V the mixture volume. The value of ΔH is positive.

A refinement (further development) of the solubility parameter concept, is that of Hansen (2007), specifying that the solubility parameter consists of the three main forces, keeping hydrocarbon molecules together

$$\delta^2 = \delta_d^2 + \delta_p^2 + \delta_h^2$$

where the subscripts stand for *dispersive*, *polar* (dipole-dipole) and *hydrogen* bonding interactions. Dispersive forces are temporary van der Waals attractive forces, the weakest intermolecular force. The strength of the inter-molecular forces can be shown by the inequality

$$F_d < F_p < F_h$$

Not included in the inequality expression are the ionic bond forces, because they are not involved in hydrocarbons. Ionic bond forces are stronger than the hydrogen bond forces. Inter-molecular (between) forces keep molecules together and are responsible for their *physical* properties. Intra-molecular (internal) forces keep molecules together and are the basis for their *chemical* properties.

The derivation of the *Flory-Huggins Theory* is summarized in the following. For the change in entropy and the change in enthalpy, the subscript 1 (=one) stands for solvent and subscript 2 (=two) for solute. Two-dimensional lattice theory was used by the original authors (Flory, 1942; Huggins, 1942). The change in entropy on mixing is

$$\Delta S_M = -R(n_1 \ln \phi_1 + n_2 \ln \phi_2)$$

and the change in enthalpy on mixing is

$$\Delta H_M = RT \chi_1 n_1 \phi_2$$

The symbols stand for number of moles (n) and volume fractions (ϕ). The symbol χ stands for the *Interaction Parameter*, sometimes called the free energy parameter, defined by

$$\chi = \frac{v(\delta_a - \delta_o)^2}{RT}$$

where v is the molar volume (mol/m^3) of the mixture and the subscripts a (=2) and o (=1) stand for asphaltene and oil. Perhaps more correctly, the o stands for the solubility parameter of the maltenes fraction of the crude oil; maltenes are the saturates, the aromatics and the resins (all but asphaltenes).

Combining the change in entropy and the change in enthalpy equations gives

$$\Delta G_M = RT(\chi_1 n_1 \phi_2 + n_1 \ln \phi_1 + n_2 \ln \phi_2)$$

The *Gibbs Free Energy* can be related to chemical potential (see *Appendix T – Chemical Potential*). The above model/theory was simplified by Hirschberg *et al.* (1984), when applying it to asphaltenes in crude oil.

The presence of a solute lowers the chemical potential of a solvent relative to its value as pure solvent. The above equation for ΔG_M can be partially differentiated with respect to number of moles, n, to give the change in chemical potential between the actual/effective and the ideal (at standard conditions) as

$$\mu - \mu^o = RT \left[\ln(1 - \phi_2) + \left(1 - \frac{1}{x}\right)\phi_2 + \chi \phi_2^2 \right]$$

The molar volume ratio is given by

$$x = \frac{v_1}{v_2}$$

The molar volumes ϕ_1 and ϕ_2 are that of the solvent and solute, respectively, in a dissolution process. In a precipitation process, the volume fractions need to be *reversed* such that

$$\mu - \mu^o = RT \left[\ln(1 - \phi_1) + \left(1 - \frac{1}{x}\right)\phi_1 + \chi \phi_1^2 \right]$$

Knowing that $(1 - \phi_1) = \phi_2$ and assuming that the solvent volume fraction is much, much larger than the solute fraction, the following approximation results

$$\frac{\Delta \mu}{RT} \simeq \ln \phi_2 + 1 - \frac{1}{x} + \chi$$

Because the molar volume ratio x can be replaced by v_2/v_1 the above equation can be written

$$\frac{\Delta \mu}{RT} \simeq \ln \phi_2 + 1 - \frac{v_1}{v_2} + \chi$$

This equation/relationship is the same as illustrated by Mohammadi and Richon (2007) and used by Hirschberg *et al.* (1984). It is the *Flory-Huggins Theory* applied to asphaltene precipitation from crude oil.

Viscosity and activation energy

The resistance of a fluid to flow is due to intermolecular friction. This resistance is called viscosity. The most commonly used viscosity is the *dynamic viscosity*, also called absolute viscosity, expressed by μ mPa·s. A greater force is required to flow a high-viscosity fluid than a low-viscosity fluid. The term *kinematic viscosity* is sometimes encountered. It is simply the dynamic viscosity divided by the fluid density, expressed by $\eta = \mu/\rho$ m^2/s. The general term viscosity, refers most commonly to the absolute/dynamic viscosity.

Because the viscosity of fluids depends on intermolecular friction/interactions, their viscosity characteristics can be very different. Fluids have been characterized/classified, as Newtonian and non-Newtonian. Newtonian fluids respond linearly to shear strain while non-Newtonian fluids respond non-linearly to shear. Three non-Newtonian fluids are commonly identified: shear thinning, shear thickening and plastic. The viscosity of the first two either decreases or increases with increasing shear strain, while the plastic fluid is characterized by a solid behaviour at low shear strain and fluid behaviour a high shear strain. Other terms and definitions are also used in the literature. Paraffinic oils may exhibit plastic behaviour at low temperatures closed to the PPT (pour point temperature).

The viscosity of pure liquids at oil production and ambient temperatures is generally Newtonian. Upon cooling, some liquids take on non-Newtonian characteristics. The major challenge in oil and gas production is when oil and water mix to make an emulsion, and when solids precipitate to make a slurry. Emulsion is a liquid-liquid mixture and a slurry is a liquid-solid mixture. Aspects of emulsions affecting flow assurance solids are discussed in several chapters. Hydrate particles in water and diesel oil affect measured viscosity in laminar flow, but not in turbulent flow (*Section 5.8 – Prevention by Cold Flow*). The relative viscosity of oil-water and water-oil emulsions may affect the deposition of inorganic solids (*Section 6.9 – Carbonate Scale*). The formation of naphthenate solids occurs at oil-water interfaces (*Section 7.6 – Interface Processes*). Viscosity plays a role in all flow assurance challenges.

The density of an *emulsion* and a *slurry* is a physical property that can be calculated directly from the general equation

$$\rho = \rho_A \alpha + (1 - \alpha)\rho_B$$

The volume fraction α is used because density has the unit mass per volume. Volume fraction in multiphase flow (gas-liquid and liquid-liquid) is most commonly called void fraction. In other instances, the symbol ϕ is used for volume fraction. To illustrate this point, the specific volume v m³/kg of an emulsion and a slurry, can be calculated from the general expression

$$v = v_A x + (1 - x)v_B$$

where x is mass fraction. In the case of an emulsion, the A can stand for the oil phase and the B for the water phase. In the case of a slurry, the A can stand for the oil phase and the B for the solid-wax phase.

The viscosity of an *emulsion* and a *slurry* is a physical property that *cannot* be calculated directly from a fundamental equation. As presented in *Appendix Y1 – Two-Phase Flow Variable and Equations*, the following empirical relationship is commonly used for the viscosity of homogeneous *gas-liquid* and *liquid-liquid* mixtures

$$\frac{1}{\mu} = \frac{x}{\mu_A} + \frac{(1 - x)}{\mu_B}$$

Viscosity has the unit Pa·s which is equivalent to kg/m·s. Therefore, the use of mass fraction x in the above empirical relationship seems most logical. This particular viscosity relationship is often identified as the *McAdams Equation*. The relationship can be used for emulsions, but it *cannot* be used for slurries. Solid particles in a slurry do not have a viscosity, *per se*. Alternative ways are needed calculate/estimate the effective viscosity of slurries, examples being those of Abdel-Waly (1997) and Tiwary and Mehrotra (2004), for wax particles in model hydrocarbon liquids.

The viscosity of liquids decreases with increasing temperature (the opposite occurs for gases). In the case of liquids at petroleum production and processing pressures, the effect of pressure on viscosity is negligible (not so for gases, of course). Because liquid viscosity is intermolecular in nature, higher temperature promotes the weakening (breaking-up) of molecular bonding when subjected to shear forces. The semi-theoretical equation

$$\mu = A \exp\left(\frac{E_a}{RT}\right)$$

is commonly used to correlate viscosity data with temperature. The symbol A is pre-exponential factor (related to the molar volume) and the RT-term is the gas constant and absolute temperature. The variable E_a kJ/mol is the activation energy for viscous flow. The above equation is one form of an equation derived/proposed by Eyring (1935). The *Eyring Equation* (Henry Eyring, 1901–1981) has the same form as the *Arrhenius Equation* (Svante A. Arrhenius, 1859–1927). However, there is one big difference, the exponential term in the viscosity equation is positive, not negative. The activation energy for viscous flow is constant for a particular liquid at normal temperatures and pressures, so is the value of the pre-exponential factor. The *Eyring Equation* shows that liquid viscosity decreases with increasing temperature. The theoretical aspects of the equation and its limitations were discussed by Avramov (2007). The natural logarithm of the *Eyring Equation*

$$\ln \mu = \ln A + \left(\frac{E_a}{R}\right)\frac{1}{T}$$

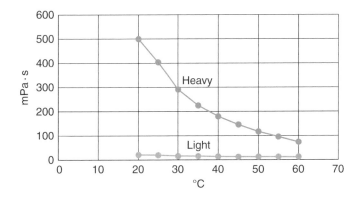

Figure M1.1 Dynamic viscosity μ mPa·s versus temperature °C for 21.93° API heavy oil and 42.41° API light oil (Singh *et al.*, 2012).

can be used to plot $\ln \mu$ versus $1/T$ to obtain the gradient/slope, and hence the activation energy of viscous flow and the pre-exponential factor (an intercept in the plot).

The viscous activation energy in hydrocarbon fluids was studied by Singh *et al.* (2012). Two crude oils from Oman were used, a heavy oil 21.93° API and a light oil 42.41° API gravity. The density and viscosity of the two oils were accurately measured from 20°C to 60°C. The viscosity values for the temperature range are shown in Figure M1.1. The viscosity of the light oil decreased from about 20 to 10 mPa·s. The viscosity of the heavy oil decreased exponentially from 500 to 75 mPa·s. A plotting of the natural logarithm of the viscosity versus the inverse of the absolute temperature, gave the following activation energies: heavy oil 38.5 kJ/mol and light oil 13.5 kJ/mol.

It has been inferred in the literature (for example, Kinker, 2000), that the activation energy of viscous flow is typically one-third to one-fourth of the latent heat of vaporization. One a different topic, the latent heat of vaporization is used in the definition of the *Hildebrand Solubility Parameter*, presented in *Appendix L2 – Regular Solution Theory*. The heat of vaporization of pure/individual hydrocarbons is found in tables in handbooks, including Griesbaum *et al.* (2012) and Riazi (2005). Correlations of the heat of vaporization of crude and refined oils are not readily available in the literature. The standard practice is to estimate the heat of vaporization from a summation based on the molar composition, for example

$$\Delta H_{fg} = \Delta H_{LV} = \sum_{1}^{n} x_i \Delta H_i$$

The subscripts *fg* (fluid-gas) and *LV* (liquid-vapour) indicate heat of vaporization, change from liquid to vapour. For simplicity, the subscripts are not used in the summation. In some publications, the subscripts v and V are used for the heat of vaporization. Instead of ΔH, capital L is sometimes used for heat of vaporization. The kind of summation shown above, is standard in commercial PVT software.

Table M1.1   Latent heat of vaporization of common liquid hydrocarbons, including API gravity and boiling point temperature (Cragoe, 1929).

Product	API	T °C	ΔH kJ/kg
Gasoline	60°	138	270
Naphtha	50°	171	240
Kerosene	40°	227	200
Fuel Oil	30°	304	156

The latent heat of vaporization of common liquid hydrocarbons are shown in Table M1.1. The values were converted to SI units from a table in engineering units (Btu, lb, °F), given by Cragoe (1929). The table values are from an old publication and may not be as up-to-date as desired. However, the table is useful because it shows both API gravity and heat of vaporization. The values can be used for estimation purposes (approximate calculations).

The data in Table M1.1 were plotted and the latent heat of vaporization (at the boiling point) found to follow the empirical correlation (based on linear curve-fitting)

$$\Delta H_{fg} = 44.6 + 3.82 \text{API}$$

Based on this approximate correlation, the heat of vaporization of the light crude (API 42.41°) was calculated 207 kJ/kg and that of the heavy crude (API 21.93°) was calculated 128 kJ/kg. These values seem counter intuitive (higher molecular weight has lower heat of vaporization), but can be explained by considering the boiling point temperatures. The heavier liquids boil at a higher temperature than the lighter liquids. At the higher temperature, the molecules have already received greater thermal energy such that vaporization requires less energy. The heat of vaporization of all liquids approaches zero when the temperature reaches the critical temperature.

The heat of vaporization in Table M1.1 has the unit kJ/kg, while the unit of the activation energy of viscous flow has the unit kJ/mol. The API gravity and molecular weight of six crude oils are given in Table 3.2 (Fan *et al.*, 2002). In API classification terminology, four of the oils are medium and two of the oils are light (see Table X1.1). A plot and curve fit made of the relationship between molecular weight kg/kmol and API gravity, resulted in the following correlation

$$M = 320 - 2.6 \text{API}$$

The correlation was tested against two light crude oils reported by Goual and Firoozabadi (2002): (1) API 33° and M = 242 and (2) API 35° and M = 206. The empirical correlation gives 234 kg/kmol and 229 kg/kmol respectively, correspondingly 3.3% lower and 11.2% higher. Even though few data points were used in the curve fitting, the empirical correlation seems to give reasonable values. The molecular weight of the light and heavy crude oils of Singh *et al.* (2012) were calculated using the correlation and are shown in Table M1.2.

Table M1.2 Properties of two crude oils, first four columns from Singh et al. (2002), the last three columns estimated in present text. The enthalpy values are the heat of vaporization.

Crude	API	ρ kg/m^3	E_a kJ/mol	M kg/kmol	ΔH kJ/kg	ΔH kJ/mol
Light	42.41°	811	13.5	210	207	43.5
Heavy	21.93°	919	38.5	263	128	33.7

The ratio of the molar heat of vaporization to the activation of viscous flow, can now be determined for the two crude oils in Table M1.2. For the light oil, 43.5/13.5 = 3.2 and for the heavy oil 337/38.5 = 0.9. Therefore, the one-third to one-fourth rule-of-thumb is only applicable for the light oil, not the heavy oil. The rule-of-thumb may be more appropriate for pure hydrocarbons than for mixtures of hydrocarbons.

Wax-in-kerosene deposition model

Flow-loop data for paraffin wax in kerosene deposition, were presented and analysed by Gudmundsson (1977). Paraffin wax was dissolved in kerosene and passed through counter-current pipe-in-pipe heat exchangers. Most of the laboratory data exhibited an exponential buildup to an asymptotic value. The properties of the wax-in-kerosene solutions used and the experimental flowrates and temperatures employed, are found in Bott and Gudmundsson (1976). The data analysis was based on the deposition-release methodology, given by the expression

$$\frac{dx}{dt} = \psi_d - \psi_r$$

The term dx/dt gives the gradient of the deposition profile curve for thickness versus time. The right-hand terms are the *deposition function* and the *release function*, as indicated by the subscripts d and r. Instead of deposit thickness, other parameters can be used in the data analysis, depending on the purpose of the work. In the oil and gas industry, the effect of deposition on pipeline pressure drop may be in focus, hence the use of deposit thickness. In the process industry, the fouling of heat exchangers may be in focus, hence the use of fouling resistance.

The main focus of the data analysis was the fouling resistance of heat exchangers, resulting in the main relationship

$$\frac{dR_f}{dt} = k_1 s J - k_2 \left(\frac{x}{u^*}\right) \tau_w$$

expressed in terms of the fouling resistance R_f (inverse of heat transfer coefficient). The shear stress at the wall τ_w and the friction velocity u^* are used in the relationship (see *Appendix H – Universal Velocity Profile*). The deposition function

$$\psi_d = k_1 s J$$

was taken to depend on the mass flux to the solid surface J kg/s $\cdot$ m^2 and a *stickability* give by

$$s = \frac{k_3 \Delta T^+}{\tau_w}$$

The parameters k_1, k_2 and k_3 can be considered as coefficients or constants. The dimensionless deposition temperature driving force was defined by

$$\Delta T^+ = \frac{T_c - T_w}{T_b}$$

where the subscripts c, w and b stand for cloud (WAT), wall and bulk, respectively. The bulk temperature is the inlet temperature of the wax-in-kerosene solution. The dimensionless temperature is for systems where the solubility of the solid phase increases with temperature.

Paraffin mass flux was expressed by

$$J = hc$$

where h m/s is the mass transfer coefficient and c kg/m^3 the concentration of paraffin wax in the solvent. The mass transfer coefficient for *molecules* in solution can be obtained from standard convective mass transfer correlations. The mass transfer coefficient of *particles* of paraffin wax was based on the approximation

$$h = k_4 u^*$$

where k_4 is a parameter/constant and u^* the friction velocity

$$u^* = \sqrt{\frac{\tau_w}{\rho}}$$

The release function included deposit thickness x to give the exponential form of the deposition-release model

$$\psi_r = k_2 \left(\frac{x}{u^*}\right) \tau_w$$

Taking into account the definition of wall shear stress, the release function becomes directly proportional to the friction velocity.

The experimental results were not extensive enough to make it possible to quantify the parameters/coefficients/constants. However, it was possible to find the two following proportionalities

$$\frac{dR_f}{dt} \propto \frac{c\Delta T^+}{\rho u^*}$$

$$R_f^\infty \propto \frac{c\Delta T^+}{(\rho u^*)^2}$$

The concentration-temperature driving force term $c\Delta T^+$ appears in both proportionalities. Furthermore, that the asymptotic deposit thickness (fouling resistance) is more dependent on the fluid flow velocity than is the rate of deposit build-up. The model does not take aging into account. Note that ρu^* has the unit of mass flux, kg/s · m^2, tangentially in the flow direction.

Pigging model for wax

Mechanical scrapers are used to remove paraffin wax deposits in crude oil and gas condensate pipelines. To plan and carry out optimal scraping/pigging operations, it is desirable to know the wax deposit profile along the pipeline and the mechanics of wax removal. The present model addresses/considers the forces required to remove wax deposits using a bypass scraper, assuming single phase liquid flow (oil or gas condensate). The model proposed is quite basic/simple and can be used to illustrate how the main parameters affect the pressure required to remove paraffin wax from a pipeline wall. Mechanical scrapers are but one of many types of gauges used in pipeline pigging operations.

The force that moves a scraper in a pipeline, is the pressure difference (upstream and downstream)

$$F_p = \Delta p \frac{\pi d^2}{4}$$

where d is the pipeline inner diameter (ID). In non-horizontal situations, the gravitational force on the scraper (upward or downward) needs to be included.

The opposing forces are the frictional force (between scraper and pipe wall) and the wax removal force

$$F_f + F_{wax}$$

REMOVAL FORCE

In the model proposed, deposit wax will be removed when its compressive stress is exceeded. The scraper pig pushes on the deposited wax such that it breaks up and loosens from the pipe wall. It assumes that the adhesive force between the paraffin wax and the pipe wall is included in the removal force. Axial stress is given by the general relationship

$$\sigma = \frac{F}{A}$$

having the unit N/m^2, the same as pressure. The compressive strength of a hard paraffin wax has been reported by Hossain *et al.* (2009) to be 658.4 kPa. Other property values reported by the same authors for the paraffin wax were density of $0.7855 \, kg/m^3$ and modulus of elasticity of $55.7 \, MPa$. Paraffin wax deposits in pipelines are a mixture of wax and oil/condensate, so the quoted compressive strength must be viewed as a maximum value. In lieu of other data, it is proposed here that the compressive strength of paraffin wax deposits in pipelines depends linearly on the oil/condensate content such that

$$\sigma_{wax} = 658(1 - \phi_{oil})$$

where ϕ_{oil} is the volume fraction of liquid hydrocarbon. This proposed/assumed linear relationship can be improved if and when more relevant data are found in the literature.

The wax removal force will be given by the expression

$$F_{wax} = \sigma_{wax} A_{wax}$$

The cross-sectional area of a clean pipe will be

$$A_{pipe} = \frac{\pi}{4} d^2$$

The cross-sectional flow area in a pipe with a paraffin wax deposit will be

$$A_{fluid} = \frac{\pi}{4} (d - 2\delta)^2$$

and the cross-sectional area of the wax will be

$$A_{wax} = \pi (\delta d + \delta^2)$$

It follows that the removal force becomes

$$F_{wax} = \sigma_{wax} \pi (\delta d - \delta^2)$$

where δ is the wax thickness. This expression can be improved by including an efficiency factor η, based on laboratory experiments and/or field experience. For simplicity, this factor will not be included in the following text.

An order of magnitude calculation can be carried out assuming a 24 inch (ca. 600 mm) pipeline diameter and a deposit thickness of 6 mm. First, assuming a hard wax deposit (no liquid content).

$$F_{wax} = 658 \cdot \pi \cdot (0.006 \cdot 0.6 - 0.006 \cdot 0.006) = 7.37 \, kN$$

For a wax deposit containing 50% volume fraction liquid hydrocarbon, the wax removal force will be one-half this value ($=3.7 \, kN$), assuming the linear relationship suggested above.

FRICTION FORCE

The friction force between the scraper and pipe wall F_f is a difficult quantity to estimate. Fundamentally, the friction force due to gravity between an object moving along a flat wall is given by

$$F_f = \mu_f Mg$$

where μ_f is the coefficient of friction between to materials, M is the mass/weight of the object and g is the gravitational constant. The material friction force is the minimum (ideal) force acting on a scraper. The real force will be larger because the diameter of the flexible discs is a bit larger than the pipeline diameter. The coefficient of friction relevant for pipeline scrapers and may lie in the range

$$0.1 < \mu_f < 1$$

Different values are reported in the literature for static and dynamic coefficients. Experiments are required to find out whether the above simple equation for F_f can be used. An order of magnitude force can be calculated assuming $\mu_f = 1$ and $M = 50\,\text{kg}$, resulting in a friction force of about 0.5 kN.

The length of the wax scraper is not used in the calculation (the weight depends on the length and diameter of the scraper and the materials of construction). The sealing disks in scrapers are typically some sort of hard rubber, resulting in a relatively high coefficient of friction, even greater than 1. Nevertheless, the friction force is much smaller than the wax removal force, in the present example.

The coefficient of friction above is a constant. In reality it may depend on velocity, just as pressure drop due to wall friction in pipes depends on velocity squared, and the friction factor decreases with Reynolds number. Furthermore, the diameter of the hard disks in a scraper pig may be a few percentages larger than the pipeline internal diameter. It can be argued that the frictional force obtained from the simple equation above is the minimum force. An additional force would be that due to the compression/bending of the hard disks.

FORCE BALANCE

The force balance for a paraffin wax deposit containing 50% volume fraction crude oil or gas condensate will be

$$F_p = F_f + F_{wax} = 0.5 + 3.7 = 4.2\,\text{kN}$$

the minimum pressure drop across the scraper (that drives it forward) will be

$$\Delta p = 4.2 \cdot \frac{4}{\pi \cdot 0.6^2} = 0.15\,\text{bar}$$

using the full diameter of the pipeline.

The friction force is likely to depend on the scraper velocity. Looking for information on such an effect, a paper by Botros and Golshan (2010), suggested that the following polynomial could be used for the friction force (based on pigging a gas pipeline)

$$F_f = c_0 + c_1 u_{pig} + c_2 u_{pig}^2$$

where $c_0 = 150\,\text{N}$, $c_1 = -4\,\text{N} \cdot \text{s/m}$ and $c_2 = 0.04\,\text{N} \cdot \text{s}^2/\text{m}^2$. Assuming a scraper velocity of 2 m/s the polynomial gives the friction force as $F_f = 142\,\text{N} = 0.142\,\text{kN}$. This value is much lower than the 0.5 kN value calculated/estimated above.

An inspection of the polynomial shows that the c_1 coefficient is negative such that the friction force decreases with velocity, assuming a reasonable velocity range. This seems counter intuitive and needs to be looked at. A model based on fluid flow in a thin layer between a scraper and a pipe wall, will depend on both the fluid friction factor and fluid/scraper velocity. The fluid friction factor decreases with velocity (Reynolds number), so perhaps the above polynomial is alright.

BYPASS SCRAPER

A 600 mm diameter pipeline flowing 750 kg/m^3 oil/condensate at 2 m/s (upstream of scraper) has a mass flow rate of 425 kg/s, based on

$$m = \rho u A$$

An orifice-type flow restriction can be modelled by

$$m = C A_2 \sqrt{2 \rho \Delta p}$$

where C is a coefficient (discharge) and A_2 the restricted flow area. The coefficient has typically a value between 0.5 and 0.7. Assuming a mass flow rate 10% of the total flow rate, a pressure drop of 15 kPa (estimated above) and $C = 0.6$ the effective flow area of the restriction will be 0.0149 m^2 and the diameter 138 mm. This diameter is 23% of the clean pipeline diameter, which seems an order-of-magnitude too high. A more reasonable value would seem be 1–5% of the clean pipe diameter.

Criterion for what percentage by-pass flow would be required for successful scraping is not known. Knowledge about the paraffin wax deposition profile would be an important input in selecting/developing a launching a useful criterion for a scraper. One possible criterion could be that the volume fraction wax in the flow downstream of the scraping pig should not be higher than 10% volume fraction. Slurry flow studies may aid in finding a good criterion.

A bypass scraper should have a non-return valve to make it possible to move it by pressurizing the downstream side; for example, in case the scraper gets stuck. A bypass scraper for paraffin wax removal should also have a plate of some sort at the downstream end to distribute the fluid directly to the wall (radial fluid jet to aid in removing deposits). A non-return valve and fluid jet distribution plate would add some pressure drop through the by-pass pig.

PRESSURE GRADIENT

A bypass flow of 42.5 kg/s may ensure that the scarped-off paraffin wax is carried downstream at greater velocity than that of the scarping pig (this is an empirical assumption). The average fluid velocity downstream the scarper is 10% greater than 2 m/s (=2.2 m/s). The pressure gradient in the pipeline will not be the same upstream and downstream of the scraper, based on the *Darcy-Weisbach Equation*

$$\Delta p = \frac{f}{2}\frac{L}{d}\rho u^2$$

Upstream of the scraper, the diameter is the full pipeline diameter. Downstream, the diameter is reduced by the paraffin wax deposit. The flowing densities will be slightly different, but not by much because the densities of oil/condensate and paraffin wax are about the same.

The wall friction factor upstream and downstream will be different. Experience indicates that the effective wall roughness of a paraffin wax deposit is significantly higher than that of a bare steel pipe. It is proposed here that the effective roughness used in pressure drop/gradient calculations with paraffin wax deposits should be π-times higher than the thickness δ. This proposal has some merit in the fundamentals presented in *Appendix F – Friction Factor of Structured Deposits* and in field testing reported by Stevenson *et al.* (2015).

Other variable of relevance are temperature and viscosity. Temperature in subsea pipelines decreases exponentially with length according to

$$T_2 = T_o + (T_1 - T_o)\exp\left[\frac{-U\pi d}{mC_p}L\right]$$

The surrounding temperature is T_o (o = outside) and U W/m$^2 \cdot$ K the overall heat transfer coefficient (ranges from 15 to 25 for subsea pipelines). The temperature subscripts 1 and 2 signify inlet and outlet conditions, respectively. The temperature-with-distance equation is derived in *Appendix A – Temperature in Pipelines*.

SLURRY FLOW

The oil/condensate and scraped-off paraffin wax will flow as slurry downstream of the scraper. In a unit length (=1 m) of an example pipeline, the volume of the scraped-off wax will be

$$A_{wax} = \frac{\pi}{4}(d - 2\delta)^2$$

$$A_{wax} = \pi(\delta d + \delta^2)$$

$$A_{wax} = \pi(0.006 \cdot 0.6 + 0.006 \cdot 0.006) = 0.0114\,\text{m}^2$$

$$V_{wax} = 0.0114 \cdot 1 = 0.0114\,\text{m}^3$$

The cross-sectional area of fluid (flowing oil/condensate) in a pipeline with a paraffin deposit will be

$$A_{fluid} = \frac{\pi(d - 2\delta)^2}{4}$$

$$A_{fluid} = \frac{\pi(0.6 - 2 \cdot 0.006)^2}{4} = 0.272\,\text{m}^2$$

Checking the numbers

$$A_{pipe} = \frac{\pi}{4}0.6^2 = 0.283\,\text{m}^2$$

$$A_{fluid} + A_{wax} = 0.272 + 0.0114 = 0.283\,\text{m}^2$$

Volume fraction paraffin wax in oil/condensate will therefore be

$$\phi_{wax} = \frac{V_{wax}}{V_{wax} + V_{flow}}$$

$$\phi_{wax} = \frac{0.0114}{0.0114 + 0.272} = 0.040$$

In other words, the wax volume percentage is about 4%. This is quite small so the viscosity of the slurry can be approximated by the *Einstein Equation*

$$\mu = \mu_o(1 - 2.5\phi_{wax})$$

where μ_o is the viscosity of the oil/condensate (same temperature and pressure), without paraffin wax particles.

Water vapour in natural gas

Methods to estimate the amount of water vapour in natural gas have been reviewed by Carroll (2003) and Mohammadi *et al.* (2005). To illustrate how pressure, temperature, sour gases, gas gravity and water salinity affect the water content in natural gas, the semi-empirical correlations developed and presented by Mohammadi *et al.* (2005) can be used. They are complete and easy to use. Semi-empirical equations can be used in the absence of full-scale PVT analysis offered in commercial computer software. A thesis by Chapoy (2004), can be consulted for numerous details of water in hydrocarbon mixtures.

The following expression was suggested by Mohammadi *et al.* (2005) to estimate the mole fraction of water in natural gases (methane data)

$$y_w = \left(\frac{p_w}{p}\right) \exp\left(a_1 \frac{p^{a2}}{T}\right)$$

where p_w MPa is the saturation pressure of water, p MPa the system pressure, $a_1 = 11.81479$, $a_2 = 0.92951$ and T K the system temperature. The constants a_1 and a_2 were obtained from methane data from 273.15 K to 477.59 K, equivalent to 0°C to 204.44°C. The methane data ranged in pressure from atmospheric to 14.40 MPa, equivalent to 144.0 bara. The average absolute deviation between the correlation and the data used ranged from 1.29% to 7.92%.

The saturation pressure of water, can be found from *Steam Tables* and empirical correlations. Mohammadi *et al.* (2005) recommend the correlation

$$p_w = 10^{-6} \cdot \exp\left[73.649 - \frac{7258.2}{T} - 7.3037 \cdot \ln(T) + 4.1653 \cdot 10^{-6} \cdot T^2\right]$$

where the pressure unit is MPa and the temperature K.

The above equations were used to calculate the mole fraction of water vapour in methane at temperatures from 20°C to 150°C and at system pressures of 10, 20 and 30 MPa. The results are shown in Figure N.1. The water content increases exponentially with temperature and is greater at low pressure than high pressure.

Mohammadi *et al.* (2005) stated that the correlation based on methane data could be used for natural gas with methane content above 70%. Such a methane concentration will cover most of the natural gases encountered in the oil and gas industry.

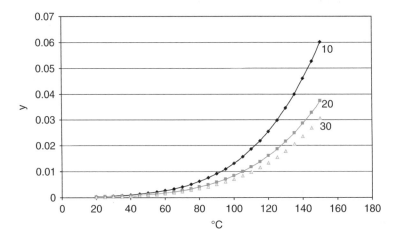

Figure N.1 Mole fraction of water vapour in methane, top line 10 MPa, middle line 20 MPa and bottom line 30 MPa.

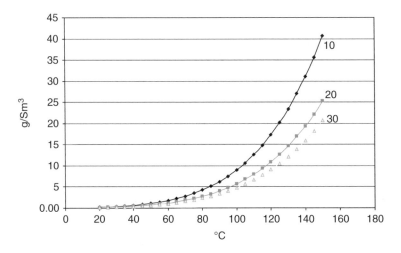

Figure N.2 Concentration of water vapour at standard conditions against temperature and pressure. Top line 10 MPa, middle line 20 MPa and bottom line 30 MPa.

The data in Figure N.1 can be represented as water concentration c mg/Sm^3 by assuming a methane gas density of about $0.6767 \, \text{kg/m}^3$ at standard conditions ($1.01325 \cdot 10^5$ Pa and 15°C, equivalent to 288.15 K). The concentration of water vapour at standard conditions c g/Sm^3 (note g and not mg) against temperature and pressure is shown in Figure N.2. The concentration is referenced to standard conditions, *not* the temperatures and pressures in the figure.

Assuming a natural gas reservoir at 90°C and 290 bara, Figure N.2 shows that the water concentration will be about 4000 mg/Sm^3. Using the well know graph Water Content of Hydrocarbon Gas in the Engineering Data Book of the Gas

Processors (Suppliers) Association (1998), the water content was estimated to be about 3500 mg/Sm^3. Similarly, using a similar graph in Rojey et al. (1997), the water content was estimated to be about 3100 mg/Sm^3. The graph in Figure N.2 gives a higher water content and can therefore be said to be conservative (gives a higher estimate and hence greater need for drying, hence avoiding designing equipment/process too small). The highest pressure in the methane data based on Mohammadi et al. (2005) was 14.4 MPa.

The presence of sour gases increases slightly the amount of water vapour in natural gas. Natural gas is called sweet when the concentration of CO_2 is below 2% vol. and the concentration of H_2S is below 1% vol. These are not absolute percentages, but guideline percentages (vary in commercial sales contracts)

A correction factor F_{sour} is used to correct for the effect of sour gases on the water vapour content, through the relationship

$$y_{w,sour} = F_{sour} y_{w,sweet}$$

where $y_{w,sweet}$ is the same as y_w above and y_{ideal} below. Mohammadi et al. (2005) suggested the following correlation for the correction factor

$$F_{sour} = 1 - y_{effective} \left[c_1 \left(\frac{T}{T_o} \right) + c_2 \left(\frac{T}{T_o} \right) \left(\frac{p}{p_o} \right) + c_3 \left(\frac{p}{p_o} \right) \right]$$

The subscript o refers to 0°C and atmospheric pressure absolute (273.15 K and $1.01325 \cdot 10^5$ Pa). Note that the reference temperature is not 15°C. The constants have the values $c_1 = 0.03185$, $c_2 = 0.01538$ and $c_3 = -0.02772$ (note that the last constant is negative). The temperature T and pressure p are the system (reservoir) values.

Instead of accounting for the effect of the different sour gases on water content separately, it is customary to use the following expression for the equivalent mole fraction of sour gases

$$y_{effective} = y_{H_2S} + 0.75 y_{CO_2}$$

The sour gas correlations were used here to calculate the increase in water vapour in methane gas (natural gas) assuming 5% vol. H_2S and 10% vol. CO_2. The corresponding mole fractions are 0.05 and 0.1. The results are shown in Figure N.3 for 10 MPa system (reservoir) pressure. The correction factor ranges from 1.134 at 20°C to 1.042 at 150°C. The volume fractions of sour gases used are relatively high, selected here for illustration purposes. The sweet gas water content at 90°C was 0.0092 and the sour gas water content 0.0100 mole fraction, an increase of about 8%. Mohammadi et al. (2005) claim the correlations can be used with good accuracy for volume fractions of sour gases up to about 0.07.

Increase in the gravity of natural gas reduces the amount of water vapour. Mohammadi et al. (2005) suggested the following correction factor to take into account the effect of gas gravity with temperature

$$F_{gravity} = 1 + b_1(\gamma - 0.554) + b_2(\gamma - 0.554) \left(\frac{T}{T_o} \right) + b_3(\gamma - 0.552)^2 \left(\frac{T}{T_o} \right)^2$$

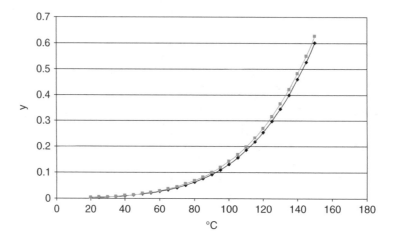

Figure N.3 Mole fraction water vapour in sweet gas (lower line) and sour gas (upper line) at 10 MPa system (reservoir) pressure. Based on 5% vol. H_2S and 10% vol. CO_2.

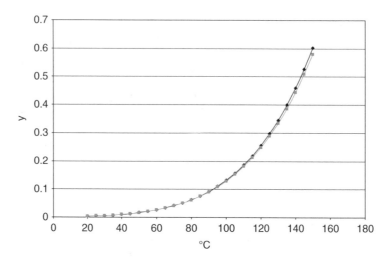

Figure N.4 Mole fraction of water vapour in sweet gas with gravity 0.7 and sweet gas with gravity 0.9, both at 10 MPa.

where γ is gas gravity and the temperatures as above. The constants have the values $b_1 = 0.17006$, $b_2 = -0.15241$ (note that the constant is negative) and $b_3 = -0.04515$ (note that the constant is negative). The effect of gravity is illustrated in Figure N.4. The upper line is for gas gravity 0.7 and the lower line for gas gravity 0.9, both at 10 MPa system (reservoir) pressure. At 90°C the mole fraction of water decreases by about 2% when the gas gravity increases from 0.7 to 0.9.

Dissolved solids (salts) in water reduce the vapour pressure of the water (aqueous solution). Therefore, the presence of salts reduces the water content of natural

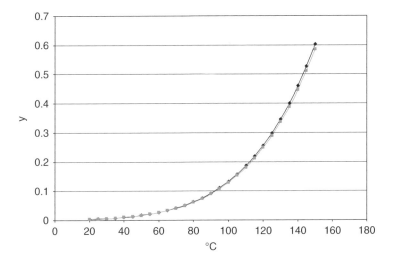

Figure N.5 Mole fraction of water vapour in natural gas in contact with water having 5% wt. dissolved salts at 10 MPa system (reservoir) pressure. Upper curve gas in contact with water, lower curve gas in contact with 5% wt. saline water.

gases. Mohammadi *et al.* (2005) suggested the following correction factor to take into account the effect of salt

$$F_{salt} = 1 - 4.920 \cdot 10^{-3} \cdot w_{salt} - 1.7672 \cdot 10^{-4} \cdot w_{salt}^2$$

where $w\%$ wt. is the weight percent of salt in brine. Seawater contains 3–5% wt. dissolved solids. At 90°C the mole fraction of water was reduced from 0.0092 to 0.0090 corresponding to about 2%.

The water content of natural gas is strongly dependent on temperature and also dependent on pressure, but to a lesser extent. The water content is affected by a small extent by the presence of sour gases (increase water content), gas gravity (decreases water content) and the liquid water phase salinity (decreases water content). Considering the mole fraction of water vapour in sweet gas (mostly methane) in contact with pure liquid water as ideal, the real mole fraction of water can be expressed by the compound relationship

$$y_{real} = F_{sour} \cdot F_{gravity} \cdot F_{salt} \cdot y_{ideal}$$

Based on the values above the real mole fraction of water vapour at 10 MPa and 90°C becomes

$$y_{real} = 1.0844 \cdot 0.9792 \cdot 0.9710 \cdot 0.009243 = 0.009530$$

an increase of about 3% from ideal to real. The presence of sour gases increases the mole fraction about 8% while the increase in gas gravity and dissolved salts decrease the mole fraction by less than 3% each.

Appendix O

Hydrate dissociation pressure

Correction factors for effects of carbon dioxide and nitrogen on the dissociation pressure of natural gas hydrate were developed by Østergaard et al. (2000). The correction factors are given by the expression

$$B_k = (a_{0,k} F_m + a_{1,k}) f_k + 1.000$$

where the subscript k stands for CO_2 or N_2 and f_k the mole fraction of the same, in the fluid (natural gas mixture). F_m is a molar ration of non-hydrate over hydrate forming molecules (see Section 5.4 – Equilibrium Lines). The parameters $a_{0,k}$ and $a_{1,k}$ are polynomial temperature correction functions given by

$$a_{0,k} = d_1(T - T_0)^3 + d_2(t - T_0)^3 + d_3(T - T_0) + d_4$$

$$a_{1,k} = d_1(T - T_0)^3 + d_2(t - T_0)^3 + d_3(T - T_0) + d_4$$

where d_1 to d_4 are constants. The equations have the same functional form. $T_0 = 273.15$ such that $(T - T_0)$ is the temperature $T°C$. There are four sets of constants as shown in Table O.1, two sets with subscripts 0 and 1 for each of the non-hydrocarbon components.

The application of the non-hydrocarbon correction factors was illustrated by Østergaard et al. (2000), by using the gas composition shown in Table O.2.

Table O.1 Constants in temperature correction function.

a	d_1	d_2	d_3	d_4
0, N_2	$1.1374 \cdot 10^{-4}$	$2.61 \cdot 10^{-4}$	$1.26 \cdot 10^{-2}$	1.123
1, N_2	$4.335 \cdot 10^{-5}$	$-7.7 \cdot 10^{-5}$	$4.0 \cdot 10^{-3}$	1.048
0, CO_2	$-2.0943 \cdot 10^{-4}$	$3.809 \cdot 10^{-3}$	$-2.42 \cdot 10^{-2}$	0.423
1, CO_2	$2.3498 \cdot 10^{-4}$	$-2.086 \cdot 10^{-3}$	$1.63 \cdot 10^{-2}$	0.650

Table O.2 Composition of a natural gas mixture.

Component	Mole %	Component	Mole %
C_1	73.95	n-C_5	0.74
C_2	7.51	C_6	0.89
C_3	4.08	C_{7+}	7.18
i-C_4	0.61	CO_2	2.38
n-C_4	1.58	N_2	0.58
i-C_5	0.50		

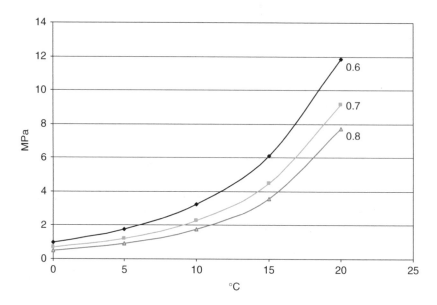

Figure O.1 Equilibrium lines for natural gas with gravity 0.6 (top line), 0.7 (middle line) and 0.8 (bottom line). Based on the Makogon (1988) correlation.

The molecular weight of the hydrate forming components (methane, ethane, propane, butane) was calculated to be 17.19 kg/kmol, representing an overall mole fraction of 0.8773. The specific gravity of the hydrate forming components was 0.6766. The correlations were used to illustrate the effect of the two non-hydrocarbon gases on the dissociation pressure of gas hydrate from 0°C to 30°C. The results are shown in Figure 5.10 in the main text.

An empirical correlation by Makogon (1988), based on gas gravity, was reported by Katz and Lee (1990). Assuming that the correlation applies to sweet gas; that is, without the presence of carbon dioxide and hydrogen sulphide (note that in the petroleum literature, sweet gas is often called dry gas). The equilibrium pressure of natural gas mixtures was expressed by the equations

$$\ln(p) = 2.3026\beta + 0.1144(T + \kappa T^2)$$

$$\beta = 2.681 - 3.811\gamma + 1.679\gamma^2$$

$$\kappa = -0.006 + 0.011\gamma + 0.011\gamma^2$$

The units used in the correlation were p kg/cm^2 and T °C. The correlation was used to calculate the equilibrium line for natural gas mixtures having gravities 0.6, 0.7 and 0.8, shown in Figure O.1 (kg/cm^2 converted to MPa by multiplying by 0.09804). The lighter the natural gas mixture, the higher the equilibrium pressure. The correlations were reported to be most accurate for the temperature range 0–15°C. The calculations shown in Figure O.1, however, have been extended to 20°C because that is the most common temperature at which gas hydrate forms in subsea pipelines (here shown to range from about 8–12 MPa).

Boiling point elevation of seawater

Increasing salinity of seawater increases its boiling point. A simple correlation by Bromley *et al.* (1974), was recently compared to the most up-to-date methods and data, and found to be one of the best (Sharqawy *et al.*, 2010). The Bromley *et al.* (1974), correlation expresses the boiling point elevation by

$$\text{BPE} = \frac{xT}{a_1}\left[1 + a_2 T + a_3 T \sqrt{x} + a_4 x - a_5 xT\left(\frac{T - a_6}{T - a_7}\right) - \frac{a_8 x(1 - x)}{T}\right]$$

where x is the weight fraction of salt and T K the temperature. Note that 5% wt. of salt gives $x = 0.05$. The constants are given in Table P.1. In the Sharqawy *et al.* (2010) paper, the value of a_8 was erroneously given as 2.583.

The Bromley *et al.* (1974), correlation was used to calculate the boiling point elevation of seawater containing 2 to 12% wt. salt (mass fraction 0.02 to 0.12). The values obtained agreed with the values given by Bromley *et al.* (1974). Using the a_8 value given by Sharqawy *et al.* (2010), gave wrong values.

The boiling point elevation (BPE) of seawater containing 5, 15 and 25% wt. salt (seawater salt) from 0 to 30°C is shown in Figure P.1. Note that seawater has salinity of 3–5% wt. The Bromley *et al.* (1974), correlation was based on salinity up to about 12% wt.

Table P.1 Constants in the Bromley *et al.* (1974) correlation.

Constant	Value
a_1	13832
a_2	0.001373
a_3	0.00272
a_4	17.86
a_5	0.0152
a_6	225.9
a_7	236
a_8	2583

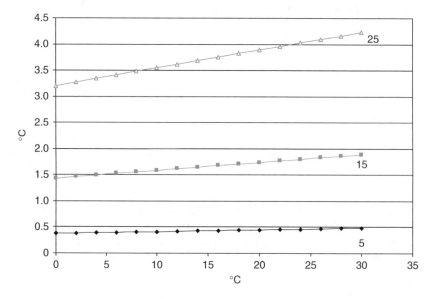

Figure P.1 BPE of seawater containing 5, 15 and 25% wt. salinity.

Solubility of gases in water

The solubility of gases in water is an essential part of the precipitation of some of the big-five flow assurance solids. Natural gas hydrates are formed from light hydrocarbons dissolved in water. Inorganic scaling depends directly on the dissolution of carbon dioxide in produced waters. The formation of naphthenates occurs because carboxylic acids dissociate and combine/react with ionic species in solution (water/brine). In vapour-liquid equilibria (VLE) and equations of state (EOS), the solubility of gases in water and liquid hydrocarbons, are similarly important.

The practical aspects of using *Henry's Law* have been discussed by Carroll (1999) and Smith and Harvey (2007). Henry's constants (and partition coefficients) for 14 gaseous solutes in water, have been presented by Fernández-Prini *et al.* (2003). A comprehensive thermodynamic model for calculating methane solubility in aqueous fluids, including the effect of dissolved NaCl, was presented by Duan and Mao (2006). The variation in *Henry's Law* constant for CO_2 as a function of temperature and salinity, was correlated by Wilkinson (2001). The concentration of NaCl ranged from 0.5–2.0 moles.

SOLUBILITY LAWS

The solubility of gas in water/brine depends on temperature, pressure and salinity. *Henry's Law* (William Henry, 1774–1836) expresses the solubility of a gas in water by the relationship

$$p_i = k_H m_i$$

where p_i is the partial pressure of a particular gas in the gas-phase, k_H is the Henry's constant of the gas and m_i is the gas concentration in the liquid-phase expressed in molality (Henry, 1803). The total pressure of a gas mixture is expressed by *Dalton's Law*

$$p = p_A + p_B + p_C + \cdots + p_N$$

where the A, B and C and N stand for the partial pressure of the particular gases (in the gas-phase). Henry's Law can also be written

$$p_i = k_H x_i$$

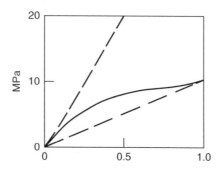

Figure Q1.1 Pressure versus mole fraction, illustrating Henry's Law (upper dashed line), Raoult's Law (lower dashed line) and non-ideal (=real) solution (middle curve).

where the x_i stands for the mole fraction of a particular gas molecule dissolved in water. The terminology used is that the dissolved gas is the *solute* and the liquid water is the *solvent*. The units of the Henry's constant depend on the units used for pressure and concentration (molality, mole fraction and others). Furthermore, the constants given in some publications, may be the *reciprocal* of the Henry's constant defined above.

   Henry's Law applies to dilute solutions, as given by the limiting expression

$$\lim_{x \to 0} \left(\frac{p}{x} \right) = k_H$$

Another law that expresses the relationship between partial pressure of dissolved gas (gas in solution) is *Raoult's Law*

$$p_i = p_i^* x_i$$

where the symbols are the same as above. However, Henry's constant has been replaced by the saturation pressure of the particular gas molecule. Meaning, the saturation pressure of the pure gas/component, named p_{pure} in *Section 4.4 – Thermodynamics of Precipitation*. The saturation pressure is not at standard conditions, but at the local (*in situ*) temperature and pressure. Raoult's Law applies in non-dilute solutions, as given by the limiting expression

$$\lim_{x \to 1} \left(\frac{p}{x} \right) = p^*$$

The subscripts have been dropped in both of the limiting expressions, just for simplicity.

   The gas solution laws have been discussed by Carroll (1999). A figure commonly used to emphasize the difference in *Henry's Law* and *Raoult's Law*, is shown in Figure Q1.1. The figure is schematic and shows the partial pressure versus the mole fraction of a particular gas molecule dissolved in water: from zero, where *Henry's Law* applies; from unity, where *Raoult's Law* applies. The solid line between the two dotted lines, is the *effective partial pressure* of the particular gas molecule.

The dotted and solid lines in Figure Q1.1 are hypothetical. The vapour pressure of the pure gas is shown to be about 10 MPa, and the Henry's constant is shown to be 40 MPa, based on mole fraction. The figure shows that the solid curve (effective partial pressure) approaches each of the laws asymptotically. Here and in general, the two laws give reasonable values for *Henry's Law* for $x < 0.1$ and for *Raoult's Law* for $x > 0.9$. Hypothetical values were used in Figure Q1.1 to illustrate the transition between the two laws. For the gases of interest in oil and gas operations, $k_H >> p^*$. For carbon dioxide, for example, $k_H \simeq 166$ MPa at 25°C and 100 kPa pressure, while the saturation pressure $p^* \simeq 5.7$ MPa at 20°C.

SOLUBILITY CONSTANTS

The solubility of gases in aqueous solutions depends on temperature, pressure and salinity. Extensive measurements have been made to determine *Henry's Law* constant for different gases at different salinities, at temperature and pressure of interest in oil and gas operations. One such study is that of Fernández-Prini *et al.* (2003), who collected experimental from the previous 60 years. For carbon dioxide, 80 data points were used; for methane, 45 data points were used. Data for 14 gases were collected and correlated. The gases were five noble gases, three diatomic gases, carbon monoxide and carbon dioxide, hydrogen sulphide, methane and ethane and sulphur hexafluoride. Henry's constants for the gases was correlated using the relationship

$$\ln\left(\frac{k_H}{p_w^*}\right) = \frac{A}{T_r} + \frac{B}{T_r}(1 - T_r)^{0.355} + \frac{C}{T_r^{0.41}} \exp(1 - T_r)$$

where A, B and C are constants specific for each of the 14 gases correlated. The pressure p_w^*, is the saturation pressure of water and $T_r = T/T_c$ the *reduced temperature* of water (the solvent). The critical temperature of water $T_c = 374.15$°C. In the correlation, the saturation pressure of water has the unit bara ($=0.1$ MPa).

Henry's constants for carbon dioxide and hydrogen sulphide are shown in Figure Q1.2. The figure shows the constants versus temperature. The pressure is atmospheric pressure below 100°C and the saturation pressure of water at higher temperatures. The value of Henry's constant at atmospheric pressure and 20°C is shown in Table Q1.1. The higher the Henry's constant, the lower the solubility; the lower the Henry's constant, the higher the solubility.

Henry's constants for methane and ethane are shown in Figure Q1.3. The figure shows the constants versus temperature, based on the saturation pressure of water. The curves for methane and ethane are similar. The Henry's constant values are orders of magnitude larger than the constants for the sour gases, CO_2 and H_2S. It means that the solubility of the light hydrocarbons are orders of magnitude lower than the solubility of the sour gases.

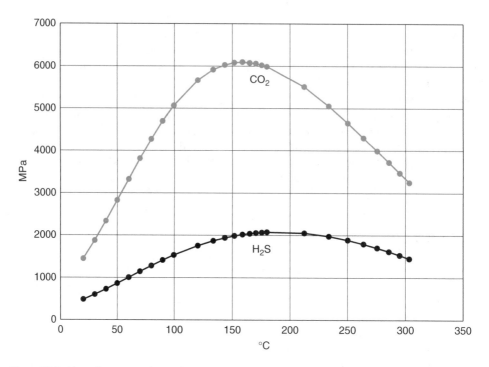

Figure Q1.2 Henry's constant for carbon dioxide (upper curve) and hydrogen sulphide (lower curve). MPa versus °C.

Table Q1.1 Henry's constant at atmospheric pressure and 20°C.

Gas	k_H MPa
CO_2	145
H_2S	48
CH_2	3610
C_2H_6	2590

CONCENTRATION OF GASES

The gas solubility law of Henry (1803) can be used to calculate the concentration of gases in aqueous solutions, meaning produced water in oil and gas operations. Taking into account the non-ideality of real gas solutions, *Henry's Law* can be expressed by

$$f_i = k_H x_i$$

where the partial pressure of the gas has been replaced by the fugacity of the gas. The fugacity represents the effective partial pressure of gas molecules, given by

$$f_i = \varphi_i p_i$$

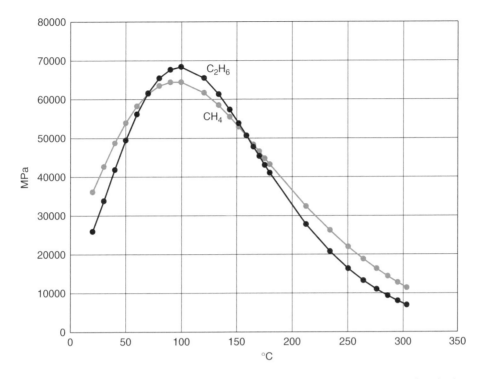

Figure Q1.3 Henry's constant for methane (upper curve at low and high temperatures) and ethane. MPa versus °C.

The symbol φ_i is the fugacity coefficient (the symbol ϕ_i is also used) of the gas, as presented in *Appendix T – Chemical Potential*. For a *pure gas only* (not gas mixtures), the coefficient is related to the compressibility factor by

$$\varphi_i = \frac{1}{z_i}$$

where z_i is the compressibility factor (z-factor) of the indicated gas component.

The z-factor of pure methane gas is shown in Figure Q1.4 against pressure, from atmospheric to 20 MPa. Two curves are shown, the lower curve for 25°C and the upper curve for 50°C. The saturation pressure of water at 180°C is 1 MPa. The z-factor in Figure Q1.4 at pressures below 1 MPa are less than 0.98, such that the loss of accuracy will be insignificant in using this $\simeq 1$ value.

The solubility of pure methane in water is shown in Figure Q1.5 versus temperature at atmospheric pressure (1 bara = 0.1 MPa). The mole fraction was calculated directly from $1/k_H$ (the spreadsheet data was in bara). The mole fraction at 20°C was found to be 2.77×10^{-5}. Converting to mass per litre water, the value was 25 mg/L, which agrees reasonably well with solubility values in the literature.

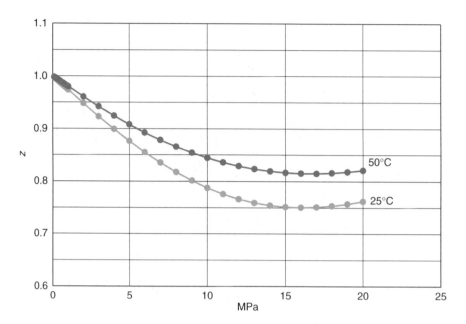

Figure Q1.4 Compressibility factor (z-factor) of pure methane versus pressure in MPa, at 25°C (lower curve) and 50°C (upper curve).

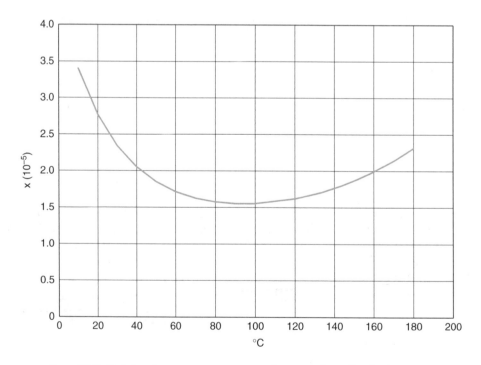

Figure Q1.5 Mole fraction versus temperature of pure methane dissolved in water.

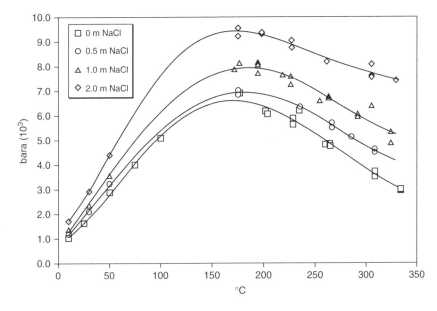

Figure Q1.6 Henry's Law constant k_H bara for carbon dioxide with temperature °C and NaCl molar concentration (Wilkinson, 2001). NaCl mol/kg from bottom and up: 0.0, 0.5, 1.0 and 2.0. Concentration of dissolved CO_2 based on mole fraction.

EFFECT OF SALINITY

The solubility of gases in aqueous solutions (water/brine) is affected by salinity. The same applies to the effect of salinity on the water vapour of natural gas (see *Appendix N – Water Vapour in Natural Gas*). The presence of salt lowers the vapour pressure of water. Salinity is a measure of the amount of salts dissolved in water. Standard seawater has a salinity of 35 g/kg, equivalent to 3.5% wt. The composition of seawater and produced water are shown in *Section 6.2 – Produced Water*. The presence of dissolved salts reduces the chemical activity of inorganic species in solution (see *Appendix R – Ionic Strength and Activity Coefficient*).

The effect of dissolved NaCl on Henry's constant for CO_2 was presented by Wilkinson (2001). The experimental data of Ellis and Golding (1963), were correlated using a fifth-degree polynomial in temperature. Four sets of coefficients were presented, for zero NaCl, 0.5 mole NaCl, 1.0 mole NaCl and 2.0 mole NaCl. For reference to the salinity of seawater, one mole NaCl corresponds to 58 g/kg, equivalent to 5.8% wt. The CO_2 Henry's constant versus temperature, for the NaCl concentrations correlated, are shown in Figure Q1.6. The unit of the constant is bara (=0.1 MPa) and the concentrations are mole fractions. Although not stated directly, it can be assumed that the pressure was the saturation pressure of water at the given temperatures. The Henry's constant values for pure water, agree reasonably well with the values shown in Figure Q1.2 above. The polynomial correlations of Wilkinson (2001) are practical and simple to use. The comprehensive thermodynamic model of Duan and Mao (2006), applies only to the effect of NaCl on the solubility of methane.

Chemical equilibrium

A typical chemical reaction can be expressed by the *Law of Mass Action*, given by the relationship

$$aA + bB \leftrightarrow cC + dD$$

where the lower-case letters show the number of atoms or molecules (called stoichiometric coefficient) and the upper-case letters the chemical symbol for the atoms or molecules. The two-ways arrow shows that the reaction can proceed from left-to-right (forward reaction) and from right-to-left (backward reaction). Given time, ranging from milliseconds to years, a chemical reaction will attain equilibrium, when the forward and backward reactions balance each other; that is, the *Gibbs Free Energy* is zero.

To quantify the amounts of reactants (left-hand-side) and products (right-hand-side) in a chemical reaction, the semi-empirical equilibrium constant concept is used

$$K_c = \frac{c_C^c c_D^d}{c_A^a c_B^b}$$

The equilibrium constant has the subscript c to indicate that it is based on the concentration of the species involved. It is common to use the subscript sp (Solubility Product) in inorganic scaling, involving solids and ions.

Interestingly, for Norwegian readers, two professors of chemistry at the now University of Oslo, are credited with the discovery that the forward reaction is given by the product of the concentration of the reactants and vice-versa for the backward reaction. Now called the *Law of Mass Action*. They also hinted that the concentrations should have an exponent. One of the original references is Waage and Guldberg (1864).

The equilibrium constant is further defined by the ratio expressing the products divided by the reactants, using effective concentrations in terms of activity

$$K_{sp} = \frac{\{\gamma_C c_C\}^c \{\gamma_D c_D\}^d}{\{\gamma_A c_A\}^a \{\gamma_B c_B\}^b}$$

The activity of the reactant A is given by

$$a_A = \gamma_A c_A$$

and similar for the other reactant B and the products C and D. The activity of a chemical species is given by the activity coefficient γ and the concentration, discussed in *Appendix R – Ionic Strength and Activity Coefficient*. The concentration c can have different units: molarity M mol/L and molality m mol/kg. The equilibrium constant for different concentration expressions will be numerically different. The molality concentration is generally preferred.

The *fugacity* of a gas molecule was defined by Lewis (1901) as

$$f = \varphi p$$

where φ is the fugacity coefficient (dimensionless) and p pressure (partial pressure in mixtures). The definition of fugacity is analogous to the definition of effective concentration of species in solution, by the use of the activity coefficient. The concept of fugacity expresses the deviation of gas from ideal gas behaviour (effective gas pressure). In situations, where all the reactants and products are gases, the gas fugacity replaces the liquid activity in the equilibrium equation above.

The real gas equation

$$pV = znRT$$

uses the symbols p for pressure, V for volume, z for gas factor (also called compressibility factor), n for number of moles, R for gas constant and T for absolute temperature. The z-factor expresses the deviation of a real gas from that of an ideal gas. Equations of state (EOS) are used to determine the z-factor for gas mixtures. The different EOS will not be discussed here because they are widely described in the literature. The ideal (without the z-factor) and real gas equations, are examples of EOS.

The fugacity of a gas is used in the expression for the chemical potential (see *Appendix T – Chemical Potential*) of real gases

$$\mu = \mu^o + RT \ln\left(\frac{f}{p^o}\right)$$

The symbols μ and μ^o stand for the chemical potential of a gas mixture and the chemical potential of the mixture at standard conditions, respectively. The unit of the chemical potential is specific energy J/mol $=$ kJ/kmol, an intensive property Extensive properties are additive, such as mass, volume and number of molecules, while intensive properties are not, such as temperature and pressure.

The equilibrium constant can be used for all kinds of chemical reactions. However, when the species (atoms or molecules) are solids or liquid water, their activity is generally assumed unity ($=1$) in the equilibrium constant expression. The assumption being that the amounts of solids or liquid water taking part in a chemical reaction, are substantially larger than that of other species, and hence effectively constant. Another way to consider this assumption, is that the activity of a solid or liquid water is already included in the equilibrium constant. The term generally assumed, means that there are special situations where the activity of liquid water needs to be included. An example being the water vapour equilibrium between natural gas and water containing dissolved salts, as in saline produced waters.

The activity (effective concentration) of an atom or molecule (specie) in a chemical reaction is given by the relationship

$$a_i = \gamma_i M_i$$

when using molarity and

$$a_i = \gamma_i m_i$$

when using molality. The activity is given by the activity coefficient γ_i for chemical species i. The activity coefficients based on molarity and molality are different (the use of the same symbol immediately above is a simplification).

Molarity changes with temperature (volume of liquid changes with temperature) while molality is independent of temperature. In dilute solutions, both molality and molarity are proportional to the mole fraction. The molality M and molality m are directly measurable quantities (extensive properties).

Chemical reactions from reactants to products may include intermediate species not shown in the overall reaction. The effects of the intermediate species are included in the overall equilibrium constant.

The question can be asked why the effective concentrations are expressed to the power of the stoichiometric coefficients. Consider an arbitrary chemical reaction of the form

$$A + B \leftrightarrow C + 3D$$

The equilibrium constant based on concentration can be written as

$$K_c = \frac{c_C c_D c_D c_D}{c_A c_B} = \frac{c_C c_D^3}{c_A c_B}$$

Ionic strength and activity coefficient

The precipitation of barium sulphate, calcium carbonate and other mineral scales, depend on the chemical equilibrium between the ionic species involved. The quantitative extent of a chemical reaction depends on the concentration of each of the species. However, the presence of surrounding ionic species affects the effective concentrations.

Imagine a positively charge species (cation) surrounded by negatively charged species (anions). In real solutions, the effective concentration of a species is lowered by the presence of neighbouring ionic species. It means that the *effective concentration* decreases with salinity. The ionic strength of an aqueous solution (Lewis and Randall, 1921) is given by the relationship

$$I = \frac{1}{2} \sum M_i z_i^2$$

where M mg/L is the molarity and z_i the ionic charge ($\pm 1, 2, 3$) The unit of I will be the same as that of the concentration used. Typical ionic strengths are 0.001 to 0.2 for potable (drinkable) groundwater, 0.7 for seawater and >5 for oilfield brines (Keetch, 2005).

The solubility of common salt (mineral halide NaCl) in water can be stated by following chemical balance

$$NaCl = Na^+ + Cl^-$$

and the solubility product constant

$$K_{sp} = \{a_{Na^+}\}\{a_{Cl^-}\} = \{\gamma_{Na^+} m_{Na^+}\}\{\gamma_{Cl^-} m_{Cl^-}\}$$

The symbols stand for activity a, activity coefficient γ and molal concentration m mol/kg. The molal concentration is more commonly used than the molar concentration M mol/L. Both of the dissolved ions have an activity coefficient. For example, if the activity coefficient of each of the halide ions is 0.5, the overall effect will be $0.5 \times 0.5 = 0.25$. The solubility product constant remains constant, irrespective of the activity coefficients. It means that the *actual concentration* of sodium chloride in solution must increase.

The seawater and formation water compositions in Table 6.1 were used to calculate their ionic strength. The results are shown in Tables R.1 and R.2 for seawater and

Table R.1 Ionic strength of seawater. The symbol M is used for both molecular weight and molar concentration.

Ions (+ & −)	M g/mol	c mg/L	M mol/L	I mol/L
Na^+	23.0	11150	0.4848	0.2424
K^+	39.1	420	0.0107	0.0054
Mg^{2+}	24.3	1410	0.0580	0.1160
Ca^{2+}	40.1	435	0.0108	0.0217
St^{2+}	87.6	6	0.0001	0.0001
Ba^{2+}	137.3	0	0.0000	0.0000
HCO_3^-	61.0	150	0.0025	0.0012
SO_4^{2-}	96.1	2800	0.0291	0.0583
Cl^-	35.5	20310	0.5721	0.2861
				0.7312

Table R.2 Ionic strength of formation water. The symbol M is used for both molecular weight and molar concentration

Ions (+ & −)	M g/mol	c mg/L	M mol/L	I mol/L
Na^+	23.0	17465	0.7593	0.3797
K^+	39.1	325	0.0083	0.0042
Mg^{2+}	24.3	124	0.0051	0.0102
Ca^{2+}	40.1	719	0.0179	0.0359
St^{2+}	87.6	198	0.0023	0.0045
Ba^{2+}	137.3	428	0.0031	0.0062
HCO_3^-	61.0	834	0.0137	0.0068
SO_4^{2-}	96.1	0.1	0.0000	0.0000
Cl^-	35.5	28756	0.8100	0.4050
				0.8525

formation/produced water, respectively. The calculated ionic strengths were $I_{SW} = 0.73$ and $I_{FW} = 0.85$. These values are slightly higher than for groundwater but much lower than given by Keetch (2005) for brine.

Models/theories have been developed to express the activity coefficient as a function of ionic strength. The models are widely available in the literature, including Keetch (2005), who gives Drever (1997) as the main reference. The best known is the Debye-Hückel (1923) theory

$$\log \gamma_\pm = -0.5 z^2 \sqrt{I}$$

The constant 0.5 is an approximation for liquid water at 25°C. The symbol z is the ionic charge of the specie for which the activity coefficient $\gamma_\pm$ is being calculated. Three model assumptions are the following: (1) Electrolytes completely dissociated into ions in solution, (2) Solutions of electrolytes are very dilute, ionic strength less than 0.01

Table R.3 Activity coefficient $\gamma_\pm$ estimated for monovalent ($1\pm$) and divalent ($2\pm$) ions using three different interaction models (see text).

I	$\sqrt{I}$	DHL (1)	DHE (1)	PM (1)	DHL (2)	DHE (2)	PM (2)
0.001	0.0316	0.96	0.96	0.96	0.86	0.86	0.86
0.010	0.1000	0.88	0.89	0.90	0.63	0.68	0.65
0.100	0.3162	0.70	0.78	0.79	0.25	0.40	0.38
1.000	1.0000	0.30	0.60	0.69	0.00	0.20	0.20

and (3) Each ion is surrounded by ions of the opposite charge, on average. Due to the limitation of the basic Debye-Hückel model, it was extended to

$$\log \gamma_\pm = -0.5z^2 \frac{\sqrt{I}}{1 + 0.33a\sqrt{I}}$$

where the constant 0.33 is an approximation for water at 25°C. The symbol a Å is an adjustable parameter corresponding to the size of the ion (ionic radius) for which the activity coefficient is being calculated. The extended Debye-Hückel model can be used for ionic strength less than 0.1. The first of the Debye-Hückel models above is called the *Limiting Law* while the second model is called the *Extended Law*.

The activity coefficient model used by Kaasa (1998) was that of Pitzer (1973). The Pitzer (1973) ionic interaction model is widely used. It is considered the best model for ionic solutions. After the original publication, a number of improvements and extensions have been made to the model. The equations and details of the model will not be presented here.

Figures illustrating the activity coefficient with ionic strength have been presented by Reeves (2009), for monovalent and divalent ions. The figures were used to read-off values for three models: (1) Debye-Hückel Limiting Law (DHL), (2) Debye-Hückel Extended Law (DHE) and (3) Pitzer's Model (PM). The read-off values are shown in Table R.3.

The activity coefficient versus $\sqrt{I}$ for monovalent ions is shown in Figure R.1. The three lines, from bottom and up, represent DHL (1), DHE (1) and PM (1). The activity coefficient versus $\sqrt{I}$ for divalent ions is shown in Figure R.2. The square root of 0.01 is 0.1 and the square root of 0.01 is 0.1. Considering the figures, R.1 and R.2, the square root values are the limiting values for the Deye-Hückle Limited Law and Debye-Hückle Extended Law. The square root of the ionic strength of seawater and formation water is 0.85 and 0.92, respectively. It means that the Limited Law cannot be used for seawater and formation water. The Extended Law and the Pitzer's Model indicated similar values in the figures and should therefore be preferred for seawater and formation water.

The two figures below are for monovalent and divalent ions only. The ionic strength shown is much lower than found in produced brines in the oil and gas industry. Seawater and formation water have both monovalent and divalent ions. It is not clear in the present work what ionic radius should be used in the Extended Law for real aqueous solutions (seawater and formation water). The two figures with the three are presented here for illustration purposes; to show how the activity coefficient

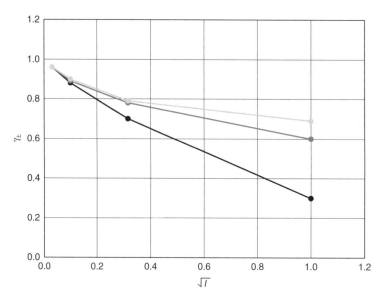

Figure R.1 Activity coefficient $\gamma_\pm$ versus the square root of ionic strength $\sqrt{I}$ for monovalent ions. The three lines, from bottom and up, represent DHL (1), DHE (1) and PM (1).

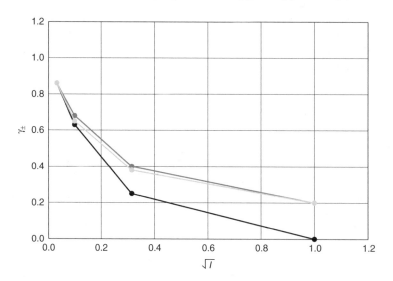

Figure R.2 Activity coefficient $\gamma_\pm$ versus the square root of ionic strength $\sqrt{I}$ for divalent ions. The three lines, from bottom and up, represent DHL (2), DHE (2) and PM (2).

decreases profoundly with increasing ionic strength, especially for divalent ions. The Pitzer Model, being more complicated to calculate that the other models, is nevertheless the model used in most software packages, both academic and commercial. The Pitzer Model is considered the most accurate of the three models presented in the present text.

Appendix S

Gibbs free energy

Thermodynamics is a field of study dealing with the conversion of energy in processes, described by extensive and intensive properties. The bedrocks of thermodynamics are the conservation of energy and the definition of entropy. Conservation means that energy cannot be created nor destroyed. Entropy expresses the transformation from an ordered to a disordered system. The value of entropy approaches zero at zero absolute temperature. Extensive properties are additive and include mass, volume, enthalpy and entropy. Intensive properties are non-additive and include pressure, temperature, density and concentration.

A number of thermodynamic formulations have been developed to facilitate the solution of practical problems. One of the important formulations is Gibbs Free Energy, defined by the relationship

$$G \equiv H - TS$$

where enthalpy is defined by the relationship

$$H \equiv U + pV$$

and internal energy by the summation of the relevant forms of energy, for example

$$U \equiv \sum (E_{thermal} + E_{mechanical} + E_{chemical})$$

Mechanical energy consists of potential energy (energy of position) and kinetic energy (energy of motion). An energy balance will in practice also include work, which is the form of energy a system exchanges with the surroundings. Not included above are the other forms of energy, such as magnetic, electric and nuclear.

Entropy can be defined by the relationship

$$dS \equiv \frac{\delta Q}{T}$$

where δQ stands for an infinitesimal change in heat (thermal energy), in a reversible process. The unit of entropy is energy per absolute temperature, joule/kelvin J/K. An infinitesimal change in heat will occur at constant temperature (isothermal process), making dS an exact differential, such that entropy is a property of a system. Energy can also be expressed by specific energy, per mass J/kg or per mole J/mol. It follows that entropy can also be expressed as specific entropy.

Heat flows from a hot object to a cold object. The change in entropy can be used to describe any process going from a concentrated state to a dispersed state. The change in entropy goes only in one direction and can be described in terms of time by the inequality

$$\frac{dS}{dt} > 0$$

expressing that entropy always increases in any process.

Helmholtz Free Energy is defined by the relationship

$$F \equiv U - TS$$

Instead of the symbol F, the symbol A is also used in the literature. In the present text, F is also used for *force* and A for *area*; it should be clear from the context, which is which. For changes in energy from state-to-state, the fundamental Gibbs Free Energy and Helmholtz Free Energy relationships can be written

$$\Delta G = \Delta H - \Delta(TS)$$

$$\Delta F = \Delta U - \Delta(TS)$$

The purpose of the Helmholtz Free Energy relationship (equation) is to express a thermodynamic process carried out at constant temperature and constant volume. For a process carried out at *constant temperature* the above relationships reduce to

$$\Delta G = \Delta H - T\Delta S$$

$$\Delta F = \Delta U - T\Delta S$$

Similarly, for change in enthalpy in a *constant pressure* process, the change in enthalpy can be written

$$\Delta H = \Delta U + p\Delta V$$

Combining the above three equations gives

$$\Delta G = \Delta F + p\Delta V$$

for a process at *constant temperature* and *constant pressure*.

For a thermodynamically reversible process at constant temperature, based on the definition of entropy

$$\Delta S = \frac{Q}{T}$$

the Helmholtz Free Energy equation (at constant temperature and constant volume) can be written as

$$\Delta F = \Delta U - Q$$

The conservation of energy can be expressed by the relationship

$$\Delta U = Q - W$$

The Q is the thermal energy absorbed from the surroundings, and the W the work lost to the surroundings (external work). It follows that

$$W = -\Delta F$$

expressing the maximum available work of a reversible process, at constant temperature and constant pressure. Maximum available work is also called exergy, becoming zero at equilibrium. It is worth noting that F depends on T and V, while G depends on T and p. Pressure is more readily measured than V, making Gibbs Free Energy more easily used than Helmholtz Free Energy.

Consider the Gibbs Free Energy expressed for two special cases: (1) constant pressure process and (2) constant temperature process. The following general expression can be written

$$G = U + pV - TS$$

and in differential form

$$dG = dU + pdV + Vdp - TdS - SdT$$

From energy conservation

$$\delta Q = dU + \delta W$$

The δ's an infinitesimal change, while the d is an operator on a state variable (extensive property). Assuming that the infinitesimal work is due to volume change only, the following can be written

$$\delta Q = dU + pdV$$

The term for work is that for an idea gas. From the definition of entropy, results in

$$TdS = dU + pdV$$

Combining with the *Gibbs Free Energy* equation in differential form gives

$$dG = Vdp - SdT$$

For the special case of *constant pressure*

$$dG = -SdT$$

and

$$\left(\frac{\partial G}{\partial T}\right)_p = -S$$

For the special case of *constant temperature*

$$dG = Vdp$$

and

$$\left(\frac{\partial G}{\partial p}\right)_T = V$$

For an appreciable process (not infinitesimal) taking place at *constant temperature*, the following integral can be written

$$\Delta G = \int_{p_1}^{p_2} Vdp$$

Using the ideal gas law

$$pV = nRT$$

and inserting into the integral, gives

$$\Delta G = nRT \int_{p_1}^{p_2} \frac{dp}{p}$$

and

$$\frac{\Delta G}{n} = RT \ln\left(\frac{p_2}{p_1}\right)$$

It may be of interest to note that if the above equation is multiplied by the mass rate kg/s and divided by the molecular weight kg/mol, it expresses the *isothermal compression power* W for an ideal gas.

The *Gibbs-Helmholtz Equation*, expresses the temperature dependence of Gibbs Free Energy in a *constant pressure* process.

$$\Delta G = \Delta H + T\left(\frac{\partial(\Delta G)}{\partial T}\right)_p$$

The derivation of the *Gibbs-Helmholtz Equation* starts with the basic Gibbs Free Energy equation

$$G = H - TS$$

$$dG = -SdT + Vdp$$

From above, for a constant pressure process

$$\left(\frac{\partial G}{\partial T}\right)_p = -S$$

By substitution from the Gibbs Free Energy equation, the following can be written

$$G = H + T\left(\frac{\partial G}{\partial T}\right)_p$$

and

$$\frac{G}{T} = \frac{H}{T} + \left(\frac{\partial G}{\partial T}\right)_p$$

Taking the partial derivative

$$\left(\frac{\partial G/T}{\partial T}\right)_p = -\frac{G}{T^2} + \frac{1}{T}\left(\frac{\partial G}{\partial T}\right)_p$$

and

$$\left(\frac{\partial G/T}{\partial T}\right)_p = \frac{1}{T}\left[-\frac{G}{T} + \left(\frac{\partial G}{\partial T}\right)_p\right]$$

From the expression for G/T above, substituting into the above equation, the following can be written

$$\left(\frac{\partial G/T}{\partial T}\right)_p = \frac{1}{T}\left\{-\left[\frac{H}{T} + \left(\frac{\partial G}{\partial T}\right)_p\right] + \left(\frac{\partial G}{\partial T}\right)_p\right\}$$

and

$$\left(\frac{\partial (G/T)}{\partial T}\right)_p = -\frac{H}{T^2}$$

This is one form of the *Gibbs-Helmholtz Equation*. A more practical form is

$$\left(\frac{\partial \left(\frac{\Delta G}{T}\right)}{\partial \left(\frac{1}{T}\right)}\right)_p = \Delta H$$

For endothermic reactions $\Delta H^o > 0$ and for exothermic reactions $\Delta H^o < 0$.

The *derivation* of the *van't Hoff Equation*, uses the *Gibbs-Helmholtz Equation* as a starting point at standard conditions (superscript o).

$$\left(\frac{\partial(\Delta G^{\circ}/T)}{\partial(1/T)}\right)_p = \Delta H^{\circ}$$

In *Appendix T – Chemical Potential* the equations

$$\Delta G^{\circ} = -RT \ln K$$

$$K = \exp\left(\frac{-\Delta G^{\circ}}{RT}\right)$$

were developed. In some texts, the relationship between the standard change in Gibbs Free Energy and the equilibrium constant, is called the *Lewis Equation*. The subscript p was used in *Appendix T – Chemical Potential*, to signify that the equilibrium constant for ideal gas was based on pressure. Note therefore, the subscript p does not signify a constant pressure process. It has been noted in the main text and appendices, that the equilibrium constant subscript depends on the intensive properties used (pressure/fugacity and concentration/activity).

A combination of the equations above, gives the *van't Hoff Equation*

$$\left(\frac{\partial \ln K}{\partial(\frac{1}{T})}\right)_p = -\frac{\Delta H^{\circ}}{RT^2}$$

Expressed in terms of normal differentials instead of partial differentials

$$\frac{d \ln K}{dT} = -\frac{\Delta H^{\circ}}{RT^2}$$

can be written

$$d \ln K = -\frac{\Delta H^{\circ}}{R}d\left(\frac{1}{T}\right)$$

$$\ln K_2 - \ln K_1 = \ln\frac{K_2}{K_1} - \frac{\Delta H^{\circ}}{R}\left(\frac{1}{T_2} - \frac{1}{T_1}\right)$$

$K_2 = K(T_2)$ and $K_1 = K(T_1)$. A plot of $\ln K$ on the y-axis and $1/T$ on the x-axis, gives the slope $-\Delta H^{\circ}/R$. For an endothermic reaction, the slope is negative, less than zero (<0) and for an exothermic reaction, the slope is positive (>0). When the slope is less than zero, the equilibrium constant increases with temperature, and when the slope is larger than zero, the equilibrium constant decreases with temperature. *Le Châtelier's Principle* is illustrated in the above equation: when a system at equilibrium is subject to change in concentration, temperature, volume or pressure, then the system readjusts itself to counteract the effect of the applied change, and a new equilibrium is established. Simply stated, whenever a system in equilibrium is disturbed, the system will adjust itself in such a way that the effect of the change will be eliminated.

Chemical potential

Chemical potential expresses the tendency of a substance to transfer from one thermodynamic state to another. The states involved can be vapour and liquid phases in equilibrium. The states can also involve chemical reactants and products in equilibrium. The chemical potential has a function analogous to temperature and pressure. A temperature difference causes heat to flow; a pressure difference causes bulk matter to flow. In general, the chemical potential can be determined from an equation of state (EOS).

Chemical potential for a *pure substance* is defined by the relationship

$$\mu \equiv \left(\frac{\partial G}{\partial n} \right)_{T,p}$$

where G is the Gibbs Free Energy and n the number of moles. The definition refers to constant temperature T and constant pressure p. In the case of a mixture, the chemical potential can be expressed by the relationship

$$\mu_i = \left(\frac{\partial G}{\partial n_i} \right)_{T,p,n_{j \neq i}}$$

where $n_{j \neq i}$ refers to the substances in the mixture, other than the substance n_i.

Because chemical potential does not have an absolute value, it needs to be related to some reference state. The reference state used is the chemical potential at standard conditions; temperature 25°C and pressure 1 bara ($=100\,\text{kPa}$).

In thermodynamics, it is common to derive expression assuming constant temperature and pressure. Expressions based on a constant pressure process, allow the temperature to vary, and vice versa. It is also common to use the ideal gas law in the derivation of expressions, thereafter to make provisions for the same expressions to apply to non-ideal gases, ideal and non-ideal liquid solutions and even solids. For non-ideal gases, the pressure is replaced by *fugacity*, and in non-ideal solutions, the molality is replaced by *activity*. Both involve coefficients, the fugacity coefficient and the activity coefficient, respectively.

The fugacity coefficient ϕ and the activity coefficient γ are defined for pure components by the expressions

$$f = \phi p$$

$$a = \gamma m$$

where f is the fugacity, ϕ the fugacity coefficient and p the pressure. The symbol a stands for activity, γ the activity coefficient and m for molality. The expressions above are for a pure substance. In mixtures, the expressions should have subscripts, to indicate which of the components they represent (A, B, C, ...). The fugacity has the unit of pressure and the activity has the unit of concentration, here molality mol/kg. Originally, the fugacity was used to express the non-ideal behaviour of gases. Subsequently, is has also been used for the non-ideal behaviour of liquids and even solids (Lira-Galeana et al., 1996).

The general relationship for the change in *Gibbs Free Energy*, for an ideal gas in a constant temperature process, can be expressed by

$$\Delta G = nRT \ln\left(\frac{p_2}{p_1}\right)$$

as shown in *Appendix S – Gibbs Free Energy*. In a constant temperature process, the pressure changes from one state to another state. Assuming that the pressure in the denominator is the standard pressure, and the pressure in the numerator is the process pressure

$$\Delta G = nRT \ln\left(\frac{p}{p^o}\right)$$

$$\Delta G = nRT (\ln p - \ln p^o)$$

Furthermore, assuming that the pressure in the numerator is the partial pressure of gas/substance A, the overall *Gibbs Free Energy* equation can be written

$$G_A = G_A^o + n_A RT \ln p_A$$

Two additional assumptions were used to arrive at this equation. First, that the standard pressure is 1 bara (=100 kPa) and second, that the process pressure can be replaced by the partial pressure of gas/substance. When $p^o = 1$, then $\ln p^o = 0$. Standard pressure is nowadays 1 bara, but was previously 1 atm (=101.325 kPa).

The unit of the *Gibbs Free Energy* above is joule (=J). When divided by the number of moles, it becomes the chemical potential μ J/mol, the molar *Gibbs Free Energy*. The relationships below represent the various ideal and real (non-ideal) substances. The chemical potential at standard pressure in the relationships, although expressed by the same chemical potential symbol, are based on the property expressing the concentrations (p, f, m and a).

Chemical potential of ideal gases

$$\mu = \mu^o + RT \ln(p)$$

Chemical potential of real gases

$$\mu = \mu^o + RT \ln(f)$$

Chemical potential of an ideal solution

$$\mu = \mu^o + RT \ln(m)$$

Chemical potential of a real solution

$$\mu = \mu^o + RT \ln(a)$$

The *Gibbs Free Energy* can be related to the equilibrium constant of chemical reactions. Consider the typical reaction

$$A + B \leftrightarrow C + D$$

The change in free energy of a reaction, is the sum of the free energies of the products, minus the free energies of the reactants

$$\Delta G = G_C + G_D - G_A - G_B$$

and

$$\Delta G = (G_C^o + RT \ln p_C) + (G_D^o + RT \ln p_C) - (G_A^o + RT \ln p_A) - (G_B^o + RT \ln p_B)$$

The individual molar free energy terms can be expressed collectively such that

$$\Delta G = \Delta G^o + RT \ln \frac{p_C p_D}{p_A p_B}$$

In may texts, the collective partial pressure terms are called the reaction quotient

$$Q = \frac{p_C p_D}{p_A p_B} = K_p$$

At equilibrium, the reaction quotient is equivalent to the equilibrium constant. The subscript p means that the constant is expressed using pressure (here, partial pressure). At equilibrium, the change in *Gibbs Free Energy* is zero such that

$$\Delta G^o = -RT \ln K_p$$

and

$$K_p = \exp\left(\frac{-\Delta G^o}{RT}\right)$$

while the subscript p stands for pressure, the equilibrium constant can also be based on other expressions of concentration: K_m for molality and K_x for mole fraction. For ionic solutions K_a is used when the concentration is expressed in terms of activity. In inorganic scaling, the most common subscript is *sp* for solubility product.

Appendix U

Solubility product constants

Values of the solubility product constant at standard conditions (standard state), are readily available in the literature, both in scientific/technical articles and handbooks (Perry *et al.*, 1998; Lide, 2004). However, the values from different sources are not always consistent. Care has to be taken when selecting values to use in calculations. For example, in Table 6.7 the solubility product constants K_{sp} for anhydrite and gypsum are 9.1×10^{-6} and 2.4×10^{-5}. Other solubility constants found in the literature are 4.9×10^{-5} and 3.1×10^{-5} for anhydrite and gypsum, respectively. Perhaps not a large difference, but because the exponential function is used in some calculations, small errors can have large effects. A potential inherent problem is the sparsity of information about the pertinent chemical and thermodynamic conditions for which the solubility product constants are reported.

In olden days, standard conditions (standard state) were 25°C and 1 atmospheric pressure (=101.325 kPa). Nowadays, the standard pressure is 1 bara (=100 kPa), which is 1.3% lower. In engineering calculations, this difference is of negligence consequence. An additional specification, is that many K_{sp} tables are based on a 1 M solution, meaning 1 mol/L, despite the fact that molality m mol/kg is more commonly used.

It is important to calculate the solubility product constant at temperatures of relevance in oil and gas operations. Literature data have been compiled by Kaasa (1998), in the appropriate temperature ranges. The constants are *thermodynamic solubility constants*, meaning that they include the effect of non-ideality, as expressed by the activity coefficient. The data for the minerals of interest were fitted to an empirical polynomial function of the form

$$\ln K_{sp} = \frac{\alpha_1}{T} + \alpha_2 + \alpha_3 \ln T + \alpha_4 T + \frac{\alpha_5}{T^2}$$

The coefficients α_1 to α_5 were determined from the available data and are shown in Table U.1, along with the valid temperature range in Celsius T °C. The solubility product constant K_{sp} is dimensionless. The temperature used in the Kaasa-correlations *must be* in absolute temperature T K.

The correlations are for *constant pressure*. If the standard pressure of 100 kPa is used, the water will boil above 100°C. In lieu of an unambiguous explanation, it must be assumed that the effective pressure was at least the vapour pressure of water. The correlations are utmost practical, because they can be used in computer programs, including software for calculations of mineral equilibria in aqueous solutions. The data

Table U.1 Coefficients in empirical correlations of the natural logarithm of *Solubility Product Constant* with temperature (Kaasa, 1998). Atmospheric pressure at temperatures below the boiling point of water; saturation pressure of water at higher temperatures.

ln K_{sp}	α_1	α_2	α_3	α_4	α_5	T °C
$CaSO_4$	−2234.4	11.6592	0	−0.048231	0	0–200
$BaSO_4$	37588	−747.61	119.28	−0.16283	−2.88 × 10^6	0–200
$CaCO_3$	−170410	3044.27	−475.704	0.376097	9.4195 × 10^6	0–300
$SrSO_4$	46550	−434.76	55.44	−0.062931	−4.4962 × 10^6	0–300
$CaSO_4 \cdot 2H_2O$	−26309.9	815.978	−138.361	0.1678630	18.6143	0–100
FeS	−6813.35	50.19636	−3.36	−0.055377	0	25–200
$FeCO_3$	129.97	−50.205	7.3134	−0.052913	0	25–300

for NaCl was not correlatable, using the same form of equation show above (Kaasa, 1998). Instead, an ordinary fourth order polynomial was used

$$K_{sp}(NaCl) = -814.18 + 7.4685T - 2.3262 \times 10^{-2}T^2$$
$$+ 3.0536 \times 10^{-5}T^3 - 1.4573 \times 10^{-8}T^4$$

For physical properties that change with temperature and marginally with pressure, it is common to correlate experimental data with an empirical relationship of the form

$$C_p = \alpha + \beta T + \gamma T^2$$

for example, for heat capacity C_p with the units kJ/kg·K or kJ/mol·K. The Greek letters are coefficients specific for the fluid (liquid or gas) in question; their numerical values can be found in literature articles and handbooks. Note that heat capacity has the same unit as the gas constant R. The above type of equation is appropriate because heat capacity is strongly dependent on temperature and almost independent of pressure (the subscript p stands for constant pressure). The heat capacity can be used to determine the heat of formation/reaction based on the simple relationship

$$\Delta H = C_p \times \Delta T$$

At temperatures other than 25°C, the enthalpy of the formation/reaction of compounds in solutions has two parts: The enthalpy at standard conditions and an integral of the heat capacity for temperatures from T K to some higher/lower temperature T K. The heat capacity and entropy of most components depends on temperature, but insignificantly on pressure.

$$\Delta H = \Delta H^\circ + \int_{T^o}^{T} C_p dT$$

$$\Delta S = \Delta S^\circ + \int_{T^o}^{T} \frac{C_p}{T} dT$$

Consider the effect of *pressure* on the solubility product constant K_{sp}. The effect is quite small, compared to the effect of temperature. For a constant temperature process (see *Appendix S – Gibbs Free Energy*) the partial derivative of *Gibbs Free Energy* was derived as

$$\left(\frac{\partial G}{\partial p}\right)_T = V$$

and in *Appendix T – Chemical Potential*, the standard state *Gibbs Free Energy* at equilibrium was derived as

$$\Delta G^o = -RT \ln K$$

The partial derivative above can be recast as

$$\left(\frac{\partial \Delta G_m^o}{\partial p}\right)_T = \Delta V_m^o$$

where the superscript indicates standard state/conditions and the subscript m indicates molar conditions. V_m^o stands for *molar volume* at standard conditions. In many publications the molar volume is expressed by the lower case v to indicated a partial/specific property. The relevant unit is L/mol or some equivalent such as m³/mol. The partial volume of an aqueous solution with several species, A and B etc., can be written as

$$V_m^o = n_A V_m^o(A) + n_B V_m^o(B) + \cdots$$

Combining the two *Gibbs Free Energy* equations above at standard conditions gives

$$d \ln K = \frac{-\Delta V_m^o}{RT} dp$$

which upon integration results in

$$\ln\left(\frac{K_p}{K_{p^o}}\right)_T = \frac{-\Delta V_m^o}{RT}(p - p_o)$$

Molar volume m³/mol is the inverse of molar density mol/m³. Specific volume m³/kg and specific density kg/m³ are related properties.

The isothermal (constant temperature) elastic bulk modulus (bulk modulus of elasticity) is defined from

$$B_T = -V\left(\frac{\partial p}{\partial V}\right)_T$$

The modulus unit B_T Pa and is the inverse of compressibility κ Pa^{-1}. In differential form the term $\Delta V/V$ expresses the relative change in volume when pressure is applied. The bulk modulus can also be expressed in terms of specific volume

$$B_T = -v\left(\frac{\partial p}{\partial v}\right)_T$$

Table U.2 Properties of selected substances at 15°C. The Weight is the molecular weight and Modulus is the bulk elastic modulus.

Substance	Formula	Density kg/m³	Weight kg/kmol	Modulus Pa
Ethylene glycol	$C_2H_6O_2$	1110	62.07	3.03×10^9
Methanol	CH_3OH	791	32.04	0.952×10^9
Water	H_2O	999	18.02	2.22×10^9

Table U.3 Molar volume at 1 bara and calculated molar volume at 100 bara and liquid density at 100 bara.

Substance	Volume m³/kmol	Volume @ 100 bara	Density @ 100 bara	Difference kg/m³
Ethylene glycol	0.0559	0.0557	1112	2 (0.8%)
Methanol	0.0405	0.0401	799	8 (1.0%)
Water	0.0180	0.0180	1003	4 (0.4%)

Rearranging and using the differential form

$$\frac{1}{v} dv = -\frac{1}{B_T} dp$$

Upon integration and rearranging

$$v = v_o \exp\left[\frac{-1}{B_T}(p - p_o)\right]$$

The Kaasa (1998) solubility product constant correlations above, show changes with temperature at constant pressure. Furthermore, molality m mol/kg is used for concentrations because liquid density ρ kg/m³, for practical purposes, is independent of pressure. To test/examine the constant pressure assumption, the equation immediately above can be used. Selected properties are shown in Table U.2, relevant for hydrate inhibition (see *Chapter 5 – Natural Gas Hydrate*). The values in Table U.2 are for 15°C because the bulk modulus of elasticity in a table by Martínez (2016), was reported at this temperature. The bulk modulus B is the inverse of compressibility κ.

Based on the properties in Table U.2, and the above equation for the molar volume v m³/kmol, the molar volume and liquid density was calculated at 100 bara and 15°C. The calculation results are shown in Table U.3. The liquid density increases slightly when the pressure increases from 1 bara to 100 bara. The increase amounts to 0.8%, 1.0% and 0.4%. Clearly, the assumption of constant pressure is reasonable in most instances.

Solubility and nucleation

The formation of solids from solutions occurs by crystallization, for example hydrate, scale and wax. The overall process can be divided into three main steps. First, the solution must become supersaturated (inherent driving force for crystallization). Second, nucleation must take place (birth of new crystals). Third, the crystals must grow in solution (bulk) and/or on already deposited crystals.

The formation and growth of crystals is a three-dimensional process. The process is not instantaneous and requires removal of thermal energy (heat of crystallization). A solution must be cooled below the solubility limit to become supersaturated, as illustrated in Figure V.1 (sometimes called Ostwald-Miers diagram). New crystals can be formed when the solution has been cooled below the nucleation line. The zone between the solubility line and nucleation line is called the metastable zone. Crystal growth takes place in both the metastable zone (on existing crystals) and the supersaturated zone (also called labile zone).

The difference in temperature between the solubility line and crystallization line depends on the system in question, the materials (hydrate, scale, wax), presence of impurities (sand, silt, corrosion products) and flow conditions (degree of turbulence). For example, in clean laboratory systems, hydrate will start to form 4–6°C below

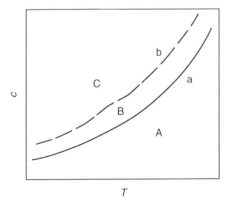

Figure V.1 Concentration with temperature for crystallization from solution (Ostwald-Miers Diagram). A, stable undersaturated state. B, metastable zone. C, labile supersaturated state. a, solubility line. b, nucleation line.

the equilibrium line. In dirty systems, as in oil and gas production, this temperature difference will be smaller.

Asphaltenes are amorphous; that is, non-crystalline. Nevertheless, a certain super-saturation is required for asphaltenes to precipitate from crude oil (de Boer *et al.*, 1995). And precipitation takes time (also in SARA analysis for asphaltenes). It means that precipitation and deposition will not necessarily occur locally where supersaturation first occurs.

The reverse process to crystallization, the solution of precipitated solids, occurs at the solubility line, not at the nucleation line. The formation and dissolution of hydrates is an example of the crystallization and dissolution limits illustrated in Figure V.1. In the case of batch experiments for hydrates the y-axis will not be concentration but pressure (dissociation pressure).

Calcite and silica chemistry

Calcite and silica are both pH-dependent scales. It means that the concentration/activity of the constituent species changes with changes in acidity. The chemical balances for both calcite and silica include three main species and the hydrogen ion. The distribution of the species against pH depend of the equilibrium constants, also known at solubility product constants. The mole fraction of the calcium and silica species based on the constants at standard state, 25°C and 100 kPa. The value of the equilibrium/solubility product constants at other temperatures can be estimated using the *van't Hoff Equation*.

Carbon dioxide is a key component in carbonate chemistry. The solubility of calcium dioxide in water, can be expressed by *Henry's Law* as presented in *Section 6.3 – Scaling Minerals*. The solution and hydration of carbon dioxide in water can be expressed by

$$CO_2\,(g) \leftrightarrow CO_2\,(aq)$$
$$CO_2\,(aq) + H_2O \leftrightarrow H_2CO_3$$

where H_2CO_3 is carbonic acid. It may be noted that the heat of dissolution of carbon dioxide in water is about 20 kJ/mol. Carbonic acid is classified as a weak acid. The hydration equilibrium constant is given by

$$K_h = \frac{[H_2CO_3]}{[CO_2]}$$

Reported values of $K_h = 1.70 \times 10^{-3}$ at 25 °C and $K_h = 1.3 \times 10^{-3}$. As always, equilibrium constant values reported in the literature do vary. The equilibrium constants herein are based on near-zero ionic strength. Above and below the specification (*aq*) is conveniently left out because it should be obvious that the species are dissolved in water. Furthermore, the present equilibrium constants should have the superscript indicating standard state conditions, namely K°.

The *first* dissociation of carbonic acid to bicarbonate is given by

$$H_2CO_3 \leftrightarrow HCO_3^- + H^+$$

and

$$K_1 = \frac{[HCO_3^-][H^+]}{[H_2CO_3]}$$

Equilibrium constants found in the literature include $K_1 = 2.00 \times 10^{-4}$ and $K_1 = 2.50 \times 10^{-4}$.

The *second* dissociation of carbonic acid (bicarbonate ion to carbonate ion) is given by

$$HCO_3^- \leftrightarrow CO_3^{2-} + H^+$$

and

$$K_2 = \frac{[CO_3^{2-}][H^+]}{[HCO_3^-]}$$

Values found in the literature include $K_2 = 4.69 \times 10^{-11}$ and $K_2 = 5.61 \times 10^{-11}$.

Because it is not possible by chemical methods to distinguish between CO_2 (*aq*) and H_2CO_3 (*aq*), their total concentration is commonly expressed as $H_2CO_3^*$ or simply as CO_2 (*aq*). The concentration of bicarbonate dissolved in water is much lower than the concentration of dissolved carbon dioxide. It follows that the *first* dissociation of carbonic acid can also be written as

$$K_{11} = \frac{[HCO_3^-][H^+]}{[CO_2 (aq)]}$$

The equilibrium constant has been given a double subscript to distinguish it from the first equilibrium constant. Quoted values of the constant are $K_{11} = 4.45 \times 10^{-7}$ and $K_{11} = 4.6 \times 10^{-7}$.

Calcium carbonate scale results from the reaction balance between the calcium ion and the carbonate ion

$$Ca^{2+} + CO_3^{2-} \leftrightarrow CaCO_3 (s)$$

Calcium carbonate will also be dissolved in water. The same chemical relationship applies for the dissolution equilibrium

$$Ca^{2+} + CO_3^{2-} \leftrightarrow CaCO_3 (aq)$$

but now the specification (*aq*) has been added to the calcium carbonate.

The carbonates $CaCO_3$ and $FeCO_3$ are pH-dependent, as shown in Table 6.3. The hydrogen ion H^+ appears in the chemical balances for the bicarbonate and carbonate ions above. It follows that their chemical balances depend on the acid-base properties of aqueous solutions. One way to illustrate this dependence is to calculate the mole fraction of the carbonate species in a range of acidity expressed as pH. For simplicity, the carbonate ion species are arbitrarily named as follows

$$A = H_2CO_3, \quad B = HCO_3^- \quad and \quad C = CO_3^{2-}$$

The mole fraction of the A ion can then be written as

$$f_A = \frac{A}{A + B + C}$$

and similarly for the ions B and C. Using algebra the following can be shown

$$f_A = \frac{[H^+]^2}{[H^+]^2 + [H^+]K_1 + K_1 K_2}$$

$$f_B = \frac{[H^+]K_1}{[H^+]^2 + [H^+]K_1 + K_1 K_2}$$

$$f_C = \frac{K_1 K_2}{[H^+]^2 + [H^+]K_1 + K_1 K_2}$$

such that

$$f_A + f_B + f_C = 1$$

The immediately above mole fraction equations are used in the main text, showing that the A-ion dominates at low pH, the B-ion dominates in the range $7 < pH < 10$ and the C-ion at higher pH.

According to Walther (2009), the chemical equilibrium equations for silica can be represented by the following five balances

$$SiO_2\ (A) \leftrightarrow SiO_2\ (aq)$$

$$SiO_2\ (Q) \leftrightarrow SiO_2\ (aq)$$

$$SiO_2\ (aq) + 2H_2O \leftrightarrow H_4SiO_4$$

$$H_4SiO_4 \leftrightarrow H_3SiO_4^- + H^+$$

$$H_3SiO_4^- \leftrightarrow H_2SiO_4^{2-} + H^+$$

where Q stands for quartz and A stands for amorphous silica.

The first balance for the solubility of amorphous silica has $pK_1^o = 2.71$ ($K_1^o = 1.95 \times 10^{-3}$). The second balance for the solubility of quartz has $pK_2^o = 4.00$ ($K_2^o = 1.0 \times 10^{-4}$). Comparing the two solubility product constants, shows that amorphous silica is more soluble than crystalline quartz, as borne out by Figure 6.13.

The neutrally charged $SiO_2\ (aq)$ is regularly written as H_4SiO_4. Walther (2009) argued that the solubility product constant for the first reaction, could also be used for the third balance; the difference being the heat of formation of two water molecules. Therefore, for the present purpose, $K_3^o \simeq K_1^o$. The fourth balance has $K_4^o = 10^{-9.82}$ and the fifth balance $K_5^o = 10^{-13.10}$.

There exists an analogy between the components/species in calcite chemistry and silica chemistry. Both are pH-dependent. The following analogies are relevant: silicic acid H_4SiO_4 (=AA) and carbonic acid, once-dissociated silicic acid $H_3SiO_3^-$ (=BB) and the bicarbonate ion and twice-dissociated silicic acid $H_2SiO_3^{2-}$ (=CC) and the carbonate ion. One consequence of the analogy is that the mole fraction equations above, can also be used for silica. By replacing the calcium carbonate equilibrium constants with the silicic acid constants, the mole fraction with pH can be calculated.

The calculated values were used to plot the silica mole fractions with pH in the main text. The AA-ion dominates at pH below about 10, the BB ion-dominates in the range $10 < pH < 13$ and the CC-ion at higher pH-values.

Crude oil composition

The American Petroleum Institute (API) defines the gravity of crude oils and petroleum liquid products by the equation

$$API = \frac{141.5}{\gamma_o} - 131.5$$

where the specific gravity is given by

$$\gamma_o = \frac{\rho_o}{\rho_w}$$

The density values are those of oil ρ_o and water ρ_w at standard conditions. The liquid density unit used does not matter because API gravity is dimensionless. Light oils command a higher price in the marketplace; heavy oils command and a lower price. The API gravity is expressed in degrees (not temperature degrees).

The API gravity is used to classify oils as light, medium, heavy and extra heavy as shown in Table X1.1. In the case of liquid water, it relative density is one ($=1$) and the $API = 10°$. Extra heavy oil will sink in water. Trading of crude oil is conventionally based on prices relative to WTI (West Texas Intermediate) and Brent (Brent Blend). Both of these crude oils are light and sweet. Crude oil is sweet if the sulphur content is less than 0.5% wt. The API gravity of WTI and Brent are 39.6° and 38.6°, respectively. Their relative densities (gravities) are 0.827 and 0.835, also respectively. And their sulphur contents are 0.24% wt. and 0.37% wt. The exact demarcations in °API in Table X1.1 depend on the production region of each crude oil. What constitutes a light crude in a particular region depends on commodity trading in that region (there are no absolute demarcations).

API gravity is often used to correlate the physical properties of crude oil, such as density and viscosity. As an example, the density of Canadian crudes is schematically shown in Figure X1.1. Light crude has density less than 900 kg/m³ and heavy crude higher density. The oils described as bitumen, heavy, medium and light have API gravity in the ranges 0–10, 10–22, 22–30 and above 30°API, respectively. These ranges are approximately the same as in Table X1.1, indicating that the API demarcations and description used to classify crude oils can be different from region to region.

The viscosity of crude oil increases with decreasing °API. The sulphur content and the acid content lower the market price of crude oil. The viscosity of crude oils against

Table XI.1 Classification of crude oils according to °API (measured at U.S. standard conditions, temperature 60°F and atmospheric pressure).

Crude Oil	°API
Light	>31.1
Medium	22.5–31.1
Heavy	<22.3
Extra Heavy	<10.0

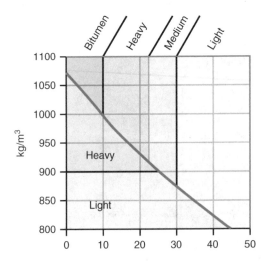

Figure XI.1 Density ρ_o kg/m³ and demarcations of Canadian crude oil against API gravity (Schmitt, 2005).

API gravity is shown in Figure X1.2 (Speight, 1999). Three API-ranges/demarcations are identified. A for conventional oil (light and medium), B for heavy oil and C for extra heavy oil. Naphthenate deposition world-wide occurs primarily in the production of heavy oils, but also in light/medium conventional oils. The viscosity in Figure X1.2 is for crude oil processes to standard conditions, for transport in an oil tanker or by a pipeline. Gas condensates are much lighter than the crudes shown in Table X1.1.

Countless correlations are found in the literature for the viscosity of oil at different conditions/states from the reservoir to surface facilities, and for different reservoirs, sedimentary basins and oil-producing regions. Undersaturated oil, saturated oil and stock-tank oil (also called dead oil) have different properties. The viscosity in Figure X1.2 is for stock-tank oil. The correlations for oil viscosity are typically expressed using logarithmic terms of the general form

$$\ln \mu_o = \alpha_1 + \frac{\alpha_2}{T} + \frac{\alpha_3}{API}$$

where the α's are correlation constants, T the temperature and API the hydrocarbon liquid gravity. The α_2/T expresses that the viscosity decreases with temperature.

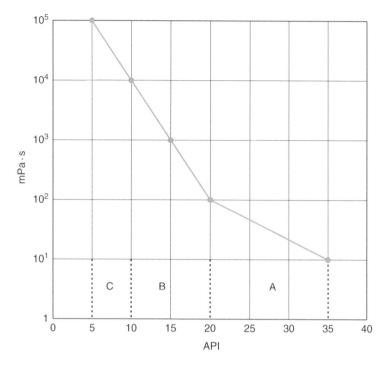

Figure X1.2 Logarithmic scale crude oil viscosity μ_o cp (=mPa · s) versus API gravity for A: Conventional crude oil (light and medium), B: Heavy oil and C: Extra heavy oil (Speight, 1999).

Table X1.2 Composition by weight of hydrocarbons in crude oil (Speight, 1999).

Hydrocarbon	Average % wt.	Range % wt.
Paraffins	30	15–60
Naphthenes	49	30–60
Aromatics	15	3–30
Asphaltics	6	

The α_3/API term expresses that the viscosity decreases with gravity. The natural logarithm in the expression relates directly with the common logarithm (base 10) in Figure X1.2.

The main four types of hydrocarbon molecules in crude oil are paraffins, naphthenates, aromatics and asphaltenes, shown in Table X1.2. The numbers in this table and numerous similar tables in the printed and online literature are predominantly based on the writings of Speight (1999) and subsequent publications.

Paraffins have a carbon to hydrogen ration C:H of 1:2 meaning that they have twice the amount of hydrogen compared to carbon. Paraffins are generally straight and branched chains, but never circular (cyclic) compounds. They are a desired content of crude oils for fuels. The shorter the paraffins are, the lighter the crude oil. *Naphthenes*

Table XI.3 Elemental composition ranges of crude oil (Speight, 1999).

Element	Weight %
Carbon	83–87
Hydrogen	10–14
Nitrogen	0.1–2
Oxygen	0.1–1.5
Sulphur	0.05–6*
Metals	<0.1

*Given as 0.5–6 in other tables.

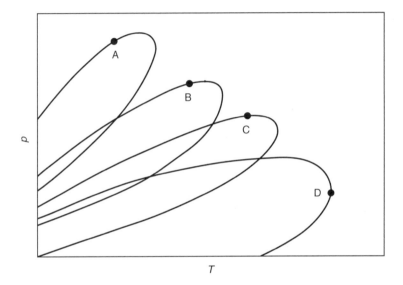

Figure XI.3 Phase diagrams of different reservoir fluids. Gas is lightest and black oil heaviest (Pedersen et al., 1989). A = gas, B = gas condensate, C = volatile oil and D = black oil.

have also a C:H ration of 1:2. They are cyclic compounds and can be described as cyclo-paraffins. They are higher in density than straight-chain paraffins and more viscous. *Aromatics* have a much less hydrogen compared to paraffins and they are more viscous. *Asphaltics* have a C:H ratio of about 1:1 making then very dense. Asphaltics are generally undesirable in crude oils. The difference in the composition of hydrocarbons in crude oils is primarily the difference in the hydrogen content. Chemical reactions in hydrocarbons involve hydrogen additions or hydrogen losses.

The elemental composition ranges of crude oil are shown in Table X1.3. The quality and price of a crude oil depends on the sulphur content and the metal elements, typically nickel and vanadium (also iron and copper). The sulphur and the metal elements create problems in the refining of crude oils. Sulphur leads to corrosion and the metal elements to detrimental performance of catalysts.

Reservoir fluids in petroleum engineering are commonly described as gas, gas condensate, volatile oil and black oil. The phase diagrams of such hydrocarbon fluids are illustrated in Figure X1.3 (Pedersen *et al.*, 1989). The solid point indicates the

Table X1.4 Typical molar percentages of petroleum reservoir fluids (Pedersen et al., 1989).

Component	Gas	Condensate	Volatile Oil	Black Oil
N_2	0.3	0.71	1.67	0.67
CO_2	1.1	8.65	2.18	2.11
C_1	90.0	70.86	60.51	34.93
C_2	4.9	8.53	7.52	7.00
C_3	1.9	4.95	4.74	7.82
C_4 $(i+n)$	1.1	2.00	4.12	5.48
C_5 $(i+n)$	0.4	0.81	2.97	3.80
C_6 $(i+n)$	6+ =0.3	0.46	1.99	3.04
C_7		0.61	2.45	4.39
C_8		0.71	2.41	4.71
C_9		0.39	1.69	3.21
C_{10}		0.28	1.42	1.79
C_{11}		0.22	1.02	1.72
C_{12}		0.15	12+ =5.31	1.74
C_{13}		0.11		1.74
C_{14}		0.10		1.35
C_{15}		0.07		1.34
C_{16}		0.05		1.06
C_{17}		17+ =0.37		1.02
C_{18}				1.00
C_{19}				0.90
C_{20}				20+ =9.18

critical point of the given fluid. Basically, if the pressure is higher than a phase diagram, the fluid is in liquid state. And if the temperature is higher than the phase diagram, the fluid is gaseous state. Inside a phase diagram the fluid is a mixture of gas and liquid.

Processed petroleum fluids have various specifications. For example, in the marketing of natural gas, carbon dioxide (CO_2) and hydrogen sulphide (H_2S), should not exceed certain limits. Natural gas is termed sweet when CO_2 is below 2% and/or H_2S is below 1%. At higher values a natural gas is termed sour.

The molar percentages of hydrocarbons in reservoir fluids are shown in Table X1.4. Also shown are the molar percentages of the common gaseous molecules nitrogen and carbon dioxide, as determined in the laboratory. Unfortunately, the hydrogen sulphide concentration is not shown. The concentrations are shown for carbon atoms from C_1 to C_{20} for use in PVT (pressure-volume-temperature) analysis of crude oils (hydrocarbon fluids) from reservoir conditions to surface facilities. The table shows the molar concentration/fractions for reservoirs producing natural gas (gas), gas condensate (condensate), volatile oil and black oil, illustrated in the schematic phase diagrams in Figure X1.3.

A typical organic chemistry classification of hydrocarbons is shown in Figure X1.4. Hydrocarbons can be classified into two distinct categories: saturated and unsaturated. Unlike saturated hydrocarbons, where all hydrogen and carbon atoms are bonded together with single bonds, unsaturated hydrocarbons have double covalent bonds

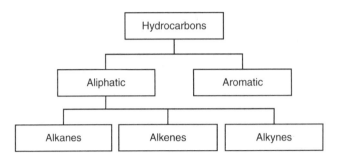

Figure XI.4 Organic chemistry classification of hydrocarbons. Alkenes and alkynes are not found in crude oil, only in refinery processes by cracking.

between the atoms (in some cases triple bonds). Unsaturated hydrocarbons have fewer hydrogen atoms bonded to carbon atoms, making them more reactive than saturated hydrocarbons.

Using the classification of organic molecules shown in Figure X1.4, the following descriptions apply:

1   Alkanes are saturated hydrocarbons having hydrogen and carbon atoms bonded to each other (*single bond*), having the chemical formula C_nH_{2n+2}, paraffin wax being an example with an integer in the range 20–40 (see *Section 4.1 – Wax in Crude Oil and Condensate*).

2   Alkenes are unsaturated hydrocarbons containing at least one carbon-to-carbon *double bond*, having the chemical formula C_nH_{2n}, ethylene being an example. Alkenes are not found in crude oil but in refinery processes.

3   Alkynes are unsaturated and contain at least one carbon-to-carbon *triple bond*, having the chemical formula C_nH_{2n-2}, acetylene being an example. Alkynes are not found in crude oil but in refinery processes.

4   Cycloalkanes (also called naphthenes) are saturated carbon and hydrogen atoms arranged in a ring structure, possibly with side chains (all the bonds are single). The chemical formula is $C_nH_{2(n+1-r)}$ where r is the number of rings. Cyclopropane C_3H_6 with one ring is an example.

5   Aromatic hydrocarbons are unsaturated and have an alternating carbon-to-carbon single and double bonds in the molecule. The chemical formula for one ring is $C_nH_{(2n-6)}$, benzene being an example. One or more of the hydrogen molecules in the ring can be replaced by a straight molecule.

Emulsions of crude oil and brine

The formation of emulsions in oil and gas production is inevitable, from the reservoir to processing. Crude oil contains all the ingredients to stabilize oil-water interfaces. The paramount challenge is the separation of emulsions into its constituent liquid phases. Separation to ensure required specifications for custody transfer of crude oil and the disposal of water/brine after processing. Crude oil should not contain too much water and disposal water/brine should not contain too much oil.

Crude oil emulsions are formed in a flow stream from a reservoir and in various pressure drop restrictions to surface facilities. The flow restrictions, provide the mixing energy to form an emulsion. A stationary emulsion will breakup into its two immiscible liquid phases, in the absence of an emulsifier. The various breakup processes of an emulsion are schematically shown in Figure X2.1, modified from Tadros (2013). An emulsion is shown in the centre of the figure. Process *A* is creaming and process *B* is sedimentation, usually resulting from gravitational or centrifugal forces. In a stationary process, lighter droplets move to the top and heavier droplets to the bottom. Process *C* is flocculation (aggregation) of droplets due to van der Waals attraction forces. Process *D* is phase inversion, that depends on the emulsifier and the volume ratio of oil and water. Process *E* is coalescence, due to the thinning and disruption of the liquid film between droplets. Two or more droplets fuse together into larger droplets; an oil layer may form. Process *F* is ripening, also called Oswald ripening. The droplet size distribution shifts (due to diffusion) to larger droplets that have lower free energy.

In oil production, after free-water knockout, the most common range of emulsified water in light crude oils (>20° API) is 5–20% vol. and in heavier crude oils (<20° API) 10–35% vol. In rare cases, water in crude oil can reach 60% vol. These percentages originate from Smith and Arnold (1987), in an early version of the Petroleum Engineering Handbook, now updated by Warren (2007) in terms of emulsion treating. In the same handbook, the fundamentals of crude oil emulsions are presented by Kokal (2007), including occurrence, types, characteristics and stability. A widely-referenced book on emulsions in the petroleum industry, is that of Schramm (1992), however not perused in the present work.

The volume fractions of oil and water/brine change during the lifetime of an oil field, the water fraction (WC = Water Cut) increasing with time. In the early life of an oil field, the oil-water mixture will be oil-continuous in most cases. However, as the volume fraction of produced water/brine increases, the emulsion will change to being water-continuous. The inversion depends on the composition of the oil, including

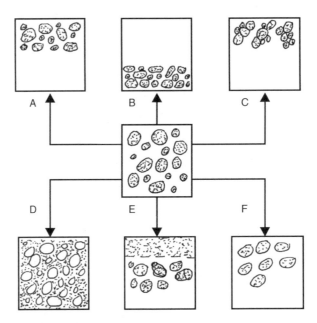

Figure X2.1 Schematic representation of various breakup processes in emulsions (see text for the A, B, C, D, E and F).

the natural emulsifiers such as big-five solids, sand and fines. In general, it can be stated that the emulsion breakup processes illustrated in Figure X2.1, are all relevant somewhere in oil and gas production and processing streams, but especially in oil-water separation. The increase in watercut in oil wells with time can be expressed by

$$\mathrm{WC}\,(t) = \frac{q_w}{q_o + q_w}$$

where the q-values are the volumetric flowrates of water and oil, identified by the subscripts w and o. The water cut at the start-up of an oil field may be small and with time it may reach as much as 90% vol. of the produced fluids. In an oil-continuous emulsion the water/brine exists as droplet surrounded by oil. The opposite is true in water-continuous emulsions.

Studies of crude oils and reservoir waters/brines show, that there is an *inversion point* from oil-continuous to water-continuous, typically at a water cut of about 60% as shown in Figure X2.2, in terms of volume fraction of water (not WC). The inversion from oil-continuous to water-continuous is illustrated by process D in Figure X2.1. On the other hand, Figure X2.2, shows the relative viscosity of arbitrary emulsions. To the left, a water-in-oil emulsion; to the right an oil-in-water emulsion. Importantly, at a volume fraction water at about 0.6, the emulsion changes from being oil-continuous to water-continuous. The figure shows the ratio of mixture viscosity divided by the viscosity of the continuous phase, at constant temperature. It is only illustrative, because

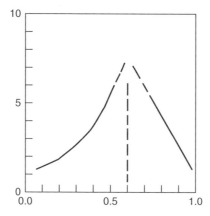

Figure X2.2 Arbitrary relative viscosity of oil-in-water and water-in-oil emulsions at constant temperature, illustrating an inversion point at water volume fraction of 0.6.

different crude oil and waters/brines will exhibit different curves, with slightly different inversion points. The viscosity of crude oil depends heavily on temperature, the viscosity decreasing with temperature (see *Appendix M1 – Viscosity and Activation Energy*). The curves in Figure X2.2 would move downward as the temperature increases. A well-known reference on the relative viscosity of suspensions is that of Thomas (1965). Suspensions include both emulsions and slurries (solids in liquids). For crude oil in water/brine emulsions, the apparent viscosity can be estimated from the expression

$$\frac{\mu}{\mu_o} = 1 + 2.5\phi + 14.1\phi^2$$

The expression was recommended by Warren (2007), in lieu of experimental values. The subscript *o* indicates the viscosity of the oil-only liquid phase, at the temperature in question. The volume fraction ϕ is that of the discontinuous phase, here the water/brine phase. If the square-term is dropped, the expression simplifies to the *Einstein Equation*.

Crude oil in water/brine emulsions are unstable because of their natural tendency to separate to reduce the interfacial area (interfacial energy). However, most emulsion in the oil industry are kinetically stable (stable over a period of time). Oilfield emulsion are classified on the basis of their stability (Kokal, 2007): *Loose emulsions* separate in a few minutes and the separated water is free water, *Medium emulsions* separate in tens of minutes and *Tight emulsions* separate in hours or even days. The droplet size distribution of petroleum emulsions is shown in Figure X2.3, in terms of a distribution function. The distribution function is given by the percentage symbol %, without specific values; the nature of the distributions should be apparent. The droplet size ranges from 1–100 μm. The crude oil and brine properties were not specified. The dominant diameters of the emulsions are round-about 7, 12 and 35 μm for the tight, medium and loose emulsions, respectively.

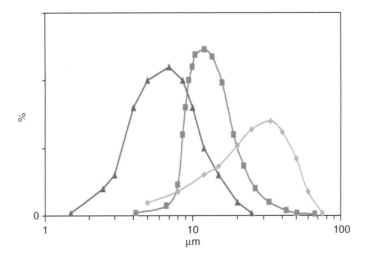

Figure X2.3 Droplet size distribution of petroleum emulsions (Kokal, 2007), from left-to-right illustrating a *tight* emulsion, a *medium* emulsion and a *loose* emulsion. Distribution function represented by %.

Emulsions are formed when two immiscible liquids, such as crude oil and formation water/brine, are vigorously mixed/stirred in the presence of emulsifiers. The greater the mixing energy and the greater the amounts of emulsifiers, the more stable the emulsion. The mystery of emulsions lies often in the emulsifiers that stabilize an emulsion. Emulsifiers include surface-active agents (surfactants) and finely divided solids. In general, surfactants are compounds that are partly soluble in both the oil-phase and the water/brine-phase. They have a hydrophobic part that has an affinity for crude oil and a hydrophilic part that has an affinity for water. Surfactants tend to concentrate at the oil/water interface, where they form interfacial films. In most cases, this leads to a lowering of the interfacial tension (see *Appendix X3 – Energy Dissipation and Bubble Diameter*), thus promoting dispersion and emulsification of the droplets. Naturally occurring emulsifiers in crude oil include such fractions as asphaltenes and resins and organic acids/bases. The big-five solids are considered natural emulsifiers. Finely divided solids act as mechanical stabilizers of interfaces. The particles must be much smaller (submicron) than the emulsion droplets. They collect at the oil-water interface and are wetted by both the crude oil and the water/brine. The two basic actions of emulsifiers, is reducing the interfacial tension and setting up a barrier between the phases.

The inversion point of oil-continuous to water-continuous emulsions, may be important for the deposition behaviour of flow assurance solids. For example, deposition may be more prevalent in oil-continuous systems than in water-continuous systems. The inversion point in Figure X2.2, is shown at water volume fraction of 0.6 (WC 60% vol.). In an example presented by Kokal (2007), the inversion point was at 0.8 (WC 80% vol.). At this juncture, it may be relevant to bring forth *Bancroft's Rule* (Wilder C. Bancroft, 1867–1953) that states: *The phase in which an emulsifier is*

Table X2.1 Physical properties of three Malaysian crude oils (Ilia Anisa et al., 2010).

Property	Crude A	Crude B	Crude C
Viscosity mPa · s	184	25	208
Density g/cm^3	0.846	0.835	0.849
API gravity	29°	34°	26°
Pour point °C	−19.5	−12	−20.4
Water % vol.	7	0.65	2
Surface tension mN/m	13.3	13.0	13.7

Table X2.2 Weight of solids (% wt.) in crude oil samples (Ilia Anisa et al., 2010).

Crude Oil	A*	Resins	Wax	Solids	R/A
A	11.7	21.6	0.42	13.89	1.85
B	2.94	32.3	9.13	9.77	11
C	8	37.7	23.8	3.27	4.71

*Asphaltenes.

more soluble constitutes the continuous phase. In practical terms, water-soluble surfactants tend to give an oil-in-water emulsion and oil-soluble surfactants tend to give a water-in-oil emulsion. The natural surfactants in crude oil are more oil-soluble than water-soluble. It follows that crude oil and brine emulsions will have an inversion point dominated by the oil-phase, as evident by inversion points occurring in the range of 0.6–0.8 volume fraction of the aqueous phase.

The phase inversion of water-in-oil emulsions for heavy and light crude oils was studied by Ilia Anisa et al. (2010). Other properties such as the effect of temperature on emulsion viscosity and mean droplet size were also studied. The properties of the three crude oils are shown in Table X2.1, and their weight composition in Table X2.2. Crudes A and C are classified as medium and crude B as light. The viscosity of the light crude is what would be expected, while the viscosities of the two medium crudes are quite high. The pour point temperatures and the water % wt. follow the viscosity values. The surface tension, on the other hand, is remarkably constant for the three crude oils.

Two of the organic big-five solids are shown in Table X2.2, expressed as % wt. The asphaltenes were separated using pentane (C_5), after dissolution in toluene to determine the insoluble solids. What the solids consisted of was not specified. The solids may have been fines and sand from the reservoir, corrosion products and perhaps inorganic scale. The composition of the formation water was not given. Silica gel was used to absorb the resins from the liquid (supernatant). The wax fraction was obtained by evaporation of the remaining liquid. The ratio R/A was quoted because sometimes it can indicate whether asphaltene precipitation and deposition will occur in wellbores, flowlines and processing facilities. It is difficult to find correspondences (relationships) between the properties of the crude oils and the measured weights of solids.

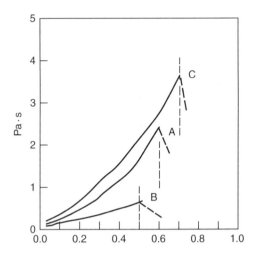

Figure X2.4 Sketch of dynamic viscosity of water-in-oil emulsions versus water volume fraction, illustrating the inversion point for three crude oils. B = 0.5, A = 0.6 and C = 0.7.

The phase inversions for the three crude oils are shown in Figure X2.4 in terms of measured viscosity (note that the scale is Pa · s, not mPa · s), against volume fraction water. It is assumed here that the measurements were taken at ambient pressure and laboratory temperatures. The curves are measured viscosity, not relative viscosity as shown in Figure X2.4 (sketch based on figure in Ilia Anisa *et al.*, 2010). The crude oils were mixed with water and vigorously stirred. The water used in the mixing was not specified. The viscosity was measured using a rotating spindle viscometer. The light crude (crude B), showed an inversion at a water volume fraction of 0.5; the lower viscosity medium crude (crude A) showed and inversion at water volume fraction of 0.6; the higher viscosity medium crude (crude C) shown an inversion at a water volume fraction of 0.7. The overall conclusion is that the inversion point of crude oils is different, depending on their bulk properties. An examination of Tables X2.1 and X2.2, points to dynamic viscosity as the most pertinent property. Although not all properties were reported for the crude oil emulsions in Figure X2.4, the figure demonstrates how the inversion point depends on the crude oil constituents.

The inversion point depends on the volume fraction of the dispersed phase, here water droplets in crude oil. As a point of reference, it may be appropriate to note the maximum packing of spheres. For random packing the volume fraction is about 0.64 and for structured packing (hexagonal) the volume fraction is about 0.74. These packing are theoretical and non-compositional.

Energy dissipation and bubble diameter

The surface tension affects greatly the diameter of a *bubble* in oil-in-water and water-in-oil emulsions and that of a *droplet* in gas-dominated flow. Turbulence theory has been used to formulate relationships for bubble/droplet diameter and energy dissipation. One such relationship was derived/suggested by Davies (1972), for maximum droplet diameter in isotropic turbulence

$$d_D = \left(\frac{4\sigma}{\rho_g} \right)^{0.6} P_E^{-0.4}$$

where σ N/m is surface tension, ρ_g kg/m^3 gas phase density (continuous phase) and P_E W/kg energy dissipation (power per kilogram). In the derivation of the equation, the densities of the continuous phase and the discontinuous phase were assumed equal. In oil-water flows, that is almost true, at least compared to oil-gas situations. Other similar equations are found in the literature; the Davies-equation is used here to illustrate the main phenomena. Davies (1986) has presented useful clarification on energy dissipation rates.

Assuming that an *analogy* exists between a droplet in gas and a bubble in liquid, the bubble diameter can be expressed by the same phenomenological equation

$$d_B = \left(\frac{4\sigma}{\rho_o} \right)^{0.6} P_E^{-0.4}$$

where ρ_o is the density of the continuous oil phase. In oil-continuous flow, the discrete phase is a water bubble; in water-continuous flow, the discrete phase is an oil bubble. In water-continuous flow the density of water ρ_w should be used. The surface/interface tension will be the same in both cases; oil-water interfacial tension. Interfacial tension is a little bit lower than surface tension. In case of no data available on interfacial tension, it can be roughly taken as the difference between the individual surface tensions (very approximate assumption).

In the bubble diameter equation, surface tension is one of the main parameters. Surface tension is defined for a single liquid surface (oil droplet in gas) while interfacial tension (IFT) is defined for the interface of two immiscible liquids (oil in water and water in oil). Surface tension and interfacial tension are essentially the same property. Interfacial tension is the *Gibbs Free Energy* per unit area J/m^2, which translates to the same unit of N/m for surface tension and interfacial tension. Surfactant molecules are preferentially found at oil-water interfaces, thereby lowering the interfacial tension. In oil and gas operations, a whole range of surface active components are found: asphaltenes, naphthenates and solid particles of paraffin wax, inorganic scale and sand (reservoir fines) and even corrosion products. An oil-water mixture/emulsion changes from water-in-oil to oil-in-water typically at watercut of about 0.6, as discussed in *Appendix X2 – Emulsions of Crude Oil and Brine*. The actual inversion point may be lower or higher, depending on the crude oil and formation brine compositions.

Many correlations are found in the literature for the surface tension of pure hydro-carbons and crude oil at conditions from reservoir to surface facilities (Buckley and Fan, 2007; Kumar, 2012; Ling and He, 2012). The countless correlations will not be presented here. For illustration purposes the simple Macleod-correlation

$$\sigma^{1/4} = C\Delta\rho$$

can be used. It shows that the surface tension depends on an experimental variable C and the density difference $(\rho_w - \rho_o)$ between water/brine and crude oil (Buckley and Fan, 2007).

Based on original work of Ramey (1973), and subsequent work by Firoozabadi and Ramey (2008), and an extensive data base for crude oils (forty-two samples), the Macleod-coefficient C was plotted against the density difference by Buckley and Fan (2007) for stock-tank oils, shown in Figure X3.1. The plot is for crude oils at atmospheric pressure, a temperature of 25°C and a distilled water acidity of pH = 6. For water density of 1 g/cm^3 and crude oil density of 0.9 g/cm^3, the coefficient in the Mcleod-equation is about C = 20. Calculating through, the interfacial tension becomes 16 mN/m, which seems reasonable. Typical crude-water values have been reported in the range 10–30 mN/m for distilled water systems.

A curve-fit was presented by Buckley and Fan (2007) based on the main variables affecting interfacial tension. They reported that the surface/interfacial tension depended primarily on the pH, TAN (Total Acid Number) and the asphaltene content (based on n-C$_7$). Furthermore, that the interfacial values correlated with crude oil viscosity. The following correlation was proposed for stock-tank oil at standard conditions

$$\sigma = 21.7 - 1.14\text{pH} + 0.754\text{A} - 1.21\text{B} + 1.15\text{C} + 0.0073\mu$$

where A stands for asphaltenes (n-C$_7$), B for total acid number (TAN), C for total base number (TBN) and μ for viscosity based on the unit cp (=mPa · s). The following values for the crude oil with the highest TAN, were selected from the forty-two samples listed by Buckley and Fan (2007): API gravity 21.9, n-C$_7$ asphaltene 0.76, TAN 3.42, TAB 2.57 and viscosity of 15.3 cp at 20°C, the interfacial tension was calculated for pH = 7 as σ = 13.3 mN/m. Using the definition of API gravity (oil density 0.922 and $\Delta\rho$ 0.078, and C = 25), and the simple $^1/_4$ correlation above, the interfacial tension was estimated to 14.5 mN/m, about 9% higher. However, if a pH of 6 was used in the correlation immediately above, the interfacial tension calculates to σ = 14.4 mN/m, which gives a good agreement. The coefficient curve in Figure X3.1 is based on pH = 6.

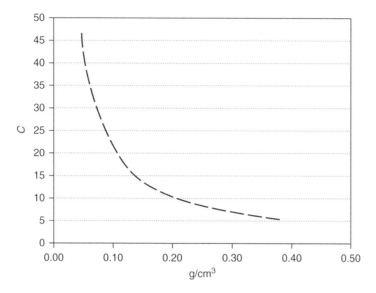

Figure X3.1 Surface/interfacial tension Macleod-coefficient *C* for immiscible oil-water emulsions against density difference $\Delta\rho$.

The effect of salinity (dissolved salts) of water/brine on interfacial tension of oil-water systems was studied by Kumar (2012). Pure hydrocarbons (n-heptane, toluene and cyclohexane) were used in experiments to represent the effects of aliphatic, aromatic and cyclic compounds, respectively. The experimental work resulted in the following correlation

$$\sigma(T,x) = \sigma_o - 0.0334T \ln(1 + 4.43x)$$

where σ_o is the interfacial tension of hydrocarbon mixed with clean water. The amount of salt (sodium chloride, calcium chloride and sodium sulphate) was expressed as mole fraction x and the temperature in T K. Interfacial tension decreases with temperature and increases marginally with pressure.

In the literature, it has been reported that the interfacial tension of hydrocarbon-water mixtures increases with salinity. However, small amounts of surfactants present in the system are reported to have the reverse effect, such that the interfacial tension decreases with salinity. The interfacial tension is reported to decrease at pH-values below 7 and also at pH-values above 7. These observations are examples of how difficult it can be to ascertain the effects of various chemical and physical properties on interfacial tension in oil-water systems. Naphthenic acids and asphaltenes are two of the most important surfactants found in crude oil (Kumar, 2012). Generally, interfacial tension is found to increase with crude oil density, viscosity and the paraffinic content.

The energy dissipation term in the bubble diameter equation, P_E, depends primarily on shear stress in the flow steam. Two situations are envisioned: flow in pipelines and wells, and flow through restrictions such as valves and bends.

In steady-state pipe flow the following force balance gives the relationship between wall shear stress and frictional pressure drop

$$\Delta p A = \tau \Delta L S$$

where A is the cross-sectional area and S the wetted circumference ($=\pi d$). For a circular flow channel with diameter d the wall shear stress, force per unit area $N/m^2 = Pa$, can be expressed as

$$\tau = \frac{A}{S}\frac{\Delta p}{\Delta L} = \frac{d}{4}\frac{\Delta p}{\Delta L}$$

Power W, defined as work per unit time, is equal to velocity times force. For pipe flow

$$P = u(A\Delta p)$$

and for unit mass W/kg

$$P_E = \frac{uA\Delta p}{A\Delta L \rho} = \frac{u\Delta p}{\Delta L \rho}$$

Frictional pressure drop in pipe flow is given by the Darcy-Weisbach Equation (see *Appendix D*)

$$\Delta p = \frac{f}{2}\frac{\Delta L}{d}\rho u^2$$

Substituting for the pressure gradient $\Delta p/\Delta L$, the energy dissipation expressed as power per unit mass W/kg ($=m^2/s^3$) becomes

$$P_E = \frac{f}{2}\frac{1}{d}u^3$$

Liquids flowing in a pipe will be subjected to the energy dissipation given by the relationship immediately above. Liquid-liquid emulsions will be subjected to at least the same energy dissipation. The real dissipation will be larger because of the more complex flow, compared to single-phase liquid flow. Also, the viscosity of emulsions is much larger than the viscosity of the individual components (oil and water).

The pressure loss in valves and other flow restrictions is much larger than in plain pipes. The following relationship, based on the kinetic energy of the flowing medium, expresses the pressure drop

$$\Delta p = K\frac{1}{2}\rho u^2$$

where K is the *pressure loss coefficient*. The average flow velocity is based on the pipe flow velocity, not the much higher velocity in the internals of a valve. The pressure loss coefficient is *small* for fully open valves and *large* for almost closed valves. Typical K-values for a gate valve and a sluice valve are shown in Figure X3.2 (Thorley, 1991). The figure shows that a 50% open gate valve has a K-factor of 2. Assuming an oil

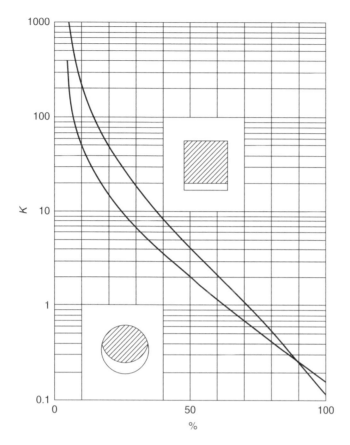

Figure X3.2 Pressure loss coefficient for gate valves (lower curve) and sluice valves (upper curve), versus percentage linear opening (Thorley, 1991).

density of 900 kg/m³ and a flow velocity of 2 m/s the pressure drop calculates to 3600 Pa or 0.036 bara. For a 10% open valve the K-factor is about 50, giving a pressure drop of 0.9 bara. For 5% open valve the K-factor is about 400 and the pressure drop 7.2 bara. Choke valves have a much higher pressure drop than gate valves. Assuming a K-factor of about 1000 the pressure drop was calculated 18 bara for the same conditions as above. Choke valves are of special design to control large pressure drops without significant metal erosion. Control valves are similarly robust.

Tables with pressure loss coefficients from different valves are found in text-books and handbooks. One such is Table X3.1 for gate valves (Blevins, 2003). The upstream/downstream nominal diameters are ½ inch, 2 inch and 6 inch, given in millimetres in the table. The valve opening is shown from 12.5% to fully open (1/8, 1/4, 3/8, 1/2, 3/4 and 1/1). The table shows that the pressure loss coefficient depends also on the upstream/downstream pipe diameter for the gate valve in question (unusual detail). In various literature citations the average K-value for gate valves are given as

Table X3.1 Pressure loss coefficient K for a gate valve, from 1/8 (=12.5%) open to 1/1 (=100%) open. Equal upstream and downstream pipe diameter *d* mm (Blevins, 2003).

Diameter	12.5%	25%	37.5%	50%	75%	100%
12.5	370	54	18.0	7.7	2.2	0.81
50.0	150	23	7.2	3.2	0.75	0.18
150.0	87	17	6.0	2.6	0.52	0.15

0.17, 0.9, 4.5 and 24 for 1/1, 3/4, 1/2, and 1/4 opening. These values are similar to the values for the 50 mm diameter gate valve in Table X3.1. The diameter is the pipe diameter, assumed the same upstream and downstream.

A method called the *Equivalent Length Method* is commonly used in industry to calculate the pressure loss due to fittings and valves. A typical value is $L/d = 8$ for a fully open gate valve. Knowing the pipe diameter, the equivalent length is added to the known length of a pipeline. The total length of a pipeline thus becomes

$$\Delta L_T = \Delta L(1 + Nd)$$

where the integer N represents the L/d of a particular fitting or valve. The equivalent length method was not used in the present text, but is mentioned here for the sake of completeness.

The K-factor method can be used to estimate the specific energy dissipation term P_E W/kg for use in the bubble diameter equation. Energy dissipation ε (symbol common in turbulence theory) in turbulent flow can be approximated by

$$\varepsilon \simeq k_E \frac{(u')^3}{l}$$

where k_E is a dimensionless constant, u' m/s the turbulence level in the main flow direction and l m a characteristic length scale (Baldi *et al.*, 2002). In pipe flow the velocity term can be assumed to be the average flow velocity. The length scale for pipe flow is commonly taken as the pipe diameter, the largest turbulent eddies being of the same order of magnitude as the pipe diameter. The phenomenology is approximate, but it gives an indication of the main variables.

The friction factor f and the pressure loss coefficient K are directly proportional to each other. The energy dissipation equation can thus take the form

$$P_E = n \frac{K}{2} \left(\frac{u^3}{d} \right)$$

where k_E has been replaced by the coefficient $nK/2$ which includes the empirical energy integer n. The value of the energy integer can be based on empirical assumptions and observations and perhaps experimental results. One consideration would be to double the integer in situations where an oil-water emulsion flows through two similar restrictions.

Two-phase flow variables and equations

The flow of two phases, liquid and gas, involves a number of specific variables. The variables are used in a number of equations necessary in two-phase flow calculations. In the oil and gas industry, the calculations involve pressure drop in pipelines and wells. Offshore, calculations are required for the cascading pressure drop from down-hole to wellhead, and further along flowlines and risers to production platforms with processing facilities. On land, the calculations required are similar. The pressure and temperature changes along wells and pipelines and processing equipment, affect the conditions governing the formation/precipitation and deposition of the whole range of flow assurance solids.

DEFINITION OF VARIABLES

Void fraction

$$\alpha = \frac{A_G}{A}$$

Hold-up

$$(1 - \alpha) = \frac{A_L}{A}$$

Velocity, u, u_G, u_L m/s
Mass flowrate, m, m_G, m_L kg/s
Volume flowrate, q, q_G, q_L m^3/s

Mass quality

$$x = \frac{m_G}{m_G + m_L}$$

$$(1 - x) = \frac{m_L}{m_G + m_L}$$

Mass velocity or mass flux

$$G = \frac{m}{A} = \rho u = \frac{u}{v} \text{ kg/s} \cdot \text{m}^2$$

Density, ρ kg/m^3
Specific volume, v m^3/kg

Average (phase) velocities

$$u_G = \frac{m_G}{\rho_G A_G} = \frac{q_G}{A_G} = \frac{Gx}{\rho_G \alpha}$$

$$u_L = \frac{m_L}{\rho_L A_L} = \frac{q_L}{A_L} = \frac{G(1-x)}{\rho_L (1-\alpha)}$$

Below, volumetric flux or superficial velocity (note that total area A is used for all three definitions)

$$U = \frac{q}{A}$$

$$U_G = \frac{q_G}{A}$$

$$U_L = \frac{q_L}{A}$$

Slip ratio is defined as

$$K = \frac{u_G}{u_L}$$

It can be shown that

$$K = \left(\frac{x}{1-x} \right) \left(\frac{\rho_L}{\rho_G} \right) \left(\frac{1-\alpha}{\alpha} \right)$$

Density (exact value) of two-phase mixture is given by

$$\rho = \alpha \rho_G + (-\alpha) \rho_L$$

It can be shown that

$$\alpha = \frac{x}{x + K(1-x)\dfrac{\rho_G}{\rho_L}}$$

If $K = 1$

$$\alpha = \frac{x}{x + (1-x)\dfrac{\rho_G}{\rho_L}}$$

The definition of GOR (Gas-Oil-Ratio) is given by

$$\text{GOR} = \frac{q_G}{q_L}$$

The void fraction equation can be arranged to

$$\frac{1}{\alpha} = 1 + \frac{q_L}{q_G} = 1 + \frac{1}{\text{GOR}}$$

such that

$$\text{GOR} = \frac{\alpha}{1 - \alpha}$$

HOMOGENEOUS FLOW

$$u = u_G = u_L$$

$$K = \frac{u_G}{u_L} = 1$$

$$x = \frac{m_G}{m_G + m_L}$$

$$\alpha = \frac{q_G}{q_G + q_L}$$

$$x = \frac{\alpha}{\alpha + (1 - \alpha)\dfrac{\rho_L}{\rho_G}}$$

$$\alpha = \frac{x}{x + (1 - x)\dfrac{\rho_G}{\rho_L}}$$

$$-\frac{dp}{dz} = \frac{dp_f}{dz} + \frac{dp_a}{dz} + \frac{dp_g}{dz}$$

$$\frac{dp_f}{dz} = \frac{fG^2}{2\rho d}$$

$$\frac{dp_a}{dz} = \frac{d(G^2/\rho)}{dz}$$

$$\frac{dp_g}{dz} = g\rho \sin \theta$$

Exact equation for mixture density

$$\rho = \alpha \rho_G + (1 - \alpha)\rho_L$$

Note that ρ has unit kg/m³; that is, mass per volume. It is therefore we use volume fraction α in the mixture density equation immediately above. The specific volume, v, which is the inverse of density ρ, can be expressed as

$$v = x v_G + (1 - x) v_L$$

Since $\rho = 1/v$ the following applies also

$$\frac{1}{\rho} = \frac{x}{\rho_G} + \frac{(1-x)}{\rho_L}$$

$$Re = \frac{Gd}{\mu}$$

Empirical equation (McAdams et al., 1942) for homogeneous mixture viscosity, commonly used in pressure drop calculations (exact equations not available) is the following

$$\frac{1}{\mu} = \frac{x}{\mu_G} + \frac{(1-x)}{\mu_L}$$

The viscosity μ has unit $Pa \cdot s$ with is equal to kg/s·m. It follows that it seems most logical to use mass fraction x in the above expression of mixture density.

The homogeneous model tends to under predict pressure drop, but can give reasonable results at high pressure and high mass flux, as in oil/gas wells, for example.

$$G = \rho u$$

$$\frac{fG^2}{2\rho d} = \frac{f\rho u^2}{2d}$$

Darcy-Weisbach equation for pressure drop in pipes

$$\Delta p = \frac{f}{2} \frac{L}{d} \rho u^2 = \frac{f}{2} \frac{L}{d} \frac{G^2}{\rho}$$

SLIP RATIO EQUATION

The slip ratio is defined as

$$K = \frac{u_G}{u_L}$$

Because

$$u_G = \frac{q_G}{A_G}$$

and

$$q_G = \frac{m_G}{\rho_G}$$

The same can be written for the liquid phase

$$K = \frac{m_G}{\rho_G A_G} \frac{\rho_L A_L}{m_L}$$

Rewriting such that

$$K = \frac{\rho_L}{\rho_G} \left(\frac{m_G A_L}{m_L A_G} \right)$$

Definition of mass fraction is

$$x = \frac{m_G}{m_G + m_L}$$

which means that

$$\frac{m_G}{m_L} = \frac{x}{1 - x}$$

Definition of void fraction is

$$\alpha = \frac{A_G}{A_G + A_L}$$

which means that

$$\frac{A_L}{A_G} = \frac{1 - \alpha}{\alpha}$$

Substitution gives

$$K = \left(\frac{\rho_L}{\rho_G} \right) \left(\frac{x}{1 - x} \right) \left(\frac{1 - \alpha}{\alpha} \right)$$

Two-phase flow regimes

In oil and gas production, multiphase flow occurs throughout the flowstream from reservoir to processing facilities. Tree main domains can be identified: flow in reservoir, flow in wellbore, flow in flowlines. The hydrocarbons entering a wellbore can be black oil, volatile oil, natural gas and gas condensate. The hydrocarbons in-place can be described as undersaturated oil, saturated oil and natural gas. For the sake of simplicity, undersaturated oil will enter a wellbore in the liquid phase, evolving to two-phase flow in the production tubing when the pressure reduces below the bubble point. Saturated oil will evolve to two-phase flow in the reservoir formation upon production due to pressure drawdown. Natural gas will enter a wellbore in the gas phase. Gas condensate evolves from natural gas in the reservoir formation, condensing out the heavy constituents due to pressure drawdown in the near-wellbore region. Superimposed on these simple hydrocarbon flows, will be the flow of produced water.

The flow in production tubing will in practice always be multiphase. Oil-water flow, oil-water-gas flow, gas-water flow and gas-condensate-water flow. The multiphase flow nature of produced fluids will continue in flowlines. Wellbores (production tubing) are near-vertical and flowlines are near-horizontal. Several types of flow regimes have been recognized in production tubing and flowlines. A multitude of models have been developed to calculate the total pressure drop in multiphase vertical flow and horizontal flow. It is far beyond the scope of the present text to delve into the voluminous literature on multiphase flow. However, a common feature of many multiphase models, is the identification of the flowing phases expressed in flow regimes. The flow regimes are often used as the basis for the semi-empirical models developed and used to calculate the total pressure drop.

The description of flow patterns tends to be more qualitative than quantitative. The observed flow patterns are ascribed different variables by different researchers. Nevertheless, the most common variable used is *volume fraction*, which ranges from a low value in bubble flow to a high value in annular flow. A simplified sketch of four flow patterns in a horizontal flowline are shown in Figure Y2.1. The figure illustrates gas-liquid flow (gas-oil, gas-oil-water, gas-water) with different volume fractions. The sketch is basically self-explanatory. In addition to volume fractions, the flow velocities trend to increase from low volume fraction to high volume fraction. Not shown in the sketch are droplets in the gas phase and gas bubbles in the liquid phase. There will be a *larger* difference (larger slip) between the gas phase velocity and the liquid phase velocity, than between the liquid phase components (oil and water). In most calculations, the liquid oil-water phase is treated as a homogeneous phase.

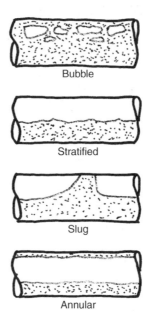

Figure Y2.1   Basic flow patterns in horizontal flow.

The flow pattern and the semi-empirical equations used to calculate the total pressure drop, are tied together by the criteria for transition from one flow pattern to another flow pattern. In the oilfield, operators may be interested in knowing what flow pattern exists in their wells and flowlines. Operators prefer continuous flow, not intermittent flow. Three of the flow regimes in Figure Y2.1 are continuous and one intermittent. Slug flow is intermittent and is definitely not desired in oil and gas operations. The liquid slugs can be short, medium and long, corresponding to liquid volume rates that cause challenges in receiving facilities. Slug catchers (large tanks) are installed in the oilfield where medium-to-long slugs are expected. Small slugs cause pressure fluctuations and mechanical vibrations that can be detrimental to the integrity of components in processing/receiving facilities. A special slugging problem is associated with platform risers, where a horizontal flowline takes a 90 degree bend upwards. Terrain slugging arises in horizontal flowlines with low-points where liquid slugs can accumulate. The pressure builds up until the liquid slug moves further along the flowline.

Flow patterns are commonly represented in flow regime maps. Various variables have been used in such maps, including superficial velocity, mass flux and Froude number. Several versions of the Froude number are found in the literature, making comparison between published studies challenging. Consistent flow regime maps for horizontal and vertical two-phase flow, were presented by Shell Global Solutions International (2007), based on a specific Froude number. The Froude number included the densities of the gas-phase and the liquid-phase (oil and water), thus capturing in-part the effect of pressure. The horizontal flow regime map is shown in Figure Y2.2. The basic Froude number expresses the ratio of inertial force to gravitational force.

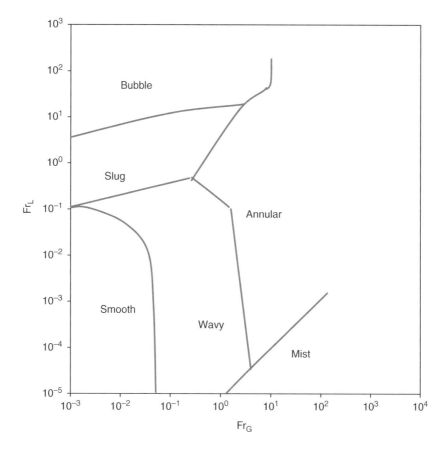

Figure Y2.2 Log-log horizontal flow regime map showing liquid Froude number (vertical axis) versus gas Froude number (Shell Global Solutions International, 2007).

The specific Froude numbers for the liquid-phase and the gas-phase for the figure above, are defined by the equations

$$Fr_L = \sqrt{\frac{\rho_L}{(\rho_L - \rho_G)gd}}$$

$$Fr_G = \sqrt{\frac{\rho_G}{(\rho_L - \rho_G)gd}}$$

where g is the gravitational constant and d the pipe diameter. The Froude number is much used for open channel flow and in ship design. In open channel flow, $Fr = 1$ is termed *critical flow*, $Fr < 1$ is termed *tranquil flow* and $Fr > 1$ is termed *rapid flow*. These basic criteria are reflected in the flow patterns illustrated in Figure Y2.2. The troublesome slug flow pattern occurs when the liquid Froude number is critical and the gas Froude number also. The stratified flow regime coincides with what is called tranquil flow, which seems appropriate. The high-velocity annular flow covers the area of the flow regime map that would be characterized as rapid flow, based on the gas-phase Froude number. The bubble flow regime is found at the top of the map. In the lower left-hand, the smooth and the wavy regimes are stratified flow regimes. The mist flow regime occurs at high volume fraction gas velocity. In the present text, the specific Froude number is a density-based Froude number.

The determination of which flow regime exists in a particular flowline, hypothetical values can be used. Oil, gas and water are assumed to flow at 25°C and 100 bar (=10 MPa) in a 150 mm ID (=0.15 m) pipeline. The nominal diameter is 6 inches. The hypothetical flowrates are shown in Table Y2.1. The oil-phase and water-phase have been assigned typical density values. The gas-phase density is based on gas composition given in Table Y2.2, resulting in a molecular weight of 18.5 g/mol (=18.5 kg/kmol).

For the flowrates and densities of the three phases, the GOR and WC were calculated, shown in Table Y2.3. Typical GOR values for volatile oil are in the range 338–570 Sm^3/Sm^3 and 20–267 Sm^3/Sm^3 for a black oil, as shown in Table 3.9. The GOR calculated for the present hypothetical cases, 230 Sm^3/Sm^3, corresponds to an upper-limit black oil. The WC was calculated 47 %, which seems reasonable (see Figure X2.4). The liquid-phase in Table Y2.3 is the *combined* oil and water, while the mixture density is that of the oil, water and gas together, at the local conditions (temperature and pressure). The superficial velocities of the gas-phase and the liquid-phase were calculated based on the total cross-sectional flow area (see *Appendix Y1 – Two-Phase Flow Variables and Equations*).

Table Y2.1 Flowrates of gas, oil and water at 10 MPa and 25°C in 0.15 m ID pipeline.

Property	Gas	Oil	Water
m kg/h	5000	20,000	20,000
ρ kg/m³	95	900	1000
q m³/h	105	22.2	20

Table Y2.2 Mole fractions (composition) of gas phase Same as gas in Table X1.4.

Molecule	y
N_2	0.003
CO_2	0.011
C_1	0.900
C_2	0.049
C_3	0.019
C_4	0.011
C_5	0.004
C_6	0.030
M	18.5 g/mol

Table Y2.3 Properties at 10 MPa and 25°C in 0.15 m ID pipeline.

Property	Value
GOR Sm^3/Sm^3	230*
WC %	47
ρ_L kg/m^3	950
ρ_G kg/m^3	95
ρ_M kg/m^3	747
U_L m/s	0.66
U_G m/s	0.83
Fr_L	0.58
Fr_G	0.23

*Standard conditions, 15°C and 100 kPa, gas density 0.98 kg/m^3

The liquid-phase and gas-phase Froude numbers place the hypothetical multiphase flow just inside the slug flow region. The slug flow region is the problem region in multiphase transport in the oil and gas industry. The only parameter that can be used to adjust the flow regime away from slug flow, is the flowline diameter. A smaller diameter will increase the superficial velocities, potentially into the annular flow regime. A larger diameter will decrease the superficial velocities, potentially into the stratified flow regime. Small diameter means greater pressure drop and larger diameter means increased pipeline costs. In other words, a typical case of finding the optimum design.

Appendix Z

Multiplier pressure drop method

The pressure drop due to wall friction in pipelines can be estimated by using the multiplier method. The two-phase pressure drop is simply the *superficial* pressure drop to either of the single-phase gas or liquid, multiplied with a factor greater than unity (=one). Lockhart and Martinelli (1949) carried out experiments in pipes 1.5–25 mm in diameter at pressures from atmospheric to 3.4 bara (=50 psia). They used air and different liquids, including water and hydrocarbons. G stands for gas and L for liquid.

In the multiplier method, the two-phase frictional pressure drop can be written

$$\frac{dp}{dz} = \phi_G^2 \left(\frac{dp}{dz}\right)_G = \phi_L^2 \left(\frac{dp}{dz}\right)_L$$

which is equivalent to writing

$$\Delta p = \phi_G^2 \Delta p_G = \phi_L^2 \Delta p_L$$

The variables ϕ_G and ϕ_L are the multipliers for gas and liquid, respectively. That the multipliers are written squared, has no particular technical significance. Apparently, the square of the multipliers was used by Lockhart and Martinelli (1949) because they used plotting paper in early times before the computer age. The square form has no practical consequence in two-phase flow pressure drop calculations.

The multiplicator method is based on the superficial pressure drop of *one* (gas or liquid) of the phases. The superficial velocities (see *Appendix Y1 – Two-Phase Flow Variables and Equations*) are given by the relationships

$$U_G = \frac{q_G}{A}$$

$$U_L = \frac{q_L}{A}$$

Expressed in another way, the single-phase pressure drops are calculated assuming each phase flows alone in the pipeline. The total flow area A is used, not the specific flow areas A_G and A_L of the individual phases.

Lockhart and Martinelli (1949) correlated their experimental data by the parameter X given by the ratio of the superficial pressure drops/gradients

$$X^2 = \frac{\left(\dfrac{dp}{dz}\right)_L}{\left(\dfrac{dp}{dz}\right)_G}$$

$$X^2 = \frac{\Delta p_L}{\Delta p_G}$$

As with the multipliers above, that the Martinelli parameter is squared, has no particular significance, other than what plotting paper was available at the time. Lockhart and Martinelli (1949) presented their experimental results where the multipliers (ϕ_G and ϕ_L) were plotted against the parameter X on a log-log paper.

The frictional pressure for liquid (L) can be calculated from the *Darcy-Weisbach Equation* (see *Appendix D – Darcy-Weisbach Equation*)

$$\Delta p_L = \frac{f}{2}\frac{L}{d}\rho_L U_L^2 = \frac{f}{2}\frac{L}{d}\frac{G_L^2}{\rho_L}$$

where the symbols have the usual meaning and G is mass flux.

$$G_L = \rho_L U_L = G x$$

and x the mass fraction.

The frictional pressure drop for gas flow (G) in a horizontal pipeline can be calculated from the equation

$$\frac{dA^2 M}{f m^2 z R T}(p_2^2 - p_1^2) - \frac{d}{f}\ln\left(\frac{p_2^2}{p_1^2}\right) + L = 0$$

derived in *Appendix G – Pressure Drop in Gas Pipelines and Wells*. It turns out for practical situations that the second term in the equation is much, much smaller than the first term (contains pressure in Pa squared). Furthermore, if average pressure and average properties are used in the equation, it simplifies to the *Darcy-Weisbach Equation*

$$\Delta p_G = \frac{f}{2}\frac{L}{d}\bar{\rho}_G U_G^2 = \frac{f}{2}\frac{L}{d}\frac{G_G^2}{\bar{\rho}_G}$$

where

$$\bar{\rho}_G = \frac{\bar{p} M}{\bar{z R T}}$$

The simplification does not affect greatly the accuracy of the frictional pressure drop in the multiplicator methodology. The reason being that the accuracy of two-phase flow calculations is much poorer than the accuracy of single-phase flow.

Table Z.1   Properties and values used in calculations.

Property	Average Value
Liquid density	950 kg/m^3
Liquid viscosity	2.5 mPa · s
Watercut	25%
Molecular weight	20 kg/kmol
Compressibility	0.7
Gas density	60 kg/m^3
Temperature	15°C
Pressure	50 bara
Pipe diameter	0.1 m

Table Z.2   Multiplicator flow calculations for individual (liquid and gas) phases.

Velocity m/s	Re$_L$	$(\Delta p/\Delta L)_L$	Re$_G$	$(\Delta p/\Delta L)_G$
1.0	37,000	105	487,800	3.54
2.0	74,000	354	975,600	12.1
3.0	111,000	720	1,463,000	24.5
4.0	148,000	1192	1,951,000	40.6
5.0	185,000	1762	2,439,000	60.0

As an example, multiplicator calculations were carried out for typical subsea flow-line conditions. The properties assumed are shown in Table Z.1. Of course, properties change with temperature and pressure. The calculation example was based on the pressure gradient, so the assumption of average properties seems reasonable. The calculation results are shown in Table Z.2. Reference is made to *Section 2.4 – Two-Phase Flow in Pipelines*, in particular the *Chisholm Equation*

$$\phi_L^2 = 1 + \frac{20}{X} + \frac{1}{X^2}$$

for turbulent-turbulent flow. The calculations gave $X = 5.42$ such that the liquid multiplier $\phi_L = 2.2$. Therefore, the two-phase liquid flow pressure gradient should be multiplied by this factor. At flow velocity 2 m/s the two-phase pressure gradient becomes 780 Pa/m and at flow velocity 4 m/s is becomes 2620 Pa/m. The pressure gradient in two-phase flow is more than double that of the single-phase pressure gradient.

Index

Note: Page numbers in **bold** indicate figures and tables